From DNA damage and stress signalling to cell death

To the previous generation, and for the next one:
Charlotte, Martin, Julien and Vincent.

G.dM

To my Parents, Children and Grandchildren.

S.S

From DNA damage and stress signalling to cell death
Poly ADP-ribosylation reactions

Edited by

Gilbert de Murcia
CNRS, Strasbourg

and

Sydney Shall
*Department of Molecular Medicine,
King's College School of Medicine and Dentistry, London*

OXFORD

UNIVERSITY PRESS

Great Clarendon Street, Oxford OX2 6DP

Oxford University Press is a department of the University of Oxford.
It furthers the University's objective of excellence in research, scholarship,
and education by publishing worldwide in
Oxford New York
Athens Auckland Bangkok Bogotá Buenos Aires Calcutta
Cape Town Chennai Dar es Salaam Delhi Florence Hong Kong Istanbul
Karachi Kuala Lumpur Madrid Melbourne Mexico City Mumbai
Nairobi Paris São Paulo Singapore Taipei Tokyo Toronto Warsaw
with associated companies in Berlin Ibadan

Oxford is a registered trade mark of Oxford University Press
in the UK and in certain other countries

Published in the United States
by Oxford University Press Inc., New York

© Oxford University Press 2000

The moral rights of the author have been asserted
Database right Oxford University Press (maker)

First published 2000

A catalogue record for this book is available from the British Library

Library of Congress Cataloguing in Publication Data

From DNA damage and stress signalling to cell death: poly-ADP-ribosylation reactions
/Gilbert de Murcia, Sydney Shall, editors.
    p. cm.
Includes bibliographical references and index.
1. ADP-ribosylation. I. Murcia, Gilbert de. II. Shall, S. (Sydney), 1932–
QP625.A29 F76 2000   572′.7921—dc21   00–058032

ISBN 0 19 850633 3

Typeset in Great Britain by
Phoenix Photosetting, Chatham, Kent
Printed in Great Britain
on acid-free paper by
Biddles Ltd, Guildford & King's Lynn

Foreword

Takashi Sugimura

This monograph on poly (ADP-ribosylation) reactions has been edited by Professors Gilbert de Murcia of Strasbourg, France and Sydney Shall of London, England, both of whom have been very industrious in this field of science. The publication of this monograph is very timely and well reflects the renewed interest in poly (ADP-ribose). This volume should speed up progress in studies of this polymer and we hope that investigators, not only in this field but also in other fields, will read it.

Poly (ADP-ribose) was discovered more than 30 years ago. The availability of the cloned genes of the enzymes involved in poly (ADP-ribose) metabolism and in the function of this polymer, as well as subsequent studies using gene-manipulated animals have opened up a new era in this field as in so many others. Basic biological phenomena such as DNA repair, differentiation, apoptosis, and important medical conditions such as ischaemia-reperfusion injury, cancer, autoimmune diseases, diabetes, and parkinsonism are now understood to be connected to the functions of this biopolymer.

It is a great honour for me to have been asked by the editors to recall the early days in this field. In 1963, Paul Mandel, and his co-authors P. Chambon and J. D. Weil, triggered the science of poly (ADP-ribose) with a paper entitled 'Nicotinamide mononucleotide activation of a new DNA-dependent polyadenylic acid synthesizing nuclear enzyme' (1).

This was the first ringing of the bell to tell us of the presence of this polymer. It was a remarkable discovery. If I may be allowed to mention my personal experience; when I learned of their finding in a chapter of the 1964 *Annual Review of Biochemistry* (2) written by Gerhard Schmidt, I was deeply impressed. In those days, the allosteric regulation of enzyme activity by low molecular weight compounds was fashionable in biochemistry. I optimistically imagined that this new regulatory mechanism might be disturbed in hepatic carcinoma. Our laboratory soon confirmed the reported phenomenon; namely, that incorporation of radioactivity from [^{14}C-adenine]-ATP into an acid-insoluble fraction was markedly enhanced by the addition of nicotinamide mononucleotide (NMN) in rat liver nuclear extracts. Naturally, we expected that the acid-insoluble product would be poly (A). Contrary to this expectation, the product was resistant to alkaline treatment. However, the product was digested by snake-venom phosphodiesterase, yielding a small amount of 5′-[^{14}C-Ade]-AMP and a large amount of an unknown compound. This compound behaved similarly on both paper and thin-layer chromatography to 5′-ADP, and to 3′,5′-ADP prepared from coenzyme A and 2′,5′-

ADP prepared from NADP, but it was not exactly the same. This new unknown compound consisted of one mole of adenine, **two** moles of ribose, and **two** moles of phosphate, and it was digested by alkaline monophosphatase. At the end of 1966, we came to the conclusion that the unknown compound was 2′[5″-phosphoribosyl]-5′-AMP and we sent a manuscript to *Biochimica Biophysica Acta,* which was published in 1967 (4). During the Christmas break, we saw a paper that had just been published in *Biochemical and Biophysical Research Communications* written by P. Chambon, J. D. Weil, J. Doly, M. T. Strosser and P. Mandel under the title of 'On the formation of a novel adenylic compound by enzymatic extracts of liver nuclei' (3). The conclusions were the same as ours (see ref. 4). We had previously presented an interim result to the Annual Meeting of the Japanese Biochemical Society (in 1967) but we had had no direct contact with Professor Mandel, although he had visited Tokyo on the occasion of the 7th International Congress of Biochemistry in 1967. The same conclusion coming quite independently both from Strasbourg and Tokyo was just a coincidence. It was a difficult problem in both laboratories to determine the exact structural chemistry of the ribose–ribose bond; however, this was eventually established in the paper by J. Doly and M. F. Petek from Paul Mandel's laboratory (5). Meanwhile, Professor Osamu Hayaishi's research group in Kyoto was studying NAD biosynthesis starting from L-tryptophan. They were also trying to determine the structure of the unknown compound, since they were aware of the presence of this compound from our presentation at the Japanese Biochemical Society Meeting. Their paper appeared in the *Journal of Biological Chemistry* in 1967 (6). Since these three laboratories published the structure of poly (ADP-ribose) at almost the same time, the presence of this unique and novel biopolymer received citizenship in the scientific community, even though it was claimed by some that the polymer was an *in vitro* artefact. Later, Professor Shall proved the presence of poly (ADP-ribose) polymerase (PARP) activity in slime mould (*Physarum polycephalum*) nuclei (7) and extended the research field to eukaryotes in general. The presence of an antibody against poly (ADP-ribose) in the sera of patients with systemic lupus erythematosus (8) suggested to us the natural occurrence of the polymer in humans. The enzyme, poly (ADP-ribose) glycohydrolase that cleaves the ribose–ribose bonds was then discovered (9) and the branched structure of the poly (ADP-ribose) was also established (10). The discovery of the inhibitor (3-aminobenzamide) against PARP (11) and the sensitization of cells to DNA damaging agents by this inhibitor (12), shed light on understanding the involvement of poly (ADP-ribose) in DNA repair. The monoclonal antibody 10H against poly (ADP-ribose) was raised in our laboratory (13) and has been widely used for the study of poly (ADP-ribose). This hybridoma is now available from RIKEN Cell Bank, Tsukuba, Japan.

The first international meeting on poly (ADP-ribose) was held in Hamburg, organized by Professor Helmuth Hilz in 1972. International meetings have been held periodically since then, in the style of club meetings: organized in Bethesda, USA, by Osamu Hayaishi and myself in 1973; in Tomakomai, Hokkaido, Japan, by myself in 1974 – that meeting, in such an exotic and beautiful place, was attended by Paul Mandel (Strasbourg, France), Osamu Hayaishi (Kyoto, Japan), Yasutomi Nishizuka (Kobe, Japan), Helmuth Hilz (Hamburg, Germany), Sydney Shall (Sussex, England), Mark Smulson (Washington, DC, USA), and D. Michael Gill (Boston, USA). In 1976, Helmuth Hilz organized a meeting in Blankenese, in Hamburg, Germany; and Mark Smulson and myself organized one in Bethesda, USA, in 1979. In Tokyo during 1982, there was a meeting organized by Masanao Miwa, Osamu Hayaishi, Sydney Shall, and myself. Then Felix R. Althaus, Helmuth Hilz, and Sydney Shall arranged a meeting in Vitznau, Switzerland, in 1984. Myron K. Jacobson and Elaine L. Jacobson held a meeting in Dallas, Texas, USA, in 1987. A meeting was held in Quebec, Canada, arranged by Guy G. Poirier and Pierre Moreau in 1990. Masanao Miwa arranged the next meeting in Tsukuba, Japan, in 1992. I still clearly remember the voice of Professor Paul Mandel, who was ill and could not join us at the Quebec meeting, being transferred through transatlantic telephone to the meeting hall in Quebec. He passed away in 1992 – we had lost a father of poly (ADP-ribose). The last two international meetings were in 1994 in Strasbourg, France, organized by Gilbert de Murcia and colleagues, and in 1997 in Cancun, Mexico, arranged by Rafael Alvarez-Gonzalez.

The cloning of the *PARP* gene was done energetically by several laboratories and now PARP knockout animals have become available. The second and third generations of scientists working on poly (ADP-ribose) are making much rapid progress these days. This monograph was prepared by Gilbert de Murcia, a flag-carrier representing the third generation, and by Sydney Shall who has witnessed the entire course of the progress through the first to third generations.

I am sure that this volume will become a most important milestone in biology and medicine, that we can now already begin to imagine developing in the near future.

I wish to express my admiration to the Editors, Contributors, and Publisher for their efforts.

References

1. Chambon, P., Weil, J. D., and Mandel, P. (1963). Nicotinamide mononucleotide activation of a new DNA-dependent polyadenylic acid synthesizing nuclear enzyme. *Biochem. Biophys. Res. Commun.*, **11**, 39.
2. Schmidt, G. (1964). Metabolism of nucleic acids. *Annu. Rev. Biochem.*, **33**, 667.

3. Chambon, P., Weil, J. D., Doly, J., Strosser, M. T., and Mandel, P. (1966). On the formation of a novel adenylic compound by enzymatic extracts of liver nuclei. *Biochem. Biophys. Res. Commun.*, **25**, 638.

4. Sugimura, T., Fujimura, S., Hasegawa, S., and Kawamura, Y. (1967). Polymerization of the adenosine 5′-diphosphate ribose moiety of NAD by rat liver nuclear enzyme. *Biochim. Biophys. Acta*, **138**, 438.

5. Doly, J. and Petek, M. F. (1966). Etude de la structure d'un composé (poly, ADP ribose) synthétisé par des extraits nucléaires de foie de poulet. *Compt. Rend.*, **263**, 1341.

6. Nishizuka, Y., Ueda, K., Nakazawa, K., and Hayaishi, O. (1967). Studies on the polymer of adenosine diphosphate ribose. I. Enzymic formation from nicotinamide adenine dinucleotide in mammalian nuclei. *J. Biol. Chem.*, **242**, 3164.

7. Brightwell, M. and Shall, S. (1971). Poly(adenosine diphosphate ribose) polymerase in *Physarum polycephalum* nuclei. *Biochem. J.*, **125**, 67.

8. Kanai, Y., Kawaminami, Y., Miwa, M., Matsushima, T., Sugimura, T., Moroi, Y., and Yokohari, R. (1977). Naturally-occurring antibodies to poly (ADP-ribose) in patients with systemic lupus erythematosus. *Nature*, **265**, 175.

9. Miwa, M. and Sugimura, T. (1971). Splitting of the ribose–ribose linkage of poly (adenosine diphosphate-ribose) by a calf thymus extract. *J. Biol. Chem.*, **246**, 6362.

10. Miwa, M., Saikawa, N., Yamaizumi, Z., Nishimura, S., and Sugimura, T. (1979). Structure of poly (adenosine diphosphate ribose): Identification of 2′-[1″-ribosyl-2″-(or 3″-)(1‴-ribosyl)]adenosine-5′,5″,5‴-tris (phosphate), as a branch linkage. *Proc. Natl Acad. Sci., USA*, **76**, 595.

11. Purnell, M. R. and Whish, J. D. (1980). Novel inhibitors of poly (ADP-ribose) synthetase. *Biochem. J.*, **185**, 775.

12. Durkacz, B. W., Omidiji, O., Gray, D. A., and Shall, S. (1980). (ADP-ribose)n participates in DNA excision repair. *Nature*, **283**, 593.

13. Kawamitsu, H., Hoshino, H., Okada, H., Miwa, M., Momoi, H., and Sugimura, T. (1984). Monoclonal antibodies to poly (adenosine diphosphate ribose) recognize different structures. *Biochemistry*, **23**, 3771.

Preface

It is now 13 years since Felix R. Althaus and Christof Richter published the last major monograph (*ADP-ribosylation of proteins*, Springer-Verlag, Berlin, 1987) on poly (ADP-ribosylation). In the last decade there have been major changes in this area, as in many other fields of biology, due to the application of 'reverse genetics', that is molecular biology. The subject as outlined in this book is very different indeed from that of 13 years ago. Some outstanding issues have been settled; for example, a clear and uncontroversial picture of the molecular properties of the enzymes has been established.

The subject of poly (ADP-ribosyl)ation has suddenly become of tremendous interest to the pharmaceutical industry, because of the important discoveries that DNA damage and DNA repair are not confined to a few relatively uncommon genetic diseases. It has emerged, and this is perhaps the most surprising and possibly the most significant new development in the last decade, that many different human diseases involve damage to DNA as a crucial element in their pathogenesis. It is now clear that in humans, cancer, ageing, autoimmune disease, cerebrovascular disease, and coronary artery disease, involve significant damage to DNA. Consequently, the study of DNA damage and its repair has a wider significance than was previously appreciated. Molecular pathologists in all these disciplines will have to become intimately acquainted with the processes of DNA damage and repair. Poly (ADP-ribose) polymerase is only involved in some pathways of DNA repair, although it seems that these pathways recur in many human diseases. This may partly be caused by oxidative damage in whose repair poly (ADP-ribose) polymerase is much involved.

Another novel aspect that may turn out to be of considerable and broad significance is the discovery of novel PARP homologues—tankyrase (PARP-5) and VPARP (PARP-4)—which are components of the processes that maintain the ends of chromosomes (telomeres) intact. There is considerable evidence that telomeres are significant in cancer development, in the life span of cells in culture, and possibly in the life span of humans.

In this book, we have endeavoured to concentrate on recent results, but at the same time to relate current developments to the older data. Divided into seven chapters, this monograph presents a broad perspective of poly (ADP-ribosylation) reactions, with particular emphasis on new areas which are expected to expand in the near future. We have tried to avoid repetition of material in different chapters. However, where authors have approached a common subject from different perspectives, or different aspects are being discussed, we have occasionally allowed the same material to be discussed in more than one chapter. The text refers to articles and reviews published until the end of November 1999.

We would like to thank the authors for their dedicated efforts to review a rapidly growing area of research, and for their patience with the editors. We are grateful to each of them for their contribution.

We wish to acknowledge people from our own laboratories; in particular, Frédéric Simonin, Gérard Gradwohl, Carlotta Trucco, Murielle Masson, Miguel Molinete, Everson Alves Miranda, Sabine Jung, Guadalupe de la Rubia, Bruno Rinaldi, Aline Huber and Eric Flatter (de Murcia laboratory) and Farzin Farzaneh, Joop Gaken, and Said Aoufouchi (Shall laboratory) for their fine contribution to various aspects of this project. We also wish to thank Shanti Natarajan and Chia Miao Mah-Becherel for their generous assistance.

Strasbourg   G. de M.
March 2000   S. S.

Contents

Contributors

EL BACHIR AFFAR
Health and Environment Unit
Laval University Medical Research
  Center, CHUQ,
and Faculty of Medecine,
Laval University
Québec
Canada

JEAN-CHRISTOPHE AMÉ
Division of Pharmaceutical Sciences
College of Pharmacy
University of Kentucky
Lexington
Kentucky 40536
USA

ANGÉLIQUE AUGUSTIN
Unité 9003 CNRS
Laboratoire conventionné avec le
  Commissariat à l'Energie Atomique
École Supérieure de Biotechnologie de
  Strasbourg
Boulevard Sébastien Brant
F-67400 Illkirch-Graffenstaden
France

ALEXANDER BÜRKLE
Deutsches Krebsforschungzentrum
Heidelberg
Germany

NICOLA J. CURTIN
Cancer Research Unit
University of Newcastle upon Tyne
Medical School
Framlington Place
Newcastle upon Tyne NE2 4HH
UK

FRANÇOISE DANTZER
Unité 9003 CNRS
Laboratoire conventionné avec le
  Commissariat à l'Energie Atomique
École Supérieure de Biotechnologie de
  Strasbourg
Boulevard Sébastien Brant
F-67400 Illkirch-Graffenstaden
France

GILBERT DE MURCIA
Unité 9003 CNRS
Laboratoire conventionné avec le
  Commissariat à l'Energie Atomique
École Supérieure de Biotechnologie de
  Strasbourg
Boulevard Sébastien Brant
F-67400 Illkirch-Graffenstaden
France

MARC GERMAIN
Health and Environment Unit
Laval University Medical Research
  Center, CHUQ
and Faculty of Medecine
Laval University
Québec
Canada

BERNARD T. GOLDING
Department of Chemistry
Bedson Building
University of Newcastle upon Tyne
Newcastle upon Tyne NE1 7RU
UK

ROGER J. GRIFFIN
Department of Chemistry
Bedson Building
University of Newcastle upon Tyne
Newcastle upon Tyne NE1 7RU
UK

ELAINE L. JACOBSON
Department of Clinical Sciences/Lucille
   P. Markey Cancer Center/
   Multidisciplinary PhD Program in
   Nutritional Sciences
University of Kentucky
Lexington
Kentucky 40536
USA

MYRON K. JACOBSON
Department of Clinical Sciences/Lucille
   P. Markey Cancer Center
University of Kentucky
Lexington
Kentucky 40536
USA

JOSIANE MÉNISSIER-DE MURCIA
Unité 9003 CNRS
Laboratoire conventionné avec le
   Commissariat à l'Energie Atomique
École Supérieure de Biotechnologie de
   Strasbourg
Boulevard Sébastien Brant
F-67400 Illkirch-Graffenstaden
France

DAVID R. NEWELL
Cancer Research Unit
University of Newcastle upon Tyne
   Medical School
Framlington Place
Newcastle upon Tyne NE2 4HH
UK

CLAUDE NIEDERGANG
Unité 9003 CNRS
Laboratoire conventionné avec le
   Commissariat à l'Energie Atomique
École Supérieure de Biotechnologie de
   Strasbourg
Boulevard Sébastien Brant
F-67400 Illkirch-Graffenstaden
France

F. JAVIER OLIVER
Unidad Mixta de Investigaciones
   Médicas
Hospital Clínico San Cecilio
Universidad de Granada
Granada
Spain

GUY G. POIRIER
Health and Environment Unit
Laval University Medical Research
   Center, CHUQ
and Faculty of Medecine
Laval University
Québec
Canada

MICHAEL J. ROBERTS
Cancer Research Unit
University of Newcastle upon Tyne
Medical School
Framlington Place
Newcastle upon Tyne NE2 4HH
UK

VÉRONIQUE ROLLI
Max-Planck-Institut für Immunbiologie
Stübeweg
D-79104 Freiburg im Breisgau
Germany

ARMIN RUF
ZHF/G A30 BASF-AG
D-67157 Ludwigshafen
Germany

VALERIE SCHREIBER
Unité 9003 CNRS
Laboratoire conventionné avec le
   Commissariat à l'Energie Atomique
École Supérieure de Biotechnologie de
   Strasbourg
Boulevard Sébastien Brant
F-67400 Illkirch-Graffenstaden
France

GEORG E. SCHULZ
Institut für Organische Chemie und
    Biochemie
Albertstrasse 21
D-79104 Freiburg im Breisgau
Germany

SYDNEY SHALL
Department of Molecular Medicine
King's College School of Medicine and
    Dentistry
The Rayne Institute
123 Coldharbour Lane
London SE5 9NU
UK

SHEILA SRINIVASAN
Department of Chemistry
Bedson Building
University of Newcastle upon Tyne
Newcastle upon Tyne NE1 7RU
UK

TAKASHI SUGIMURA
President Emeritus, National Cancer
    Center
5-1-1 Tsukiji
Chuo-ku
Tokyo 104-0045
Japan

CSABA SZABÓ
Inotek Corporation
100 Cummings Center
Beverly MA 01915
USA

ALEX W. WHITE
Department of Chemistry
Bedson Building
University of Newcastle upon Tyne
Newcastle upon Tyne NE1 7RU
UK

Abbreviations

(PR)$_2$-AMP	diphosphoribosyl-AMP
3-AB	3-aminobenzamide
3-AVA	3-aminobenzoic acid
3-MB	3-methoxybenzamide
AP	apurinic/apyrimidic
APP	*Arabidopsis thaliana* homologue of *PARP*
ART	ADP-ribosyl-transferase
AT	ataxia–telangiectsia
ATM	gene responsible for ataxia–telangiectasis (ataxia–telangiectasia mutated)
ATR	AT-related protein
BCNU	1,3-bis(2-chloroethyl)-1-nitrosourea (Carmustine)
BER	base excision repair
bNOS	constitutive, neurological NOS
bp	base pair
BRCT	*br*east *ca*ncer susceptibility protein, BRCA1, *C-t*erminus
CAD	caspase-activated DNase; carbamoyl phosphate synthetase/aspartate transcarbamylase/dihydro-orotase
carba-NAD	carbanicotinamide adenine nucleotide
CBP	CRE-binding protein
CCNU	1-(2-chloroethyl-3-cyclohexyl)-1-nitrosourea (Lomustine)
Cdk	cyclin-dependent kinase
CHO	Chinese hamster ovary (also, an aldehyde group)
CRE	cyclic AMP response-element
DBD	DNA binding domain
DEVD-CHO	Asp–Glu–Val–Asp–aldehyde
DHQ	1,5-dihydroxyisoquinoline

DNA-PK	DNA-dependent protein kinase
DNA-PKcs	DNA-PK catalytic subunit
ds	double strand
DTIC	5-(3,3-dimethyl-1-triazeno)-imidazole-4-carboxamide (Dacarbazine)
ecNOS	constitutive, endothelial NOS
EM	electron microscopy
FITC	fluorescein isothiocyanate
GR	glucocorticoid receptor
GST	glutathione-S-transferase
HIV	human immunodeficiency virus
HMG	high mobility group (HMG)
HN_2	nitrogen mustard
IC_{50}	concentration required for 50% inhibition (of activity, cell growth, etc.)
ICAD	inhibitor of CAD
ICAM-1	intercellular adhesion molecule-1
ICE	interleukin-1β-converting enzyme (caspase-1)
IFN-γ	interferon-γ
IL-1	interleukin-1
IN	integrase
INH2BP	5-iodo-6-amino-1,2-benzopyrone
iNOS	inducible NOS
kDa	kilodalton
K_i	inhibition constant
KO	knockout
LC_{50}	concentration required for 50% inhibition of survival
LDH	lactate dehydrogenase
LMG	low mobility group
LPS	(bacterial) lipopolysaccharide

LTR	long terminal repeat
m-AMSA	amsacrine or 4′(9-acridimylamino) methanesulfon-*m*-anisidine
MDM2	mouse double minute 2
MEF	mouse embryonic fibroblast
MMS	methyl methanesulfonate
MMTV	mouse mammary tumour virus
MNNG	1-methyl-3-nitro-1-nitroguanidine
MNU	N-methyl-N-nitrosourea
MPTP	1-methyl-4-phenyl-1,2,3,6-tetrahydropyridine
MRC	multiprotein DNA replication complex
MTIC	5-methyltriazenoimidazole-4-carboxamide
MVP	major Vault particle
NA	nicotinamide
NAD	nicotinamide adenine dinucleotide
NLS	nuclear location (or localization) signal
NMDA	N-methyl-D-aspartate
NMR	nuclear magnetic resonance
NOS	nitric oxide synthase
PADP-R	poly(ADP-ribose)
pADPr	poly(ADP-ribose)
PALA	N-phosphonoacetyl-L-aspartic acid
PARG	poly(ADP-ribose) glycohydrolase
PARP	poly(ADP-ribose) polymerase
PARP-CF	poly(ADP-ribose) polymerase catalytic fragment
PARS	poly(ADP-ribose) synthetase
PCNA	proliferating cell nuclear antigen
PI	phosphatidylinositol
PR-AMP	phosphoribosyl AMP
RB	retinoblastoma

ROI	reactive oxygen intermediates
SCE	sister-chromatid exchange
SCID	severe combined immunodeficiency
SIN-1	3-morpholino-sydnonimine
SLE	systemic lupus erythematosus
SNAP	S-nitroso-N-DL-penicillamine
ss	single strand
TBP	TATA-binding protein
TEF	transcription enhancer factor
TGF	transforming growth factor
TNF	tumour necrosis factor
UDG	uracil DNA glycosylase
VPARP	Vault PARP
XO	xanthine oxidase
XRCC1	X-ray repair cross-complementing factor-1
YY	Yin Yang

1

ADP-ribose polymer metabolism

Jean-Christophe Amé, Elaine L. Jacobson and
Myron K. Jacobson

1.1 Introduction to ADP-ribose transfer reactions

Nicotinamide adenine dinucleotide (NAD) plays fundamental roles in both cellular energy metabolism and cellular signalling. In energy metabolism, the chemistry of the pyridine ring allows NAD to readily accept and donate electrons in hydride transfer reactions catalysed by numerous dehydrogenases. Thus NAD is a link between the oxidation of reduced substrates and processes that result in the generation of phosphoanhydride linkages in ATP required for many energy-consuming processes. The involvement of NAD in cellular signalling is based on the high group transfer potential of the glycosylic linkage between nicotinamide and ribose, which allows the oxidized form of NAD to function as a donor of ADP-ribose to specific cellular nucleophilic acceptors.

Figure 1.1 contrasts these two different metabolic uses of NAD. In energy metabolism, NAD is rapidly interconverted between its oxidized and reduced

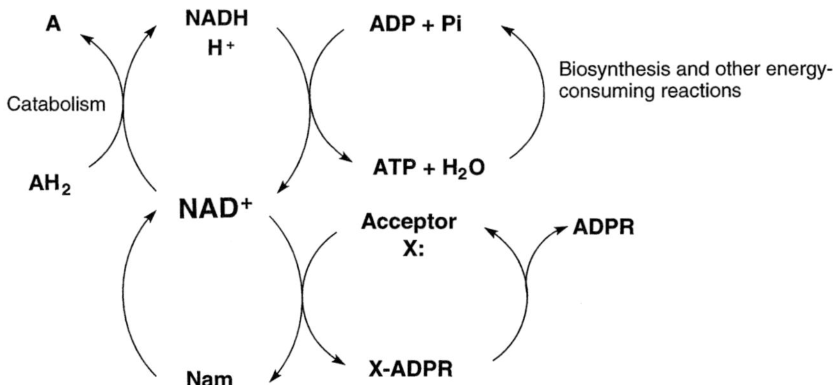

Fig. 1.1 Two distinct roles for NAD in cellular metabolism due to its involvement in hydride transfer and ADP-ribose transfer reactions. The symbol X: is used to represent an acceptor nucleophile, A is used to represent substrates of cellular dehydrogenases, Nam represents nicotinamide and ADPR represents ADP-ribose.

forms, but there is no net consumption of the nucleotide. In contrast, the utilization of NAD in ADP-ribose transfer reactions results in the release of the ADP-ribose moiety of NAD, requiring its resynthesis to avoid depletion of the NAD pool of the cell. The involvement of NAD in both hydride transfer and ADP-ribose transfer reactions provides a potentially important link that can interrelate cellular energy metabolism and cellular signalling. Since ADP-ribose transfer reactions specifically use the oxidized form of NAD, cellular signalling may be linked to the redox state of the cellular NAD pool, allowing cellular signalling in response to changes in energy status. In turn, ADP-ribose transfer reactions have the potential to alter the overall energy status of the cell via depletion of the NAD pool to the point where the NAD content of the cell could become limiting for the synthesis of ATP, which could provide a mechanism by which cellular signalling leads to cellular necrosis or apoptosis.

Although our current understanding of the function of ADP-ribose transfer reactions is incomplete, at least three different classes have been shown to be involved in cellular signalling as depicted in Fig. 1.2. Protein mono (ADP-ribosyl) transferases transfer single ADP-ribose moieties to acceptor proteins. The most widely studied mono (ADP-ribosyl) transferases are bacterial toxins that subvert the metabolism of animal cells by modifying specific target proteins. However, it is now clear that animal cells contain numerous endogenous transferases of this category that function in cell signalling (1). NAD glycohydrolases are involved in the metabolism of cyclic ADP-ribose, a potent calcium mobilizing agent that appears to function as an endogenous second messenger by moduling membrane calcium channels (2). A third class of ADP-ribose transfer reactions is readily distinguished from the other classes because they involve the generation of polymers of ADP-ribose. The metabolism of polymers of ADP-ribose is the focus of this monograph.

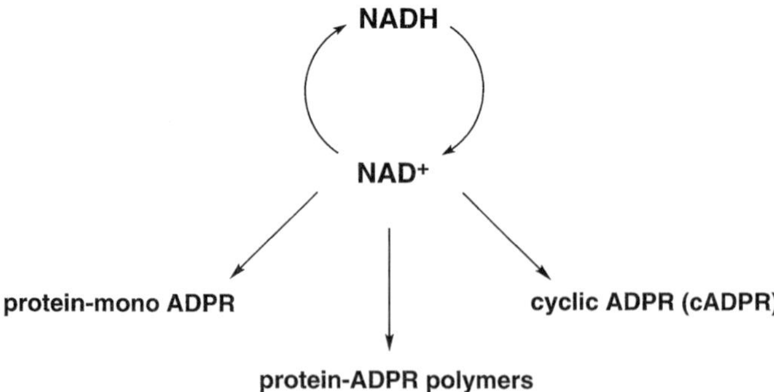

Fig. 1.2 Utilization of the oxidized form of NAD as the donor of ADP-ribose by three different classes of ADP-ribosyl transferases, resulting in the modification of cellular proteins by monomers of ADP-ribose, polymers of ADP-ribose, and the generation of cyclic ADP-ribose.

1.2 Poly (ADP-ribose)

1.2.1 Introduction—historical background

The first report on poly (ADP-ribose) (pADPR) formation was published by P. Chambon, J. Weil, and P. Mandel in 1963 (3). They found that the incorporation of [^{14}C]adenine-labelled ATP into an acid-insoluble fraction of a nuclear preparation from chicken liver was enhanced 1000-fold by nicotinamide mononucleotide (NMN). In 1966, Mandel's group postulated the structure of pADPR and the major hydrolysis product, named φ-ADP-ribose (4) where the site of linkage between adjacent ADPR moieties was described (5). During the same period, studies on pADPR were performed by Sugimura and co-workers (6) and Hayaishi's group (7). using a nuclear preparation from rat liver, they confirmed the observation made previously that NMN greatly enhanced the formation of acid-insoluble material from [^{14}C-adenine]ATP. The reaction product was hydrolysed by snake venom phosphodiesterase. However, it was resistant to alkali, which should have hydrolysed poly(A) to adenosine 3′-monophosphate (6,8,9). Furthermore, the chromatographic mobility of the product formed by snake venom phosphodiesterase was different from that of 5′-AMP but similar to that of 5′-ADP. The major hydrolysis product contained one mole of adenine, two moles of ribose, and two moles of phosphate per mole of nucleotide. The phosphates were removed by alkaline phosphomonoesterase, indicating that they were not involved in a phosphodiester or pyrophosphate bond. The major hydrolysis product was deduced to be 2′ (or 3′)-(5′-phosphoribosyl)-5′-AMP, which was named 'phosphoribosyl AMP (PR-AMP)' (10) and this product corresponded to φ-ADP-ribose (4). From these findings, it became clear that NAD was formed from NMN and ATP by NAD pyrophosphorylase in nuclei, and that the ADP-ribose moiety of NAD was then converted to poly (ADP-ribose) with the concomitant release of nicotinamide.

1.2.2 Structure of poly (ADP-ribose)

Figure 1.3 shows the basic structural features of poly (ADP-ribose) and the location of attack by enzymes that can utilize the polymers as substrates. The size of poly (ADP-ribose) synthesized both *in vitro* and *in vivo* ranges from a few to more than 200 residues (11, 12). In contrast to DNA or RNA, this polymer has a ribose (1″ → 2′) ribose–phosphate–phosphate backbone (Fig. 1.3). The anomeric carbon of one ADP-ribose moiety was found to be bound to the adenosine moiety of the adjacent one via a 1″ → 2′ glycosidic linkage, which on subsequent nuclear magnetic resonance (NMR) analysis (13–15) was identified as an alpha anomeric linkage.

Miwa *et al.* first reported that the molecular weight of isolated polymers

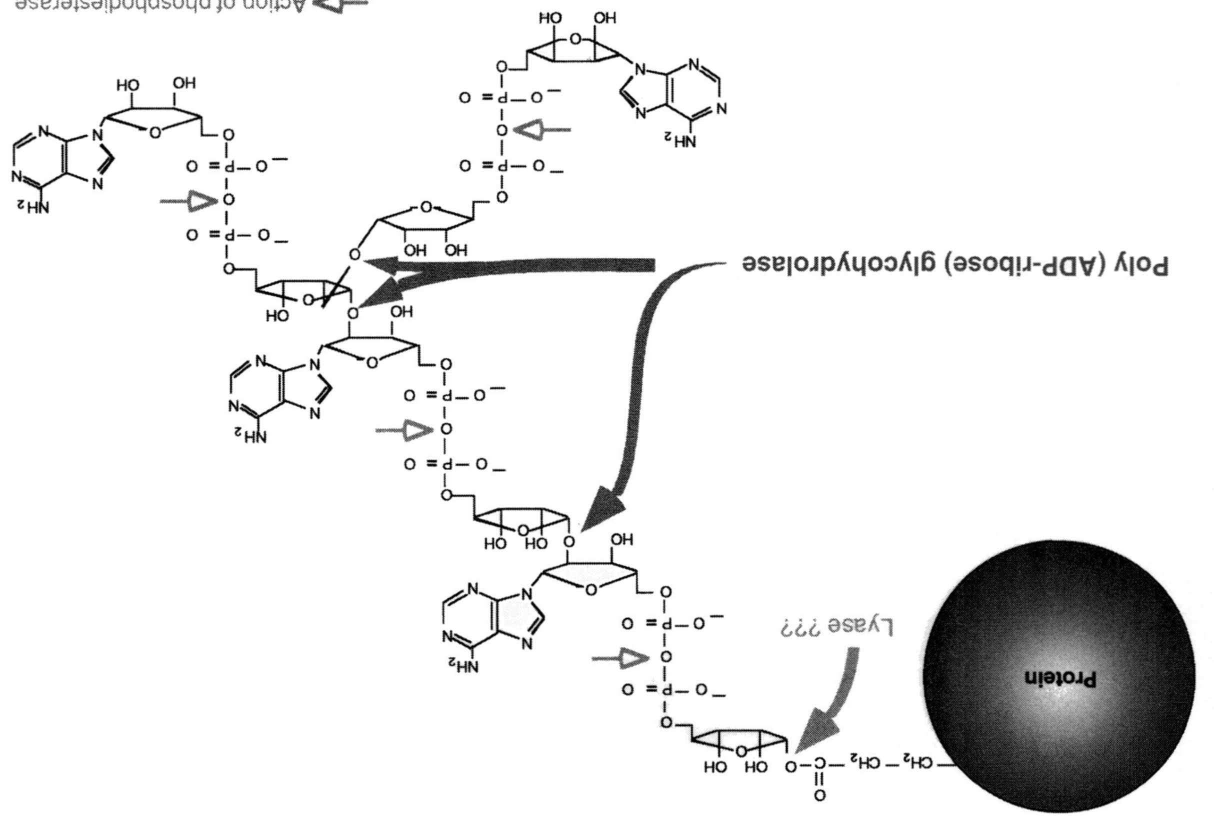

Fig. 1.3 Structure of poly (ADP-ribose).

calculated from the ratio of PR-AMP to 5′-AMP was not consistent with the sedimentation coefficient value determined by centrifugation (16). This suggested aggregation of poly (ADP-ribose) chains or a branched structure, as in the case of polysaccharides. Identification of the nucleotide, *O*-α-D-ribofuranosyl-(1‴ → 2″)-*O*-α-D-ribofuranosyl-(1″ → 2′)-adenosine-5′, 5″, 5‴-tris(phosphate), abbreviated as (PR)$_2$-AMP, in phosphodiesterase digests of polymers confirmed the presence of branching in the polymer (17–19). The content of branching residues was estimated to range from 2% to 3% of the total ADP-ribose residues. Thus, more than 6.6 branching points have been identified in purified polymers of an average size of approximately 250 residues (11), and the number of branches per polymer increases with polymer size (20).

Polymer size can be estimated by determining the relative amount of nucleotides released after digestion with snake venom phosphodiesterase as described by Miwa and Sugimura (21). They established the following relationships (22):

$$\text{Polymer size} = \frac{[\text{AMP}] + [\text{PRAMP}] + [(\text{PR})_2\text{AMP}]}{[\text{AMP}] - [(\text{PR})_2\text{AMP}]}.$$

The number of points of branching can be estimated from the following equation:

$$\text{Average number of branching points} = \frac{[(\text{PR})_2\text{AMP})]}{[\text{AMP}] - [(\text{PR})_2\text{AMP}]}.$$

1.2.3 Detection and quantification of poly (ADP-ribose)

For a long time the lack of accurate methods for the detection of poly (ADP-ribose) in biological material was a major obstacle to understanding the functions of ADP-ribose polymer metabolism. All methods used to determine ADP-ribose polymers in intact cells and tissues must be very sensitive and selective, since the content of ADP-ribose polymers is very low relative to other adenine-containing nuclear polymers such as DNA and RNA. While the quantification of ADP-ribose polymers is still not easy, a number of chemical and immunological methods are now available for studying ADP-ribose polymer metabolism.

Chromatographic methods

Major progress towards the determination of ADP-ribose polymer content in cultured cells and animal tissues was made when immobilized dihydroxyboronyl groups for affinity adsorption of ADP-ribose polymers were introduced (23–30). These resins have been used in combination with both chemical and immunological detection methods to quantify ADP-ribose polymer residues *in vivo* in animal tissues and cultured cells. For example, one approach allowing quantification of total polymer residues is based on the conversion of boronate affinity-purified polymers to nucleosides, followed by conversion of the nucleosides to highly fluorescent 1, N^6-etheno derivatives (31) which can be separated by reversed-phase HPLC and quantified at the picomole level using a fluorimeter (30). More recently, the use of [^{3}H]adenine as a radioactive precursor to NAD has further simplified polymer measurements in cultured cells (32). Additionally, the combination of boronate chromatography with high-resolution anion-exchange HPLC has allowed the preparation and separation of individual species of ADP-ribose oligomers up to the 50-mer and multibranched polymers (20). Another chromatographic method involving two-dimensional thin-layer chromatography on cellulose plates (33) has been used to separate and quantify the three diagnostic nucleotides of poly (ADP-ribose) generated by phosphodiesterase digestion: AMP, PRAMP, and (PR)$_2$AMP.

Immunodetection of poly (ADP-ribose)

Four types of antibodies have been prepared and applied to detect ADP-ribose polymers in cultured cells and tissues. Anti-poly (ADP-ribose) sera (34–37), anti-PR-AMP sera (38), antibodies specific for 5′-AMP (39, 40), and antibodies against an analogue of ADP-ribose (41) have been generated. These antisera have permitted the detection and the quantitative estimation of poly (ADP-ribose) by a variety of different approaches.

The first specific antibodies against poly (ADP-ribose) were raised in rabbits by injecting a mixture of poly (ADP-ribose) and methylated bovine serum albumin (34). The antisera obtained did not bind poly(A) or other related nucleotides, nor yeast RNA or calf thymus DNA, but bound poly (ADP-ribose) and, to a lesser degree, ADP-ribose and PR-AMP. However, a cross-reaction with double-stranded RNA, poly(A)–poly(U), or poly(I)–poly(C) duplexes was observed (42, 43). Furthermore, the reactivity of these antibodies against poly (ADP-ribose) was found to be dependent on polymer size, binding smaller polymers less efficiently than larger polymers (44). A radioimmunoassay (45) involving this antibody has been used to detect poly (ADP-ribose) *in vivo*, but to be quantitative the method needed a prior fractionation of the sample according to polymer size. Other radioimmunoassays employing various preparations of antibodies against poly (ADP-ribose) have been developed (35, 36, 45) and used to detect poly (ADP-ribose) in cells under various conditions, but again the methods were semi-quantitative.

Since polyclonal antibodies represent a mixture of antibodies with different specificities, they cannot be used to select a specific antigenic structure in poly (ADP-ribose) molecules. Monoclonal antibodies recognize specific antigenic determinants and thus have greater potential for quantitative immunoassays for ADP-ribose polymers. Hybridomas producing monoclonal antibodies to poly (ADP-ribose) were first reported in the early 1980s (37, 46). Characterization of two monoclonal antibody preparations demonstrated that one recognized the linear structure of ADP-ribose polymers, while the second recognized additional structures including the branched portions of the polymers. A limitation of chemical methods for the determination of ADP-ribose polymer content is that they are unsuitable for evaluating the response of individual cells within cell populations in terms of polymer synthesis. Thus, specific antibodies have proved to be very useful for the cytological detection of ADP-ribose polymers within individual cells. Antibodies highly specific for poly (ADP-ribose) can be easily visualized by coupling them to dyes such as fluorescein isothiocyanate (FITC) or by using FITC-labelled anti-IgG antibody. Cross-reactivity of the antibodies with DNA is of particular concern, because DNA is present in nuclei in much higher concentrations than poly (ADP-ribose). Most antibodies selected for immunofluorescence studies do not exhibit cross-reactivity with DNA, RNA, poly(A), ADP-ribose, or NAD when examined by double-immunodiffusion and membrane-binding assays (47). However, since the reactivity of the antibody varies markedly with the chain length of the polymer, the intensity of fluorescence observed in immunohistochemistry is therefore a function of both the amount and size of the polymer. An example of an observation of poly (ADP-ribose) by immunohistochemistry following the treatment of cells with a DNA damaging agent is shown in Plate 1 (37).

Electron microscopy

Among the physical methods used to investigate the structure of poly (ADP-ribose) as well as the effect of poly (ADP-ribosylation) on polynucleosomal ultrastructure, electron microscopy has been very useful (12, 48–52). This technique has allowed the visualization of the branched structure of the polymer (50) as shown in Fig. 1.4. It has also allowed the demonstration that poly (ADP-ribosylation) causes a decondensation of chromatin superstructure (48, 49, 51), which may be required to facilitate DNA repair.

1.3 Biosynthesis of poly (ADP-ribose)

1.3.1 Natural occurrence of poly (ADP-ribosy)lation activity

The natural occurrence of poly (ADP-ribose) is supported by the ubiquitous presence of enzymes that catalyse the synthesis (poly (ADP-ribose) polymerase, PARP) and turnover (poly (ADP-ribose) glycohydrolase, PARG) of

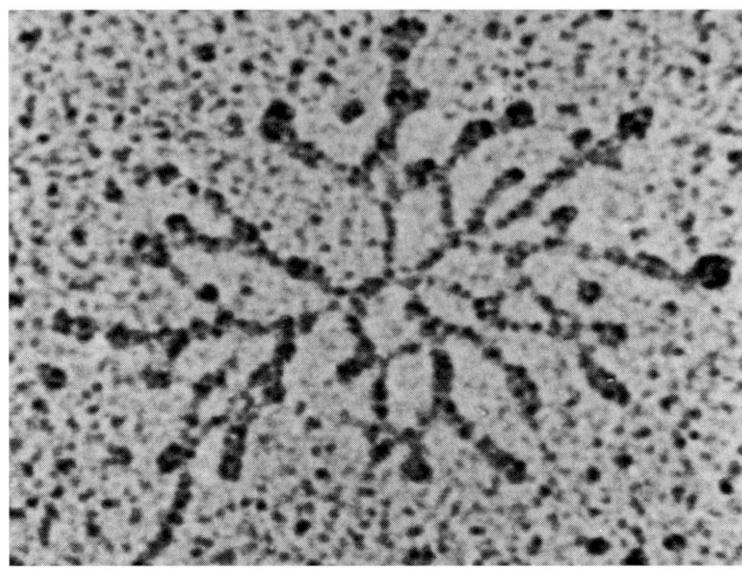

Fig. 1.4 Branched structure of poly (ADP-ribose) observed by electron microscopy (50). (20 000 ×).

the polymer. Immunohistochemical studies as well as chemical methods have demonstrated the presence of poly (ADP-ribose) in all types of nucleated mammalian cells, except granulocytes with segmented nuclei (47, 53–55). PARP activity has also been demonstrated in a number of plants (56) and lower eukaryotes including *Dictyostelium discoideum* (57) and *Cryphecodinium cohnii* (58), and recently in an archeaobacterium *Sulfolobus solfataricus* (59). Although the presence of PARP activity in yeast has been postulated (60, 61), the recent completion of the sequencing of the entire genome of the yeast *Saccharomyces cerevisiae* rules out this possibility as no significant sequence homology could be found between PARP and any protein sequences in the yeast genome.

1.3.2   Purification of poly (ADP-ribose) polymerase

PARP [EC 2.4.2.30] has been purified from a wide variety of organisms and tissues and also from cell cultures. It is a monomeric enzyme with an estimated molecular weight of about 116 kDa by polyacrylamide gel electrophoresis (62). Early purification strategies for PARP included salt extraction and the use of chromatography on DNA–cellulose, DNA–agarose, hydroxyapatite, Blue Sepharose, and nicotinamide–Sepharose. However, most of these methods are time consuming and a more effective method of PARP purification has been developed based on affinity chromatography using 3-aminobenzamide as the ligand (63). Using a column of 3-aminobenza-mide–Sepharose, PARP has been purified in a three-step protocol allowing the isolation of electrophoretically pure enzyme from human placenta (63, 64), fish (65), and *Helix pomatia* (66). Similar protocols have allowed the rapid purification of PARP from bull testis (67) and from calf thymus (68). Cloning the cDNA encoding PARP from different species has provided a new tool to investigate precisely the modular organization of the enzyme. Among the different approaches used to study the functions of the enzyme is the overexpression of the cDNA in a heterologous expression system, from which the recombinant protein can easily be purified. Different reports showed the overexpression and purification of PARP or specific domains of the enzyme in *Escherichia coli* (40-kDa catalytic domain (69), DNA binding domain (70)), or in insect cells using a baculovirus expression system (full-length human PARP (71), chicken PARP catalytic domain (72–74)). Due to high levels of overexpression in these systems the purification procedure could be greatly simplified. All the procedures are based on the use of 3-aminobenzamide affinity chromatography after protamine- and ammonium sulfate-precipitation steps. For the expression in insect cells, this protocol allowed the purification of 70 mg of pure protein per litre of cell culture (72). The purification, in the case of overexpression of *E. coli*, yielded 40 mg of protein from a 3-litre culture, but it required an extra DEAE–Sephacel chromatographic step to reach the same degree of purity (69). The purification of the DNA binding domain of PARP overexpressed in *E. coli* (70) has required a different protocol. The ability of this domain to bind DNA allowed affinity purification using a single-strand DNA–cellulose chromatographic step after hydroxylapatite adsorption chromatography–5 mg of PARP DNA binding domain per litre of bacterial culture was obtained using this procedure.

1.3.3 Enzyme structure and cloning of messenger RNA

Poly (ADP-ribose) polymerase is a multidomain protein. Early work from Shizuta's laboratory demonstrated that PARP can be divided into three functional domains (75–78). Following limited proteolysis of the purified enzyme, they characterized an M_r 54 000 fragment as the substrate binding moiety, an M_r 46 000 fragment as the DNA binding domain, and an M_r 22 000 polypeptide as the site of automodification (Fig. 1.5). The cleavage sites of these proteases have been determined precisely by amino-acid sequencing at the junction of the different domains (79–81).

The complete primary structure of PARP in various organisms was deduced following the cloning of the cDNA corresponding to the protein. This has shown that the three domains of the enzyme have molecular weights, respectively, of 55.4 kDa (NAD binding domain), 16 kDa (automodification

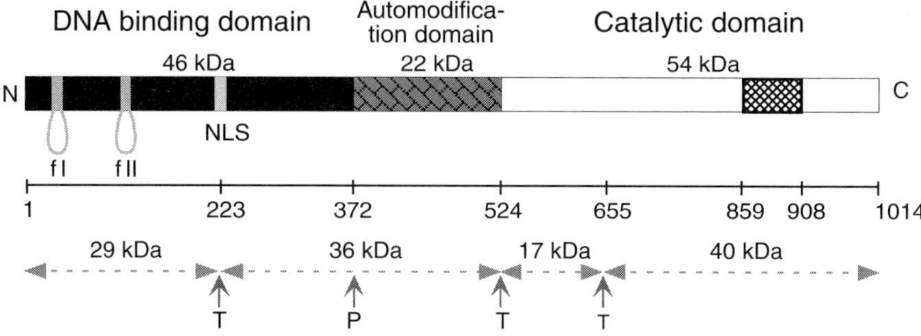

Fig. 1.5 Schematic representation of the three functional domains of the human poly (ADP-ribose) polymerase obtained after mild trypsin (T) and papain (P) digestion, according to Kameshita *et al.* (76).

domain), and 42.5 kDa (DNA binding domain) (79). The human cDNA (79, 80, 82–84) and bovine cDNA (81, 85) were initially obtained and, since then, the human cDNA has been used as a probe to isolate total or partial cDNA of PARP from rat (86), mouse (87), chicken (88), *Xenopus* (89,90), *Drosophila* (81), *Sarcophaga peregrina* (82), and the plants *Arabidopsis thaliana* (93) and *Zea mays* (94, 95).

The PARP mRNA has an open reading frame of 3042 nucleotides and codes for a protein of 113 135 Da made up of 1014 amino acids. The 5′ untranslated region of human and mouse mRNA (82) show sequences flanking the first ATG that match the sequences identified by Kozak (96) for initiation regions. Low homology has been found in this region between the two organisms, although they are both rich in guanine and cytosine (74% in human). The primary structure of PARP shows great homology at the amino-acid level between different species. Human and mouse enzymes show 92% homology (87), whereas rat (86) and bovine sequences (81, 85), respectively, show 97% and 98% homology with the human sequence. Most amino-acid changes are conservative. Thus, the enzyme structure has been highly conserved throughout evolution. Furthermore, the conserved residues are clustered into blocks of strong conservation that correspond precisely to the PARP functional domains and subdomains: the zinc fingers FI and FII, involved in DNA strand-break recognition (97, 98), the bipartite nuclear location signal (NLS) (99), and the catalytic region (69, 100, 101). The carboxy-terminal region of PARP is by far the most conserved part of the protein. The catalytic domain, in particular, contains a block of 50 amino acids (residues 859–908) that are strictly conserved among vertebrates (100%) and highly conserved among all species (92%). This region has been termed the 'PARP signature sequence'.

1.3.4 Enzymological properties of PARP

The basic scheme of the poly (ADP-ribose) polymerase reaction is as follows:

$$\text{Acceptor} + n\text{NAD} \xrightarrow[\text{poly (ADP – ribose) polymerase}]{\text{DNA}}$$
$$\text{Acceptor} - (\text{ADP} - \text{ribose})n + n\text{Nicotinamide} + n\text{H}^+.$$

Whereas numerous DNA binding proteins can serve as covalent acceptors *in vitro* (102–106), the primary acceptor *in vitro* (107,108) and *in vivo* (109, 110) is PARP itself. The automodification of PARP has been reported in bovine thymus (107), rat liver (108), calf thymus (111), and HeLa cells (112). This region of automodification contains between 15 (108) and 28 (113) amino-acid acceptor sites. Not surprisingly, this massive hyper-poly (ADP-ribosyla-tion) reaction leads to the catalytic inactivation of PARP as a result of electrostatic repulsion between the negatively charged enzyme-bound ADP-ribose polymers and DNA (111, 114).

1.3.5 Multiple enzyme activities of PARP

The carboxy-terminal region of PARP encompasses the 54-kDa catalytic domain bearing the NAD-binding pocket (Fig. 1.5). This highly conserved module is responsible for each of the following catalytic activities associated with the full-length enzyme:

- *initiation*, the attachment of an ADP-ribosyl moiety to an acceptor protein (115);
- *elongation*, whereby additional ADP-ribose moieties are attached to protein-bound ADP-ribosyl residues (116);
- *branching*, the introduction of an ADP-ribose residue via a linkage that initiates a branch along the linear portion of the polymer (17);
- *abortive NADase*, the cleavage of NAD into nicotinamide and free ADP-ribose.

While the initiation reaction occurs via a quasi-distributive mechanism (108, 117), ADP-ribose polymerization (elongation and branching reactions) appears to take place via a highly processive mechanism (118). The relative activities of initiation, linear elongation, and branching will determine the polymer size and complexity under specific conditions. It has been estimated that for every initiation step, the enzyme is capable of catalysing over 200 linear-elongation and five- to seven-branching reactions (11, 20). Thus, under many conditions, ADP-ribose chain elongation is the main catalytic function of the enzyme.

The elongation of ADP-ribose chains has been shown to occur via a pro-tein-distal elongation mechanism (119–121), where the 2′-hydroxyl group of the non-reducing end of the growing ADP-ribose is utilized as the acceptor for the covalent attachment of the newly added ADP-ribose unit. This mechanism predicts that the automodification reaction of PARP is bimolecular (intermolecular) in nature and that initial rates of auto-poly (ADP-ribosylation) should increase with the square of the enzyme concentration. In agreement with this hypothesis, studies have shown that the initial rates of automodification increase with second-order kinetics as a function of the PARP concentration at saturating concentrations of DNA and $MgCl_2$ (117). In fact, second-order kinetics were observed for both the initiation (at nM [NAD]) and elongation (at μM [NAD]) reactions (117). These results are consistent with the conclusion that PARP performs the automodification reaction as a catalytic dimer (117). Furthermore, initial rates of the automodification reaction were also observed to increase with second-order kinetics as a function of the concentration of NAD at nanomolar levels (117), suggesting that at least one molecule of NAD must be bound to each PARP molecule for efficient automodification to occur.

Experimental evidence supporting self-association of PARP was also recently obtained by molecular-sieve chromatography, non-denaturating polyacryloamide gel electrophoresis, and chemical cross-linking experiments (122). Furthermore, it has been shown that PARP from *Drosophila* contains a possible leucine zipper in the automodification domain (91). The periodic repetition of leucine residues at every seventh amino acid (leucine zipper) in the automodification domain and the basic amino acid-rich region adjacent to the leucine residues of this sequence may mediate the homo- and heterodimerization of PARP with itself and other DNA binding proteins, respectively.

1.3.6   PARP–DNA interactions

The activity of PARP *in vivo* and *in vitro* is almost totally dependent on DNA interruptions (123). DNA binding involves two zinc fingers (124) localized in the 29-kDa N-terminal domain obtained by digestion with trypsin (Fig. 1.5). The existence of this structure in PARP has been confirmed by four observations:

1. The 29-kDa domain is contained in the 46-kDa fragment (125), that is the major fragment for DNA binding.
2. Two zinc atoms have been localized on the DNA binding site (124).
3. The presence of zinc on PARP is essential to DNA binding on the 29-kDa domain (126).
4. The primary structure of this domain, which was deduced from the mRNA nucleotide sequence, shows two potential zinc-finger motifs (79, 82, 127).

The majority of proteins with zinc fingers identified to date are regulatory proteins that recognize a specific DNA sequence (128). PARP, on the other hand, does not apparently recognize a specific sequence but recognizes a break in DNA. Indeed, DNase I protection studies have shown that the 29-kDa PARP fragment binds to a 66 base-pair DNA fragment harbouring a single strand-break interruption (97, 129). The protection observed is sequence-independent, since the same result was obtained when the nick was shifted to another site in the sequence. Site-directed mutagenesis of the two zinc fingers (FI and FII, Fig. 1.5) has shown that the binding of the divalent ion to the C-terminal finger (FII) was necessary for the recognition and binding of PARP to a DNA break (98). However, a modification in the structure of FI does not modify the properties of the 29-kDa fragment regarding the recognition and binding of DNA, and it is suggested that the FI finger could: (1) stabilize FII interactions by non-specific contact with DNA; or (2) facilitate protein–protein contacts between two PARP molecules at the binding site on DNA (98). On the other hand, site-directed mutagenesis of the full-length enzyme indicates that the FI finger is important in enzyme activation by DNA containing double-strand breaks (130). These results highlight the differential role of FI and FII in the recognition of breaks and in PARP activation. The two fingers are clearly individualized as two subdomains, which is also suggestive of a specific role for each.

1.3.7 Acceptors of poly (ADP-ribose)

As discussed elsewhere in this monograph, ADP-ribose polymer metabolism clearly plays a role in the maintenance of genomic integrity. However, the molecular details of the structure–function relationship of the polymers are still poorly understood. Conceivably, the polymers can function by altering the function of covalently modified proteins, and/or by interacting with chromatin components via non-covalent interactions. Not surprisingly, all proteins shown to be covalent acceptors of ADP-ribose polymers are DNA binding proteins. The major acceptors *in vivo* are PARP and histones, although topoisomerase I and high mobility group (HMG) and low mobility group (LMG) proteins have been identified as minor, but potentially important, acceptors (62, 131–133). Polymers of ADP-ribose also have a high potential for non-covalent interactions, since each polymer residue contains two phosphate groups carrying a formal negative charge each and an adenine ring capable of base-stacking and hydrogen-bonding interactions. A number of studies now indicate that non-covalent interactions probably play important roles in polymer function (134–137).

The effects of poly (ADP-ribose) on the structure of chromatin *in vitro* has involved the use of polynucleosome preparations (138, 139). The presence of poly (ADP-ribose) on histones (mainly histone H1) induced chromatin relaxation (51, 140). Degradation of the polymers by PARG induced chromatin recondensation (52). Conditions leading to the modification of histone H2b resulted in the dissociation of the histone octamer from DNA (102, 141). The negative charges resulting from poly (ADP-ribose) synthesis appear to contribute to the establishment of a relaxed state of chromatin.

PARP may also play a direct role in chromatin structure, on the basis of its high abundance in chromatin and on its particular distribution in the nucleus. After PARP depletion by antisense RNA expression, chromatin becomes hypersensitive to treatment with either DNase I or micrococcal nuclease (142, 143), suggesting that depletion of PARP results in a more relaxed state of chromatin.

1.4   Poly (ADP-ribose) catabolism

In eukaryotic cells, two different types of enzymes are known to catalyse the hydrolysis of ADP-ribose polymers (Fig. 1.3). Poly (ADP-ribose) glycohydrolase (144–147), cleaves ribose–ribose bonds of both linear and branched portions of poly (ADP-ribose). A number of different enzymes, including snake venom phosphodiesterase, catalyse the hydrolysis of the polymer pyrophosphate linkages. While hydrolysis of the pyrophosphate linkages is useful for characterizing ADP-ribose polymers, the involvement of enzymes that catalyse hydrolysis at this linkage in the specific degradation of the polymer *in vivo* has not been demonstrated. An ADP-ribosyl protein lyase has been proposed to catalyse the removal of protein–proximal ADP-ribose monomers (148).

1.4.1   Poly (ADP-ribose) glycohydrolase

Historical

Poly (ADP-ribose) glycohydrolase (PARG) is the primary enzyme responsible for the catabolism of poly (ADP-ribose) *in vivo*. It catalyses the hydrolysis of glycosidic ($1'' \rightarrow 2'$) linkages in poly (ADP-ribose) to produce ADP-ribose (147, 149). This activity was first described by Miwa and Sugimura (146), and PARG was subsequently partially purified by Ueda *et al.* (150). A more detailed characterization led to the establishment of a primarily exoglycosidic mode of hydrolysis (147), although the enzyme also has endoglycosidase activity (151–153). Since the initial reports, numerous characterizations of the enzyme have been reported (149, 154–161). In most preparations, the K_M for ADP-ribose residues in the polymer is between 1 and 6 μM, which corresponds to the constitutive poly (ADP-ribose) concentration of most tissues if calculated per total nuclear volume. An important question debated in the literature concerns the size of PARG *in vivo*.

Numerous studies have reported that different forms of PARG exist in cells, suggesting different roles and locations within the cell. Sizes for PARG ranging from 50 kDa to 74 kDa (144, 147, 149, 154, 156–162) have been reported.

Molecular characterization of poly (ADP-ribose) glycohydrolase

The rapid synthesis of ADP-ribose polymers that occurs in response to DNA strand breaks is accompanied by very rapid polymer turnover (163–165), indicating that PARP and PARG activities are closely coordinated as cells respond to DNA damage. Therefore, knowledge of the precise mode of action of PARG on ADP-ribose polymers is important for elucidating not only the metabolism of poly (ADP-ribose) but also the biological functions of ADP-ribose polymer metabolism. Recently, Jacobson and co-workers (161) have reported the isolation and characterization of the cDNA encoding bovine PARG. An intriguing finding was that the bovine PARG cDNA clone of 4070 nucleotides codes for a protein of approximately 111 kDa (Fig. 1.6), which is nearly twice the size of the PARG protein (65 kDa) isolated from bovine thymus. Both the 65-kDa and the 111-kDa forms retain activity (161) when tested on an activity gel (162) (Fig. 1.7).

It seems likely that PARG contains a protease-sensitive site that, following proteolysis, yields a protein fragment of approximately 64 kDa which retains enzymatic activity. Several pieces of evidence favour this possibility.

1. Expression of the carboxyl-terminal portion of the cDNA resulted in enzymatic activity.
2. All the oligopeptides sequenced from the purified protein were located in the carboxyl-terminal half of the protein.
3. The only protein, other than the 65-kDa protein detected in the thymus preparation, was approximately 111 kDa.
4. The PARG activity expressed in bacteria was sensitive to proteolysis, yielding a protein of approximately 65 kDa.
5. The cleavage site in PARG is in the region of a putative nuclear location signal (NLS), and the PARP NLS is located in a protease-sensitive site (166).

Taken together with Southern analyses, the data suggest that PARG is enclosed by a single copy gene (161). To support this idea, the chromosome

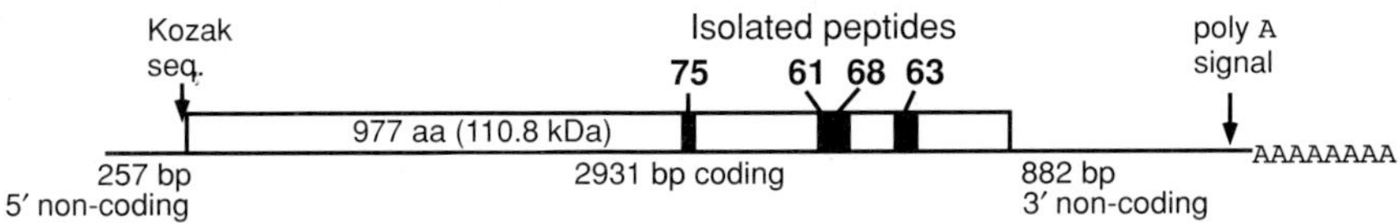

Fig. 1.6   A schematic diagram of the full-length cDNA coding for bovine PARG.

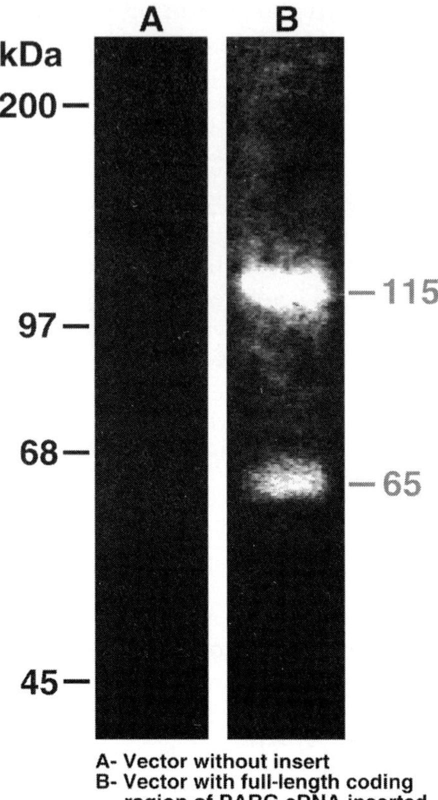

Fig. 1.7   Analysis of PARG activity following SDS-PAGE. Crude extracts (1 μl each) of cells transformed by the vector pTrcHis without an insert (lane A) or containing the full-length coding region of PARG's cDNA (lane B) were analysed by an activity gel assay. The gel was incubated following electrophoresis in 50 mM sodium phosphate buffer, pH 7.5, 50 mM NaCl, 10% glycerol, 1% Triton X-100, 10 mM β-mercaptoethanol to allow renaturation of PARG activity (162). (Taken from Lin *et al.*, 1997 (161).)

localization of the *PARG* gene has been determined at a unique location both in humans at position 10q11.23 and in the mouth at position 14B (167). Proteolysis seems likely to explain the presence of PARG activity in proteins of approximately 74 kDa and 59 kDa in bovine thymus preparations (162) and the molecular heterogeneity of different PARG preparations previously observed by various authors (144, 147, 149, 154, 156–160).

Antibodies were raised against the catalytic domain of bovine PARG (168), and studies were undertaken to assess the presence and the size of PARG in different cell and tissue extracts. Western blot analyses revealed a protein (Fig. 1.8) in the bovine tissue extracts that corresponded to a molecular weight of 65 kDa, which corresponds to the size of the purified

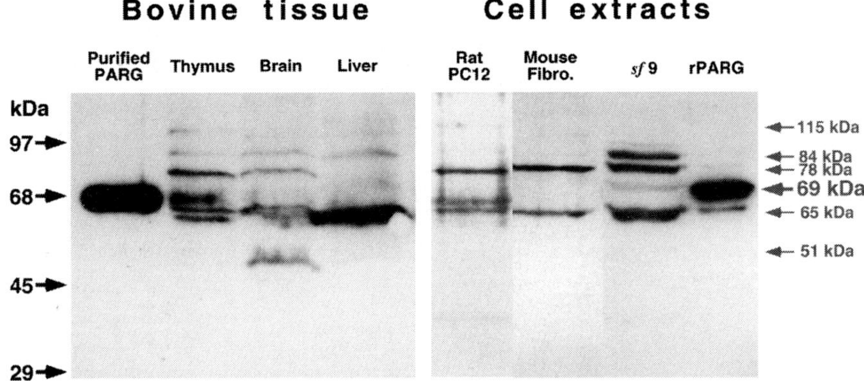

Fig. 1.8 Characterization of PARG from tissues and cells extracts by immunoblot. Different protein extracts (30 µg) from bovine tissues and cell extracts from rat, mouse, and *sf*9 insect cells were analysed by Western blot. Purified PARG from bovine thymus was loaded as controls. (Taken from ref. 168: Molecular heterogeneity and regulation of poly ADP-ribose glycohydrolase, (1999). *Mol. Cell. Biochem.*, **193**, 75, with permission.)

enzyme. Other bands of higher MW (78 kDa, 84 kDa, and 115 kDa) were also detected, but the abundance of these forms was much lower. A similar profile of bands specifically reacting with PARG antibodies was observed within extracts from rat, mouse, and insect cells, showing that PARG is a highly conserved protein. These results raise the question of the role of the N-terminal part of the protein formed by proteolysis. Also, it is not known whether the cleavages observed are specific or not.

The cDNA sequence encoding bovine PARG indicates that PARG shares little or no homology with other known sequences. A search of amino-acid sequence databanks has failed to reveal significant homology with any motif or sequences coding for known proteins. However, strong homologies have been found with partial cDNA clones in the dBEST database. Homologies were found with cDNA from various organisms including humans, *Drosophila*, mice, rats, *Caenorhabditis elegans*, *Schistosoma mansoni*, and *Toxoplasma gondii*. Subsequent cDNA library screening has allowed the characterization of the cDNAs encoding PARG from humans, *Drosophila*, mice, and *Caenorhabditis elegans*. The sequence comparisons reveal that mammalian PARGs are highly conserved. For example, human PARG has 89% identity with bovine PARG and 86% identity with murine PARG (Fig. 1.9). Sequence comparisons between mammalian and non-mammalian PARGs show high homology in the catalytic domain, but no significant homology in the putative regulatory domain.

Amino-acid sequence comparisons also reveal a putative bipartite NLS (169) that shows considerable similarity to the NLS of PARP (99). Figure 1.10 compares NLS sequences for the five known PARGs with the NLS

Fig. 1.9 Deduced amino acid sequences of bovine, human, murine, *Drosophila*, and *C. elegans*, PARGs.

                    ADP-ribose polymer metabolism

```
dPARG    (577)   ENKAS.KKKLY.DFIK..EE..LKKV.RDVP..
bPARG    (422)   ED....KRKEQCEMKHQRTE...RKIPKYIPPH
hPARG    (421)   ED....RRKEQWETKHQRTE...RKIPKYVPPH
mPARG    (413)   ED....RRKEQCEVRHQRTE...RKIPKYIPPN
CePARG    (29)   QVPTMKRRKLTEHG.NTTESLLLKEDPEEPKS

hPARP    (205)   EG....KRKGD.EVDG.VDEVA.KKKSKKEKDK
mPARP    (205)   EG....KRKGD.EVDG.TDEVA.KKKSRKETDK
bPARP    (208)   EG....KRKGD.EVDG.IDEVT.KKKSKKEKDK
aPARP    (205)   EG....KRKGE.EVDG..NVVA.KKKSRKEKEK
XlPARP   (204)   EG....KRKAD.EVDG.HSAAT.KKKIKKEKEK
DmPARP   (202)   EELPDTKRAKM.ELSDTNEEGE.KKQR......
SpPARP   (205)   EGVSSAKKAKI.EKIDEEDAASI.KELTEKIKK
```

Fig. 1.10 Comparison of amino acid sequences of a putative NLS of PARG with the NLS of PARP. Abbreviations and references for the sequences shown are as follows: dPARG, *Drosophila* PARG (175), bPARG, bovine PARG (161); hPARP, human PARG (175); mPARG, murine PARG (175); CePARG, *Caenorhabditis elegans* PARG (175, 176); hPARP, human PARP (79, 82, 127); mPARP, murine PARP (87); bPARP, bovine PARP (85); aPARP, chicken PARP (88); XlPARP, *Xenopus laevis* PARP (89); Dm-PARP, *Drosophila melanogaster* PARP (91); SpPARP, *Sarcophaga peregrina* PARP (92).

region of PARP from seven different species. The putative NLS of PARG fulfils the criteria for bipartite NLS, in that it contains conserved acidic and basic amino-acid residues at two different locations each within the region of homology to the NLS of PARP (99). The presence of an NLS is in good agreement with studies (144, 170) that found that PARG was exclusively a nuclear protein, but other studies have indicated both a nuclear and cytoplasmic location for PARG (171). However, it is possible that nuclear glycohydrolase reaches the cytoplasmic fraction during tissue homogenization, as is the case with DNA polymerases (172).

The overall organization of the mammalian PARGs is very similar (Fig. 1.11), where the NLS is located between the catalytic and putative regulatory

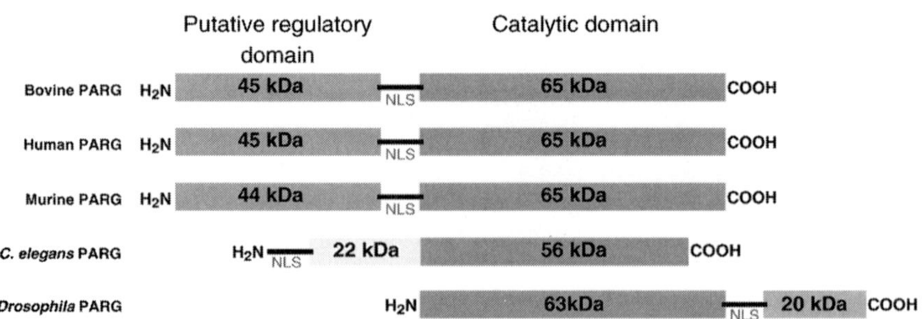

Fig. 1.11 Comparison of overall organization of PARG from five species.

domains. NLS regions normally occur on the protein surface in hinge regions between protein domains (99, 173). The NLS of mammalian PARG's is located between the 45-kDa and catalytic domains in a region very prone to proteolysis. The non-mammalian PARGs are similar in size for the catalytic domain; however, they differ in the location of the putative NLS. Furthermore, the putative regulatory domain is much smaller, and, in the case of *Drosophila*, it is located to the carboxy-terminal side of the catalytic domain rather than at the amino terminus.

Regulation of PARG expression

In the metabolism of ADP-ribose polymers, the activities of PARP and PARG are closely related. Soon after the polymer has been synthesized by PARP following DNA damage, it is extensively degraded by PARG. The net result is that the polymer has a very short half-life. The close relationship between the two proteins suggests a possible mode of regulation in which PARG expression depends on the presence of PARP. However, Western blot analyses (Fig. 1.12) show there is no significant difference in the expression of PARG in the genotypes $PARP^{+/+}$, $PARP^{+/-}$, or $PARP^{-/-}$ (168). Furthermore,

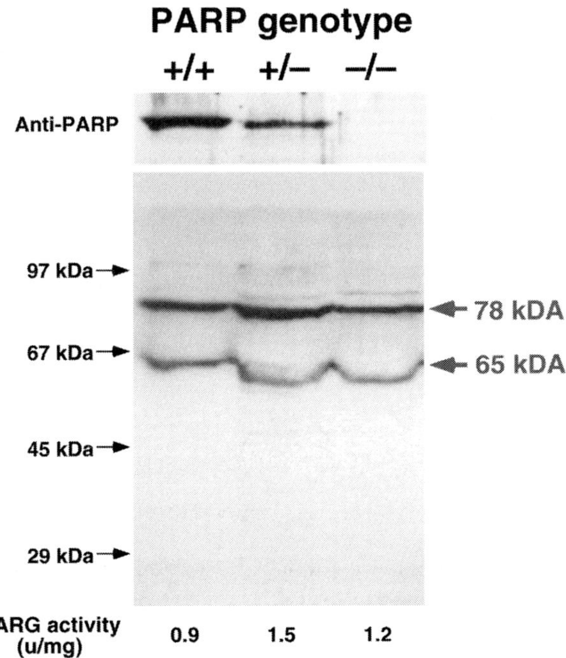

Fig. 1.12 Analysis of PARP and PARG by immunoblotting in cells containing different PARP genotypes.

PARG activity did not differ in each of these cell extracts. These data indicate that the levels of PARP and PARG protein are likely to be independently regulated.

The availability of new molecular tools for studying PARG, including the cDNA sequence, specific antibodies, stably transfected cells containing either sense or antisense PARG cDNA, and ultimately a mouse containing a disrupted *PARG* gene, will allow complete genetic modulation of the synthesis and turnover of ADP-ribose polymers. These tools should facilitate our understanding of the functions of this pathway.

1.4.2   ADP-ribosyl protein lyase

The ADP-ribosyl protein lyase activity was discovered and partially purified by Hayaishi and co-workers (174) and subsequently purified to apparent homogeneity by Oka *et al.* (148). This enzyme is unique amongst mammalian enzymes in that its reaction yields unsaturated sugars by an enzymatic elimination reaction (148). The substrate specificity of the ADP-ribosyl protein lyase is not very dependent on the protein portion, but it is highly specific for the mono (ADP-ribosyl) moiety and the carboxyl ester bond to the protein. This enzyme cleaves the last ADP-ribose residue left on the acceptor by PARG (174); however, little is known about its molecular structure.

References

1. Haag, F. and Koch-Nolte, F. (1996). Structure, function and biology of mono (ADP-ribosyl) transferases and related enzymes. In *ADP-ribosylation in animal tissues* (ed. F. Haag and F. Koch-Nolte), pp. 1–350. Plenum Press New York.
2. Lee, H. C. (1996). Modulator and messenger functions of cyclic ADP-ribose in calcium signaling. *Recent Progr. Horm. Res.*, **51**, 355.
3. Chambon, P., Weil, J. D., and Mandel, P. (1963). Nicotinamide mononucleotide activation of a new DNA-dependent polyadenylic acid synthesizing nuclear enzyme. *Biochem. Biophys. Res. Commun.*, **11**, 39.
4. Chambon, P., Weil, J. D., Doly, J., Strosser, M. T., and Mandel, P. (1966). On the formation of a novel adenylic compound by enzymatic extracts of liver nuclei. *Biochem. Biophys. Res. Commun.*, **25**, 638.
5. Dolly, J. and Petek, F. (1966). Étude de la structure d'un composé 'poly (ADP-ribose)' synthétisé par des extraits nucléaires de foie de poulet. *CR. Hebd. Seanc. Acad. Sci. Ser. D Sci. Nat.*, **263**, 1341.
6. Fujimura, S., Sugimura, T., Okabe, K., and Yoshida, T. (1965). NMN-activated poly(A) polymerase in nuclei from rat liver and hepatoma cells. *Proc. Annu. Meeting Jpn Biochem. Soc.*, **38**, 591. (In Japanese.)

7. Reeder, R. H., Ueda, K., Honjo, T., Nishizuka, Y., and Hayaishi, O. (1967). Studies on the polymer of adenosine phosphate ribose. II. Characterization of the polymer. *J. Biol. Chem.*, **242**, 3172.

8. Hasegawa, S., Fujimura, S., Shimizu, Y., Okuyama, H., and Sugimura, T. (1967). Venom phosphodiesterase hydrolysis product of ADPR polymer synthesized enzymatically from NAD$^+$. *Proc. Annu. Meeting Jpn Biochem. Soc.*, **39**, 194. (In Japanese.)

9. Fujimura, S., Hasegawa, S., and Sugimura, T. (1967). Nicotinamide mononucleotide-dependant incorporation of ATP into acid-insoluble material in rat liver nuclei preparation. *Biochim. Biophys. Acta*, **134**, 496.

10. Fujimura, S., Hasegawa, S., Shimizu, Y., and Sugimura, T. (1967). Polymerization of the adenosine 5'-diphosphate-ribose moiety of nicotinamide-adenine dinucleotide by nuclear enzyme. I. Enzymatic reactions. *Biochim. Biophys. Acta*, **145**, 247.

11. Alvarez-Gonzalez, R. and Jacobson, M. K. (1987). Characterization of polymers of adenosine diphosphate ribose generated *in vitro* and *in vivo*. *Biochemistry*, **26**, 3218.

12. Hayashi, K., Tanaka, M., Shimada, T., Miwa, M., and Sugimura, T. (1983). Size and shape of poly (ADP-ribose): examination by gel filtration, gel electrophoresis and electron-microscopy. *Biochem. Biophys. Res. Commun.*, **112**, 102.

13. Miwa, M., Saito, H., Sakura, H., Saikawa, N., Watanabe, F., Matsushima, T., and Sugimura, T. (1977). A ^{13}C NMR study of poly (adenosine diphosphate ribose) and its monomers: evidence of alpha-(1'' leads to 2') ribofuranosyl ribofuranoside residue. *Nucleic Acids Res.*, **4**, 3997.

14. Ferro, A. M. and Oppenheimer, N. J. (1978). Structure of a poly (adenosine diphosphoribose) monomer: 2'-(5''-phosphoribosyl)-5'-adenosine monophosphate. *Proc. Natl. Acad. Sci. USA*, **75**, 809.

15. Inagaki, F., Miyazawa, T., Miwa, M., Saito, H., and Sugimura, T. (1978). NMR analyses of conformation of ribosyl adenosine 5',5''-bis(phosphate) in aqueous solution. *Biochem. Biophys. Res. Commun.*, **85**, 415.

16. Miwa, M., Kato, M., Iijima, H., Tanaka, Y., Kondo, T., Kawamitsu, H., Terada, M., and Sugimura, T. (1983). Poly (ADP-ribose): structure, quantification, and biological significance. In *ADP-ribosylation, DNA repair and cancer* (ed. M. Miwa, O. Hayaishi, S. Shall, M. Smulson, and T. Sugimura), pp. 27–37. Jpn Sci. Soc. Press, Tokyo.

17. Miwa, M., Saikawa, N., Yamaizumi, Z., Nishimura, S., and Sugimura, T. (1979). Structure of poly (adenosine diphosphate ribose): identification of 2'-[1''-ribosyl-2''- (or 3''-)(1'''-ribosyl)]adenosine-5',5'',5'''-tris(phosphate) as a branch linkage. *Proc. Natl. Acad. Sci. USA*, **76**, 595.

18. Miwa, M., Ishihara, M., Takishima, S., Takasuka, N., Maeda, M., Yamaizumi, Z., Sugimura, T., Yokoyama, S., and Miyazawa, T. (1981). The branching and linear portions of poly (adenosine diphosphate ribose) have the same alpha (1 leads to 2) ribose–ribose linkage. *J. Biol. Chem.*, **256**, 2916.

19. Kanai, M., Miwa, M., Kuchino, Y, and Sugimura, T. (1982). Presence of branched portion in poly (adenosine diphosphate ribose) *in vivo*. *J. Biol. Chem.*, **257**, 6217.

20. Kiehlbauch, C. C., Aboul-Ela, N., Jacobson, E. L., Ringer, D. P., and Jacobson, M. K. (1993). High resolution fractionation and characterization of ADP-ribose polymers. *Anal. Biochem.*, **208**, 26.

21. Miwa, M. and Sugimura, T. (1982). Structure and properties of poly (ADP-ribose). In *ADP-ribosylation reactions, biology and medicine* (ed. O. Hayaishi and K. Ueda), pp. 43–62. Academic Press New York.
22. Tanaka, M., Miwa, M., Hayashi, K., Kubota, K., and Matsushima, T. (1977). Separation of oligo (adenosine diphosphate ribose) fractions with various chain lengths and terminal structures. *Biochemistry*, **16**, 1485.
23. Okayama, H., Ueda, K., and Hayaishi, O. (1978). Purification of ADP-ribosylated nuclear proteins by covalent chromatography on dihydroxyboryl polyacrylamide beads and their characterization. *Proc. Natl. Acad. Sci. USA*, **75**, 1111.
24. Minaga, T., Romaschin, A. D., Kirsten, E., and Kun, E. (1979). The *in vivo* distribution of immunoreactive larger than tetrameric polyadenosine diphosphoribose in histone and non-histone protein fractions of rat liver. *J. Biol. Chem.*, **254**, 9663.
25. Adamietz, P., Klapproth, K., and Hilz, H. (1979). Isolation and partial characterization of the ADP-ribosylated nuclear proteins from Ehrlich ascites tumour cells. *Biochem. Biophys. Res. Commun.*, **91**, 1232.
26. Alvarez-Gonzalez, R., Juarez-Salinas, H., Jacobson, E. L., and Jacobson, M. K. (1983). Evaluation of immobilized boronates for studies of adenine and pyridine nucleotide metabolism. *Anal. Biochem.*, **135**, 69.
27. Niedergang, C., Okazaki, H., and Mandel, P. (1978). A new technique for the isolation and fluorometric determination of polyadenosine diphosphate ribose *in vivo*. *Anal. Biochem.*, **88**, 20.
28. Sims, J. L., Juarez-Salinas, H., and Jacobson, M. K. (1980). A new highly sensitive and selective chemical assay for poly (ADP-ribose). *Anal. Biochem.*, **106**, 296.
29. Juarez-Salinas, H., Mendoza-Alvarez, H., Levi, V., Jacobson, M. K., and Jacobson, E. L. (1983). Simultaneous determination of linear and branched residues in poly (ADP-ribose). *Anal. Biochem.*, **131**, 410.
30. Jacobson, M. K., Payne, D. M., Alvarez-Gonzalez, R., Juarez-Salinas, H., Sims, J. L., and Jacobson, E. L. (1984). Determination of *in vivo* levels of polymeric and monomeric ADP-ribose by fluorescence methods. In *Methods in enzymology*, Vol. 106 (editors: Fimm Wold and Kivie Moldave), p. 483. Academic Press, NY.
31. Barrio, J., Secrist III, J. A., and Leonard, N. J. (1972). Fluorescent adenosine and cytidine derivatives. *Biochem. Biophys. Res. Commun.*, **46**, 597.
32. Aboul-Ela, N., Jacobson, E. L., and Jacobson, M. K. (1988). Labeling methods for the study of poly- and mono (ADP-ribose) metabolism in cultured cells. *Anal. Biochem.*, **174**, 239.
33. Keith, G., Desgres, J., and de Murcia, G. (1990). Use of two-dimensional thin-layer chromatography for the components study of poly (adenosine diphosphate ribose). *Anal. Biochem.*, **191**, 309.
34. Kanai, Y., Miwa, M., Matsushima, T., and Sugimura, T. (1974). Studies on anti-poly (adenosine diphosphate ribose) antibody. *Biochem. Biophys. Res. Commun.*, **59**, 300.
35. Kidwell, W. R. and Mage, M. G. (1976). Changes in poly (adenosine diphosphate-ribose) and poly (adenosine diphosphate-ribose) polymerase in synchronous HeLa cells. *Biochemistry*, **15**, 1213.
36. Ferro, A. M., Minaga, T., Piper, W. N., and Kun, E. (1978). Analysis of larger

than tetrameric poly (adenosine diphosphoribose) by a radioimmunoassay in nuclei separated in organic solvents. *Biochim. Biophys. Acta*, **519**, 291.

37. Kawamitsu, H., Hoshino, H., Okada, H., Miwa, M., Momoi, H., and Sugimura, T. (1984). Monoclonal antibodies to poly (adenosine diphosphate ribose) recognize different structures. *Biochemistry*, **23**, 3771.

38. Sakura, H., Miwa, M., Kanai, Y., Matsushima, T., and Sugimura, T. (1978). Formation and characterization of antibody against 2′-(5″-phosphoribosyl)-5′ AMP, the monomer form of poly (adenosine diphosphate ribose). *Nucleic Acids Res.*, **5**, 4025.

39. Bredehorst, R., Ferro, A. M., and Hilz, H. (1978). Determination of ADP-ribose and poly (ADP-ribose) by a new radioimmunoassay. *Eur. J. Biochem.*, **82**, 115.

40. Bredehorst, R., Schluter, M. M., and Hilz, H. (1981). Determination of 5-AMP in the presence of excess 3′(2′)-AMP with the aid of antibodies raised against n6-carboxymethyl-5′-AMP conjugates. Use for the quantitation of pyridine nucleotides and of protein-bound ADP-ribose. *Biochim. Biophys. Acta*, **652**, 16.

41. Meyer, T. and Hilz, H. (1986). Production of anti-(ADP-ribose) antibodies with the aid of dinucleotide-pyrophosphatase-resistant hapten and their application for the detection of mono(ADP-ribosyl)ated polypeptides. *Eur. J. Biochem.*, **155**, 157.

42. Kanai, Y, Miwa, M., Matsushima, T., and Sugimura, T. (1975). Proceedings: Anti-poly (ADP-ribose) antibody and its specificity. *J. Biochem. (Tokyo)*, **77**, 5.

43. Kanai, Y., Sugimura, T., and Matsushima, T. (1978). Induction of specific antibodies to poly (ADP-ribose) in rabbits by double-stranded RNA, poly(A)-poly(U). *Nature*, **274**, 809.

44. Kanai, Y., Miwa, M., Matsushima, T., and Sugimura, T. (1978). Comparative studies on antibody and antibody production to poly (ADP-ribose) in mice. *Immunology*, **34**, 501.

45. Sakura, H., Miwa, M., Tanaka, M., Kanai, Y., Shimada, T., Matsushima, T., and Sugimura, T. (1977). Natural occurrence of a biopolymer, poly (adenosine diphosphate ribose). *Nucleic Acids Res.*, **4**, 2903.

46. Kawamitsu, H., Hoshino, H., Miwa, M., and Sugimura, T. (1983). Monoclonal antibodies against poly (ADP-ribose) recognize different structures of poly (ADP-ribose). *Princess Takamatsu Symp.*, **13**, 41.

47. Ikai, K., Ueda, K., and Hayaishi, O. (1980). Immunohistochemical demonstration of poly (adenosine diphosphate-ribose) in nuclei of various rat tissues. *J. Histochem. Cytochem.*, **28**, 670.

48. Poirier, G. G., de Murcia, G., Jongstra-Bilen, J., Niedergang, C., and Mandel, P. (1982). Poly (ADP-ribosyl)ation of polynucleosomes causes relaxation of chromatin structure. *Proc. Natl. Acad. Sci. USA*, **79**, 3423.

49. Aubin, R. J., Frechette, A., de Murcia, G., Mandel, P., Lord, A., Grondin, G., and Poirier, G. G. (1983). Correlation between endogenous nucleosomal hyper(ADP-ribosyl)ation of histone H1 and the induction of chromatin relaxation. *EMBO J.*, **2**, 1685.

50. de Murcia, G., Jongstra-Bilen, J., Ittel, M.-E., Mandel, P., and Delain, E. (1983). Poly (ADP-ribose) polymerase auto-modification and interaction with DNA: electron microscopic visualization. *EMBO J.*, **2**, 543.

51. Fréchette, A., Huletsky, A., Aubin, R. J., de Murcia, G., Mandel, P., Lord, A.,

Grondin, G., and Poirier, G. G. (1985). Poly (ADP-ribosyl)ation of chromatin: kinetics of relaxation and its effect on chromatin solubility. *Can. J. Biochem. Cell. Biol.*, **63**, 764.

52. de Murcia, G., Huletsky, A., Lamarre, D., Gaudreau, A., Pouyet, J., Daune, M., and Poirier, G. G. (1986). Modulation of chromatin superstructure induced by poly (ADP-ribose) synthesis and degradation. *J. Biol. Chem.*, **261**, 7011.

53. Ikai, K., Ueda, K., Fukushima, M., Nakamura, T., and Hayaishi, O. (1980). Poly (ADP-ribose) synthesis, a marker of granulocyte differentiation. *Proc. Natl. Acad. Sci. USA*, **77**, 3682.

54. Ikai, K., Danno, K., Imamura, S., and Ueda, K. (1982). Immunohistochemical demonstration of poly (adenosine diphosphate-ribose) synthesis in human skin. *J. Dermatol.*, **9**, 125.

55. Ikai, K. and Ueda, K. (1983). Immunohistochemical demonstration of poly (adenosine diphosphate-ribose) synthetase in bovine tissues. *J. Histochem. Cytochem.*, **31**, 1261.

56. Chen, Y. M., Shall, S., and O'Farrell, M. (1994). Poly (ADP-ribose) polymerase in plant nuclei. *Eur. J. Biochem.*, **224**, 135.

57. Rickwood, D. and Osman, M. S. (1979). Characterization of poly (ADP-ribose) polymerase activity in nuclei from the slime mould *Dictyostelium discoideum*. *Mol. Cell. Biochem.*, **23**, 79.

58. Werner, E., Sohst, S., Gropp, F., Simon, D., Wagner, H., and Kröger, H. (1984). Presence of poly (ADP-ribose) polymerase and poly (ADP-ribose) glycohydrolase in the dinoflagellate *Cryphecodinium cohnii*. *Eur. J. Biochem.*, **139**, 81.

59. Faraone-Mennella, M. R., Gambacorta, A., Nicolaus, B., and Farina, B. (1998). Purification and biochemical characterization of a poly (ADP-ribose) polymerase-like enzyme from the thermophilic archaeon *Sulfolobus solfataricus*. *Biochem. J.*, **335**, 441.

60. Sugimura, T., Fujimura, S., Hasegawa, S., Shimizu, Y., and Okuyama, H. (1968). Polymerization of ADPR moiety of NAD$^+$ by nuclear enzyme preparation. *J. Vitaminol. (Kyoto)*, **14**, 135.

61. Kameshita, I., Yamamoto, H., Fujimoto, S., and Shizuta, Y. (1985). Antigenic determinant and interspecies cross-reactivity of a monoclonal antibody to poly (ADP-ribose) synthetase. *FEBS Lett.*, **182**, 393.

62. Althaus, F. R. (1987). Poly (ADP-ribose) biosynthesis. In ADP-ribosylation of proteins: enzymology and biological significance, (ed. F. R. Althaus and C. Richter), pp. 12–37. Springer-Verlag Berlin, Heidelberg.

63. Burtscher, H. J., Auer, B., Klocker, H., Schweiger, H., and Hirsch-Kauffmann, H. (1986). Isolation of ADP-ribosyltransferase by affinity chromatography. *Anal. Biochem.*, **152**, 285.

64. Ushiro, H., Yokoyama, Y., and Shizuta, Y. (1987). Purification and characterization of poly (ADP-ribose) synthetase from human placenta. *J. Biol. Chem.*, **262**, 2352.

65. Burtscher, H. J., Schneider, R., Klocker, H., Auer, B., Hirsch-Kauffmann, M., and Schweiger, M. (1987). ADP-ribosyltransferase is highly conserved: purification and characterization of ADP-ribosyltransferase from a fish and its comparison with the human enzyme. *J. Comp. Physiol.*, **157**, 567.

66. Burtscher, H. J., Klocker, H., Schneider, R., Auer, B., Hirsch-Kauffmann, M.,

and Schweiger, M. (1987). ADP-ribosyltransferase from *Helix pomatia*. Purification and characterization. *Biochem. J.*, **248**, 859.

67. Ruggieri, S., Gregori, L., Natalini, P., Vita, A., Emanuelli, M., Raffaelli, N., and Magni, G. (1990). Evidence for an inhibitory effect exerted by yeast NMN adenylyltransferase on poly (ADP-ribose) polymerase activity. *Biochemistry*, **29**, 2501.

68. D'Amours, D., Duriez, P. J., Orth, K., Shah, R. G., Dixit, V. M., Earnshaw, W. C., Alnemri, E. S., and Poirier, G. G. (1997). Purification of the death substrate poly (ADP-ribose) polymerase. *Anal. Biochem.*, **249**, 106.

69. Simonin, F., Hofferer, L., Panzeter, P. L., Muller, S., de Murcia, G. and Althaus, F. R. (1993). The carboxyl-terminal domain of human poly (ADP-ribose) polymerase. Overproduction in *Escherichia coli*, large scale purification, and characterization. *J. Biol. Chem.*, **268**, 13454.

70. Molinete, M., Vermeulen, W., Bürkle, A., Ménissier-de Murcia, J., Küpper, J. H., Hoeijmakers, J. H., and de Murcia, G. (1993). Overproduction of the poly (ADP-ribose) polymerase DNA-binding domain blocks alkylation-induced DNA repair synthesis in mammalian cells. *EMBO J.*, **12**, 2109.

71. Giner, H., Simonin, F., de Murcia, G., and Ménissier-de Murcia, J. (1992). Overproduction and large-scale purification of the human poly (ADP-ribose) polymerase using a baculovirus expression system. *Gene*, **114**, 279.

72. Jung, S., Miranda, E. A., Ménissier-de Murcia, J., Niedergang, C., Delarue, M., Schulz, G. E., and de Murcia, G. (1994). Crystallization and X-ray crystallographic analysis of recombinant chicken poly (ADP-ribose) polymerase catalytic domain produced in Sf9 insect cells. *J. Mol. Biol.*, **244**, 114.

73. Ruf, A., Ménissier-de Murcia, J., de Murcia, G., and Schulz, G. E. (1996). Structure of the catalytic fragment of poly (ADP-ribose) polymerase from chicken. *Proc. Natl. Acad. Sci. USA*, **93**, 7481.

74. Miranda, E. A., de Murcia, G., and Ménissier-de Murcia, J. (1997). Large-scale production and purification of recombinant protein from an insect cell/baculovirus system in Erlenmeyer flasks: application to the chicken poly (ADP-ribose) polymerase catalytic domain. *Braz. J. Med. Biol. Res.*, **30**, 923.

75. Nishikimi, M., Ogasawara, K., Kameshita, I., Taniguchi, T., and Shizuta, Y. (1982). Poly (ADP-ribose) synthetase. The DNA binding domain and the auto-modification domain. *J. Biol. Chem.*, **257**, 6102.

76. Kameshita, I., Matsuda, Z., Taniguchi, T., and Shizuta, Y. (1984). Poly (ADP-ribose) synthetase. Separation and identification of three proteolytic fragments as the substrate binding domain, the DNA binding domain and the automodification domain. *J. Biol. Chem.*, **259**, 4770.

77. Shizuta, Y., Kameshita, I., Agemori, M., Ushiro, H., Taniguchi, T., Otsuki, M., Sekimizu, K., and Natori, S. (1985). The domain structure of poly (ADP-ribose) polymerase and its role in accurate transcription initiation. In *ADP-ribosylation of proteins* (ed. F. R. Althaus, H. Hilz, and S. Shall), pp. 52–59. Springer-Verlag, Heidelberg.

78. Shizuta, Y., Kameshita, I., Ushiro, H., Matsuda, M., Suzuki, S., Mitsuuchi, Y., Yokoyama, Y., and Kurosaki, T. (1986). The domain structure and the function of poly (ADP-ribose) synthetase. *Adv. Enzyme Regul.*, **25**, 377.

79. Kurosaka, T., Ushiro, H., Mutsuuchi, Y., Suzuki, S., Matsuda, M., Matsuda, Y., Katunuma, N., Kangawa, K., Matsuo, H., and Hirose, T. (1987). Primary

structure of human poly (ADP-ribose) synthetase as deduced from cDNA sequence. *J. Biol. Chem.*, **262**, 15990.

80. Suzuki, H., Uchida, K., Shima, H., Sato, T., Okamoto, T., Kimura, T., and Miwa, M. (1987). Molecular cloning of cDNA for human poly (ADP-ribose) polymerase and expression of its gene during HL-60 cell differentiation. *Biochem. Biophys. Res. Commun.*, **146**, 403. (Published erratum appears in *Biochem. Biophys. Res. Commun.* (1987) **148**, 1549.)

81. Taniguchi, T., Yamauchi, K., Yamamoto, T., Toyoshima, K., Harada, N., Tanaka, H., Takahashi, S., Yamamoto, H., and Fujimoto, S. (1988). Depression in gene expression for poly (ADP-ribose) synthetase during the interferon-gamma-induced activation process of murine macrophage tumor cells. *Eur. J. Biochem.*, **171**, 571.

82. Cherney, B. W., McBride, O. W., Chen, D., Alkhatib, H., Bhatia, K., Hensley, P., and Smulson, M. E. (1987). cDNA sequence, protein structure, and chromosomal location of the human gene for poly (ADP-ribose) polymerase. *Proc. Natl. Acad. Sci. USA*, **84**, 8370.

83. Alkhatib, H. M., Chen, D. F., Cherney, B., Bhatia, K., Notario, V., Giri, C., Stein, G., Slattery, E., Roeder, R. G., and Smulson, M. E. (1987). Cloning and expression of cDNA for human poly (ADP-ribose) polymerase. *Proc. Natl. Acad. Sci. USA*, **84**, 1224. (Published erratum appears in *Proc. Natl. Acad. Sci. USA* (1987) **84**, 4088.)

84. Schneider, R., Auer, B., Kuhne, C., Herzog, H., Klocker, H., Burtscher, H. J., Hirsch-Kauffmann, M., Wintersberger, U., and Schweiger, M. (1987). Isolation of a cDNA clone for human NAD^+: protein ADP-ribosyltransferase. *Eur. J. Cell. Biol.*, **44**, 302.

85. Saito, I., Hatakeyama, K., Kido, T., Ohkubo, H., Nakanishi, S., and Ueda, K. (1990). Cloning of a full-length cDNA encoding bovine thymus poly (ADP-ribose) synthetase: evolutionarily conserved segments and their potential functions. *Gene*, **90**, 249.

86. Thibodeau, J., Gradwohl, G., Dumas, C., Clairoux-Moreau, S., Brunet, G., Penning, C., Poirier, G. G., and Moreau, P. (1989). Cloning of rodent cDNA coding the poly (ADP-ribose) polymerase catalytic domain and analysis of mRNA levels during the cell cycle. *Biochem. Cell Biol.*, **67**, 653.

87. Huppi, K., Bhatia, K., Siwarski, D., Klinman, D., Cherney, B., and Smulson, M. (1989). Sequence and organization of the mouse poly (ADP-ribose) polymerase gene. *Nucleic Acids Res.*, **17**, 3387.

88. Ittel, M. E., Garnier, J. M., Jeltsch, J. M., and Niedergang, C. P. (1991). Chicken poly (ADP-ribose) synthetase: complete deduced amino acid sequence and comparison with mammalian enzyme sequences. *Gene*, **102**, 157.

89. Saulier-Le Drean, B. M. (1992). Poly (ADP-ribose) polymerase in *Xenopus laevis*. PhD thesis, Université de Rennes, France.

90. Uchida, K., Uchida, M., Hanai, S., Ozawa, Y., Ami, Y., Kushida, S., and Miwa, M. (1993). Isolation of the poly (ADP-ribose) polymerase-encoding cDNA from *Xenopus laevis*: phylogenetic conservation of the functional domains. *Gene*, **137**, 293.

91. Uchida, K., Hanai, S., Ishikawa, K., Ozawa, Y., Uchida, M., Sugimura, T., and Miwa, M. (1993). Cloning of cDNA encoding *Drosophila* poly (ADP-ribose) polymerase: leucine zipper in the auto-modification domain. *Proc. Natl. Acad. Sci. USA*, **90**, 3481.

92. Masutani, M., Nozaki, T., Hitomi, M., Ikejima, M., Nagasaki, K., de Prati, A. C., Kurata, S., Natori, S., Sugimura, T., and Esumi, H. (1994). Cloning and functional expression of poly (ADP-ribose) polymerase cDNA from *Sarcophaga peregrina*. *Eur. J. Biochem.*, **220**, 607.

93. Lepiniec, L., Babiychuk, E., Kushnir, S., Van Montagu, M., and Inze, D. (1995). Characterization of an *Arabidopsis thaliana* cDNA homologue to animal poly (ADP-ribose) polymerase. *FEBS Lett.*, **364**, 103.

94. Mahajan, P. B. and Zuo, Z. (1998). Purification and cDNA cloning of maize poly (ADP)-ribose polymerase. *Plant Physiol.*, **118**, 895.

95. Babiychuk, E., Cottrill, P., Storozhenko, S., Fuangthong, M., Chen, Y., O'Farrell, M., Van Montagu, M., Inze, D., and Kushnir, S. (1998). Higher plants possess two structurally different poly (ADP-ribose) polymerases. *Plant J.*, **15**, 635.

96. Kozak, M. (1987). An analysis of 5'-noncoding regions from 699 vertebrate messenger RNAs. *Nucleic Acids Res.*, **15**, 8125.

97. Ménissier-de Murcia, J., Molinete, M., Gradwohl, G., Simonin, F., and de Murcia, G. (1989). Zinc-binding domain of poly (ADP-ribose) polymerase participates in the recognition of single strand breaks on DNA. *J. Mol. Biol.*, **210**, 229.

98. Gradwohl, G., Ménissier-de Murcia, J., Molinete, M., Simonin, F., Koken, M., Hoeijmakers, J. H., and de Murcia, G. (1990). The second zinc-finger domain of poly (ADP-ribose) polymerase determines specificity for single-stranded breaks in DNA. *Proc. Natl. Acad. Sci. USA*, **87**, 2990.

99. Schreiber, V., Molinete, M., Boeuf, H., de Murcia, G. and Ménissier-de Murcia, J. (1992). The human poly (ADP-ribose) polymerase nuclear localization signal is a bipartite element functionally separate from DNA binding and catalytic activity. *EMBO J.*, **11**, 3263.

100. Simonin, F., Ménissier-de Murcia, J., Poch, O., Muller, S., Gradwohl, G., Molinete, M., Penning, C., Keith, G., and de Murcia, G. (1990). Expression and site-directed mutagenesis of the catalytic domain of human poly (ADP-ribose) polymerase in *Escherichia coli*. Lysine 893 is critical for activity. *J. Biol. Chem.*, **265**, 19249.

101. Simonin, F., Poch, O., Delarue, M., and de Murcia, G. (1993). Identification of potential active-site residues in the human poly (ADP-ribose) polymerase. *J. Biol. Chem.*, **268**, 8529.

102. Huletsky, A., Niedergang, C., Fréchette, A., Aubin, R., Gaudreau, A., and Poirier, G. G. (1985). Sequential ADP-ribosylation pattern of nucleosomal histones. ADP-ribosylation of nucleosomal histones. *Eur. J. Biochem.*, **146**, 277.

103. Boulikas, T. (1988). At least 60 ADP-ribosylated variants histones are present in nuclei from dimethylsulphate-treated and untreated cells. *EMBO J.*, **7**, 57.

104. Boulikas, T. (1990). Poly (ADP-ribosylated) histones in chromatin replication. *J. Biol. Chem.*, **265**, 14638.

105. Yoshihara, K., Itaya, A., Tanaka, Y., Ohashi, Y., Ito, K., Teraoka, H., Tsukada, K., Matsukage, A., and Kamiya, T. (1985). Inhibition of DNA polymerase α, DNA polymerase β, terminal deoxynucleotidyl transferase, and DNA ligase II by poly (ADP-ribosyl)ation reaction *in vitro*. *Biochem. Biophys. Res. Commun.*, **128**, 61.

106. Ohahi, Y., (1986). Effect of ionic strength on chain elongation in ADP-ribosylation of various nucleases. *J. Biochem.*, **99**, 971.
107. Yoshihara, K., Hashida, T., Tanaka, Y., and Ohgushi, H. (1977). Enzyme-bound early product of purified poly (ADP-ribose) polymerase. *Biochem. Biophys. Res. Commun.*, **78**, 1281.
108. Kawaichi, M., Ueda, K., and Hayaishi, O. (1981). Multiple autopoly (ADP-ribosyl)ation of rat liver poly (ADP-ribose) synthetase. Mode of modification and properties of automodified synthetase. *J. Biol. Chem.*, **256**, 9483.
109. Adamietz, P. (1987). Poly (ADP-ribose) synthase is the major endogenous non-histone acceptor for poly (ADP-ribose) in alkylated rat hepatoma cells. *Eur. J. Biochem.*, **169**, 365.
110. Kreimeyer, A., Wielckens, K., Adamietz, P., and Hilz, H. (1984). DNA repair-associated ADP-ribosylation *in vivo*. Modification of histone H1 differs from that of the principal acceptor proteins. *J. Biol. Chem.*, **259**, 890.
111. Zahradka, P. and Ebisuzaki, K. (1982). A shuttle mechanism for DNA-protein interactions. The regulation of poly (ADP-ribose) polymerase. *Eur. J. Biochem.*, **127**, 579.
112. Jump, D. B. and Smulson, M. (1980). Purification and characterization of the major nonhistone protein acceptor for poly (adenosine diphosphate ribose) in HeLa cell nuclei. *Biochemistry*, **19**, 1024.
113. Desmarais, Y., Menard, L., Lagueux, J., and Poirier, G. G. (1991). Enzymological properties of poly (ADP-ribose) polymerase: characterization of automodification sites and NADase activity. *Biochim. Biophys. Acta*, **1078**, 179.
114. Ferro, A. M. and Olivera, B. M. (1982). Poly (ADP-ribosylation) *in vitro*. Reaction parameters and enzyme mechanism. *J. Biol. Chem.*, **257**, 7808.
115. Kawaichi, M., Ueda, K., and Hayaishi, O. (1980). Initiation of poly (ADP-ribosyl) histone synthesis by poly (ADP-ribose) synthetase. *J. Biol. Chem.*, **255**, 816.
116. Althaus, F. R. and Richter, C. (1987). ADP-ribosylation of proteins. Enzymology and biological significance. *Mol. Biol. Biochem. Biophys.*, **37**, 1.
117. Mendoza-Alvarez, H. and Alvarez-Gonzalez, R. (1993). Poly (ADP-ribose) polymerase is a catalytic dimer and the automodification reaction is intermolecular. *J. Biol. Chem.*, **2568**, 22575.
118. Naegeli, H., Loetscher, P., and Althaus, F. R. (1989). Poly ADP-ribosylation of proteins. Processivity of a post-translational modification. *J. Biol. Chem.*, **264**, 14382.
119. Ueda, K., Kawaichi, M., Okayama, H., and Hayaishi, O. (1979). Poly (ADP-ribosy)ation of nuclear proteins. Enzymatic elongation of chemically synthesized ADP-ribose-histone adducts. *J. Biol. Chem.*, **254**, 679.
120. Taniguchi, T. (1987). Reaction mechanism for automodification of poly (ADP-ribose) synthetase. *Biochem. Biophys. Res. Commun.*, **147**, 1008.
121. Alvarez-Gonzalez, R. (1988). 3′-Deoxy-NAD⁺ as a substrate for poly (ADP-ribose) polymerase and the reaction mechanism of poly (ADP-ribose) elongation. *J. Biol. Chem.*, **263**, 17690.
122. Bauer, P. I., Buki, K. G., Hakam, A., and Kun, E. (1990). Macromolecular association of ADP-ribosyltransferase and its correlation with enzymic activity. *Biochem. J.*, **270**, 17.
123. Benjamin, R. C. and Gill, D. M. (1980). Poly (ADP-ribose) synthesis *in vitro*

programmed by damaged DNA. A comparison of DNA molecules containing different types of strand breaks. *J. Biol. Chem.*, **255**, 10502.

124. Mazen, A., de Murcia, J., M., Molinete, M., Simonin, F., Gradwohl, G., Poirier, G., and de Murcia G. (1989). Poly (ADP-ribose) polymerase: a novel finger protein. *Nucleic Acids Res.*, **17**, 4689.

125. Lamarre, D., Talbot, B., Leduc, Y., Muller, S., and Poirier, G. (1986). Production and characterization of monoclonal antibodies specific for the functional domains of poly (ADP-ribose) polymerase). *Biochem. Cell Biol.*, **64**, 368.

126. Zahradka, P. and Ebisuzaki, K. (1984). Poly (ADP-ribose) polymerase is a metalloenzyme. *Eur. J. Biochem.*, **142**, 503.

127. Uchida, K., Morita, T., Sato, T., Ogura, T., Yamashita, R., Noguchi, S., Suzuki, H., Nyunoya, H., Miwa, M., and Sugimura, T. (1987). Nucleotide sequence of a full-length cDNA for human fibroblast poly (ADP-ribose) polymerase. *Biochem. Biophys. Res. Commun.*, **148**, 617.

128. Jonhson, P. F. and McKnight, S. L. (1989). Eukaryotic transcriptional regulatory proteins. *Annu. Rev. Biochem.*, **58**, 799.

129. Gradwohl, G., Ménissier-de Murcia, J., Molinete, M., Simonin, F., and de Murcia, G. (1989). Expression of functional zinc finger domain of human poly (ADP-ribose) polymerase in *E. coli. Nucleic Acids Res.*, **17**, 7112.

130. Ikejima, M., Noguchi, S., Yamashita, R., Ogura, T., Sugimura, T., Gill, D. M., and Miwa, M. (1990). The zinc fingers of human poly (ADP-ribose) polymerase are differentially required for the recognition of DNA breaks and nicks and the consequent enzyme activation. Other structures recognize intact DNA. *J. Biol. Chem.*, **265**, 21907.

131. Krupitza, G. and Cerruti, P. (1989). ADP-ribosylation of ADPR-transferase and topoisomerase I in intact mouse epidermal cells JB6. *Biochemistry*, **28**, 2034.

132. Krupitza, G. and Cerruti, P. (1989). Poly (ADP-ribosylation) of histones in intact human keratinocytes. *Biochemistry*, **28**, 4054.

133. Tanuma, S., Johnson, L. D., and Johnson, G. S. (1983). ADP-ribosylation of chromosomal proteins and mouse mammary tumour virus gene expression. *J. Biol. Chem.*, **258**, 15371.

134. Huletsky, A., de Murcia, G., Muller, S., Hengartner, M., Menard, L., Lamarre, D., and Poirier, G. G. (1989). The effect of poly (ADP-ribosyl)ation on native and H1-depleted chromatin. A role of poly (ADP-ribosyl)ation on core nucleosome structure. *J. Biol. Chem.*, **264**, 8878.

135. Mathias, G. and Althaus, F. R. (1987). Release of core DNA from nucleosomal core particles following (ADP-ribose)$_n$ modification *in vitro. Biochem. Biophys. Res. Commun.*, **143**, 1049.

136. Ohashi, Y., Ueda, K., Kawaichi, M., and Hayashi, O. (1983). Activation of DNA ligase by poly (ADP-ribose) in chromatin. *Proc. Natl. Acad. Sci. USA*, **80**, 3604.

137. Sauermann, G. and Wesierska-Gadek, J. (1986). Poly (ADP-ribose) effectively competes with DNA for histone H4 binding. *Biochem. Biophys. Res. Commun.*, **139**, 523.

138. de Murcia, G., Huletsky, A., and Poirier, G. G. (1988). Review: modulation of chromatin structure by poly (ADP-ribosyl)ation. *Biochem. Cell. Biol.*, **66**, 626.

139. Boulikas, T. (1991). Relation between carcinogenesis, chromatin structure and poly (ADP-ribosylation) (review). *Anticancer Res.*, **11**, 489.
140. Niedergang, C. P., de Murcia, G., Ittel, M. E., Pouyet, J., and Mandel, P. (1985). Time course of polynucleosome relaxation and ADP-ribosylation. Correlation between relaxation and histone H1 hyper-ADP-ribosylation. *Eur. J. Biochem.*, **146**, 185.
141. Huletsky, A. (1988). Modulation de la structure de la chromatine par la poly (ADP-ribosyl)ation. PhD thesis, Université de Sherbrooke, Québec, Canada.
142. Ding, R., Pommier, Y., Kang, V. H., and Smulson, M. (1992). Depletion of poly (ADP-ribose) polymerase by antisense RNA expression results in a delay in DNA strand break rejoining. *J. Biol. Chem.*, **267**, 12804.
143. Ding, R. and Smulson, M. (1994). Depletion of nuclear poly (ADP-ribose) polymerase by antisense RNA expression: influences on genomic stability, chromatin organization, and carcinogen cytotoxicity. *Cancer Res.*, **54**, 4627.
144. Hatakeyama, K., Nemoto, Y., Ueda, K., and Hayaishi, O. (1986). Purification and characterization of poly (ADP-ribose) glycohydrolase. Different modes of action on large and small poly (ADP-ribose). *J. Biol. Chem.*, **261**, 14902.
145. Hatakeyama, K., Nemoto, Y., Ueda, K., and Hayaishi, O. (1989). Poly (ADP-ribose) glycohydrolsae and ADP-ribosyl turnover. In *ADP-ribose transfer reactions, mechanisms and biological significance*, (ed. M. K. Jacobson and E. L. Jacobson), pp. 47–52. Springer-Verlag New York.
146. Miwa, M. and Sugimura, T. (1971). Splitting of the ribose–ribose linkage of poly (adenosine diphosphate-ribose) by a calf thymus extract. *J. Biol. Chem.*, **246**, 6362.
147. Miwa, M., Tanaka, M., Matsushima, T., and Sugimura, T. (1974). Purification and properties of glycohydrolase from calf thymus splitting ribose–ribose linkages of poly (adenosine diphosphate ribose). *J. Biol. Chem.*, **249**, 3475.
148. Oka, J., Ueda, K., Hayaishi, O., Komura, H., and Nakanishi, I. (1984). ADP-ribosyl protein lyase. Purification, properties, and identification of the product. *J. Biol. Chem.*, **259**, 986.
149. Tanaka, M., Miwa, M., Matsushima, T., Sugimura, T., and Shall, S. (1976). Poly (adenosine diphosphate ribose) glycohydrolase in *Physarum polycephalum*. *Arch. Biochem. Biophys.*, **172**, 224.
150. Ueda, K., Oka, J., Naruniya, S., Miyakawa, N., and Hayaishi, O. (1972). Poly ADP-ribose glycohydrolase from rat liver nuclei, a novel enzyme degrading the polymer. *Biochem. Biophys. Res. Commun.*, **46**, 516.
151. Brochu, G., Duchaine, C., Thibeault, L., Lagueux, J., Shah, G. M., and Poirier, G. G. (1994). Mode of action of poly (ADP-ribose) glycohydrolase. *Biochim. Biophys. Acta*, **1219**, 342.
152. Braun, S. A., Panzeter, P. L., Collinge, M. A., and Althaus, F. R. (1994). Endoglycosidic cleavage of branched polymers by poly (ADP-ribose) glycohydrolase. *Eur. J. Biochem.*, **220**, 369.
153. Desnoyers, S., Shah, G. M., Brochu, G., Hoflack, J. C., Verreault, A., and Poirier, G. G. (1995). Biochemical properties and function of poly (ADP-ribose) glycohydrolase. *Biochimie*, **77**, 433.
154. Tavassoli, M., Tavassoli, M. H., and Shall, S. (1983). Isolation and purification of poly (ADP-ribose) glycohydrolase from pig thymus. *Eur. J. Biochem.*, **135**, 449.

155. Tanuma, S., Kawashima, K., and Endo, H. (1986). Purification and properties of an (ADP-ribose)$_n$ glycohydrolase from guinea pig liver nuclei. *J. Biol. Chem.*, **261**, 965.

156. Thomassin, H., Jacobson, M. K., Guay, J., Verreault, A., Aboul-Ela, N., Ménard, L., and Poirier, G. G. (1990). An affinity matrix for the purification of poly (ADP-ribose) glycohydrolase. *Nucleic Acids Res.*, **18**, 4691.

157. Tanuma, S. and Endo, H. (1990). Purification and characterization of an (ADP-ribose)$_n$ glycohydrolase from human erythrocytes. *Eur. J. Biochem.*, **191**, 57.

158. Maruta, H., Inageda, K., Aoki, T., Nishina, H., and Tanuma, S. (1991). Characterization of two forms of poly (ADP-ribose) glycohydrolase in guinea pig liver. *Biochemistry*, **30**, 5907.

159. Uchida, K., Suzuki, H., Maruta, H., Abe, H., Aoki, K., Miwa, M., and Tanuma, S. (1993). Preferential degradation of protein-bound (ADP-ribose)$_n$ by nuclear poly (ADP-ribose) glycohydrolase from human placenta. *J. Biol. Chem.*, **268**, 3194.

160. Abe, H. and Tanuma, S. (1996). Properties of poly (ADP-ribose) glycohydrolase purified from pig testis nuclei. *Arch. Biochem. Biophys.*, **336**, 139.

161. Lin, W., Amé, J.-C., Aboul-Ela, N., Jacobson, E. L. and Jacobson, M. K. (1997). Isolation and characterization of the cDNA encoding bovine poly (ADP-ribose) glycohydrolase. *J. Biol. Chem.*, **272**, 11895.

162. Brochu, G., Shah, G. M., and Poirier, G. G. (1994). Purification of poly (ADP-ribose) glycohydrolase and detection of its isoforms by a zymogram following one- or two-dimensional electrophoresis. *Anal. Biochem.*, **218**, 265.

163. Juarez-Salinas, H., Sims, J. L., and Jacobson, M. K. (1979). Poly (ADP-ribose) levels in carcinogen-treated cells. *Nature*, **282**, 740.

164. Wielckens, K., Schmidt, A., George, E., Bredehorst, R., and Hilz, H. (1982). DNA fragmentation and NAD depletion. Their relation to the turnover of endogenous mono (ADP-ribosyl) and poly (ADP-ribosyl) proteins. *J. Biol. Chem.*, **257**, 12872.

165. Jacobson, E. L., Antol, K. M., Juarez-Salinas, H., and Jacobson, M. K. (1983). Poly (ADP-ribose) metabolism in ultraviolet irradiated human fibroblasts. *J. Biol. Chem.*, **258**, 103.

166. Lazebnik, Y. A., Kaufmann, S. H., Desnoyer, S., Poirier, G. G., and Earnshaw, W. C. (1994). Cleavage of poly (ADP-ribose) polymerase by a proteinase with properties like ICE. *Nature*, **371**, 346.

167. Amé, J. C., Jacobson, E. L., and Jacobson, M. K. (1999). Assignment of poly (ADP-ribose) glycohydrolase gene (PARG) to human chromosome 10q11.23 and mouse chromosome 14B by *in situ* hybridization. *Cytogenet. Cell. Genet.*, **85**, 269.

168. Amé, J. C., Jacobson, E. L., and Jacobson, M. K. 1999). Molecular heterogeneity and regulation of poly (ADP-ribose) glycohydrolase. *Mol. Cell. Biochem.*, **193**, 75.

169. Robbins, J., Dilworth, S. M., Laskey, R. A., and Dingwall, C. (1991). Two interdependent basic domains in nucleoplasmin nuclear targeting sequence: identification of a class of bipartite nuclear targeting sequence. *Cell*, **64**, 615.

170. Jonsson, G. G., Menard, L., Jacobson, E. L., Poirier, G. G., and Jacobson, M. K. (1988). Effect of hyperthermia on poly (adenosine diphosphate-ribose) glycohydrolase. *Cancer Res.*, **48**, 4240.

171. Tanuma, S. (1989). Purification and properties of two forms of poly (ADP-

ribose glycohydrolase). In *ADP-ribose transfer reactions: mechanisms and biological significance*, (ed. M. K. Jacobson and E. L. Jacobson), pp. 119–124. Springer-Verlag New York.

172. Weinberg, D. H., Collins, K. L., Simaneh, P., Russo, A., Wold, M. S., Virshys, D. M., and Kelly, T. J. (1990). Reconstitution of simian 40 DNA replication with purified proteins. *Proc. Natl. Acad. Sci. USA*, **87**, 8692.

173. Roberts, B. L., Richardson, W. D., and Smith, A. E. (1987). The effect of protein context on nuclear location signal function. *Cell*, **50**, 465.

174. Okayama, H., Honda, M., and Hayaishi, O. (1978). Novel enzyme from rat liver that cleaves on ADP-ribosyl histone linkage. *Proc. Natl. Acad. Sci. USA*, **75**, 2254.

175. Amé, J.-C., Huang, A., Jacobson, E. L., and Jacobson, M. K. in preparation.

176. Wilson, R., Ainscough, R., Anderson, K., *et al.* (1994). 2.2 Mb of contiguous nucleotide sequence from chromosome III of *C. elegans*. *Nature*, **368**, 32.

Poly (ADP-ribose) polymerase: structure and function

Véronique Rolli, Armin Ruf, Angélique Augustin, Georg E. Schulz, Josiane Ménissier-de Murcia, and Gilbert de Murcia

2.1 The *PARP* gene and its tissue expression

The cDNA encoding the human enzyme was cloned in 1987 by several groups simultaneously (see Chapter 1). The expressed gene encoding the human PARP is located on chromosome 1 (1) at position q41–q42 (2) while two pseudogenes are found on chromosomes 13 and 14 in bands 13q34 and 14q24 (3).The human gene consists of 23 exons contained within 43 kb of DNA (4). Each functional domain of PARP is encoded by several exons except the nuclear location signal (NLS) which is entirely contained in exon 5. The same exon–intron organization is found in the mouse in a 33-kb gene (5, 6). The *Drosophila PARP* gene spans more than 60 kb and is a single-copy gene mapped to 81F on chromosome III. It is composed of six exons (7) encoding functional domains: the automodification domain is encoded by exon 5, and the catalytic domain is almost entirely encoded by exon 6.

The 5′-flanking region of the mammalian *PARP* gene (8–11) resembles that of other 'housekeeping' genes and lacks TATA and CAAT boxes but contains two consensus sequences for Sp1 binding and three putative AP-2 binding elements (9). Schweiger and associates proposed that the promoter activity could be repressed by an excess of PARP protein (or by its DNA binding domain) interacting specifically with cruciform structures originating from two sets of inverted repeats located upstream of the ATG (5, 12). Thus, a model for autoregulation of the human *PARP* gene was proposed based on two observations. First, deletion analysis in this region demonstrated the potential role of these cruciform structures in the promoter activity (12). Second, PARP interacts preferentially with cruciform structures (13) and forms loops and crossovers with supercoiled DNA in the absence of breaks in the DNA (see Section 2.4.2) (14). Unlike mammalian *PARP* genes, the regulatory region of the *Drosophila PARP* gene contains a putative TATA box and three CCAAT boxes, as well as an octamer motif. *In situ* mRNA hybridization of *Drosophila* embryos revealed large amounts of PARP mRNA in ovaries and in an unidentified cell type dispersed around the central nervous system (7).

A direct correlation exists between cell-cycle progression and a high expression level of PARP mRNA related to the abundance of the enzyme during cell proliferation. The transcriptional induction of PARP is best documented in lymphocytes induced to proliferate with phytohaemaglutinin (15, 16) or after partial hepatectomy (17, 18). The tissue distribution of PARP activity or PARP mRNA has been examined in several organs from mouse (19) and rat (20). High levels of mRNA and of enzymatic activity were observed in testis, thymus, and spleen, while low levels were found in liver, kidney, and heart. Brain is the only example of a terminally differentiated tissue where PARP is not only abundant (19) but also seems to be inducible during glutamate-mediated neurotoxicity (21)—under genotoxic stress, no transcriptional induction can be observed in other tissues or cell types (1).

The tissue distribution of PARP protein, enzyme activity, and mRNA has been examined in several organs from rat (20) and mouse (19). Northern blot analysis indicated that the mouse *PARP* gene is expressed in testis, thymus, spleen, and brain, while low transcript levels were observed in liver and kidney (19). *In situ* hybridization is a powerful technique to precisely localize the cells expressing a defined gene in a given organ. Such an approach was employed to delineate the expression pattern of the *PARP* gene in 18.5-days postcoitum (d.p.c.) mouse fetuses (Fig. 2.1) and in mouse and rat adult tissues (17, 22–24).

In 18.5-d.p.c. mouse fetuses, PARP transcripts are strongly expressed in thymus, spleen, brain, cerebellum, spinal and cranial ganglia, intestine, colon, testis, in tooth buds, and in the nasal cavities (Fig. 2.1). Lymphoid organs, especially the thymus, express the *PARP* gene at high levels, as do the germinal centres of the spleen and Peyer's patches in the ileum, both of which are sites of B-lymphocyte maturation (25). These observations suggest that PARP could play a role in somatic hypermutation of immunoglobulin genes and in immunoglobulin class-switching. These processes were not affected in PARP-deficient mice, but it should be noted that so far, these animals have not been examined in the presence of DNA damaging conditions that are normally required to elicit a PARP response (26, 27).

The mouse *PARP* gene is also specifically expressed in the male and female gonads. In the testis, the mRNA is detected in the peripheral region of the seminiferous tubules, in which are found the spermatogonia, Sertoli cells, and spermatocytes (20) (and Fig. 2.1), whereas expression is reduced in the spermatids (24). The *PARP* gene is also expressed in ovarian follicle cells. Primary follicles display high levels of PARP transcripts, whereas expression decreases during follicle maturation, reaching after ovulation the basal level of expression observed in the ovarian stroma (25).

In the central nervous system, PARP is highly expressed in regions with a high neuronal cell density. That is, the pyramidal cells of the hippocampal

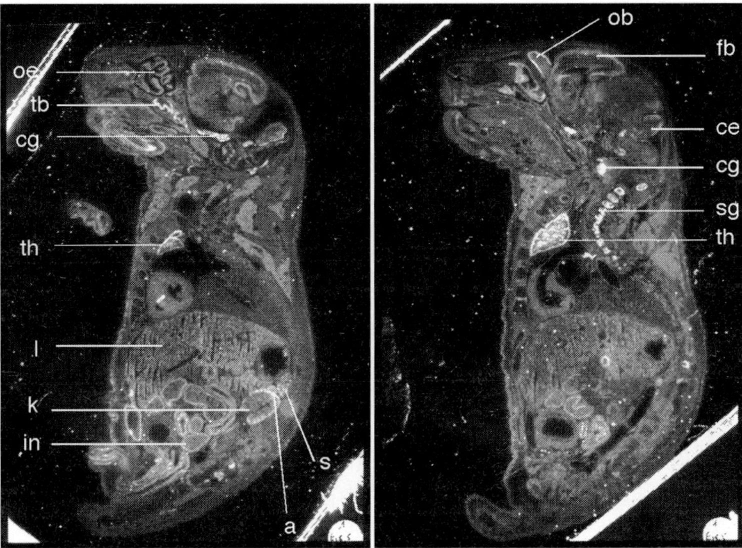

Fig. 2.1 *In situ* hybridization analysis of PARP transcript distribution in 18.5-d.p.c. mouse embryo. Two bright-field views of sagittal sections are shown. Abbreviations: a, adrenal gland; ce, cerebellum; cg, cranial ganglia; fb, forebrain; in, intestine; k, kidney; l, liver; ob, olfactory bulb; oe, olfactory epithelium; s, spleen; sp, spinal ganglia; tb, tooth buds; th, thymus. (25)

regions CA1, CA3, the granule cells of the dentate gyrus, and the Purkinje cells of the cerebellar cortex show strong expression (23, 25). The cranial and spinal ganglia, which also contain postmitotic neurones, express high levels of PARP transcripts. These observations are at odds with the common idea that PARP expression usually correlates with cell proliferation.

Interestingly, there are striking similarities and almost identical tissue expression of PARP and of other genes involved either in base excision repair (BER) such as DNA polymerase-β (19), XRCC1 (28), and DNA ligase III (29), or in the control of checkpoints and genome integrity such as p53 (30) or ATM (31).

2.2 Subcellular location of PARP

PARP activity was initially discovered in chicken liver nuclei (32) and since then this activity has been found in many other tissues and cell types using a variety of techniques. Observation of several cell lines using either conventional fluorescence microscopy (33), confocal laser scanning microscopy (34), or immunoelectron microscopy (35, 36) reveal that PARP is not homogeneously distributed in the nucleus (Plate 2).

In the nucleolus, the label essentially appears on the dense fibrillar component, a nucleolar compartment associated with transcriptional activity, and to a minor extent on the granular component (35). PARP has been shown to translate from the nucleolus to the nucleoplasm when RNA synthesis is inhibited (37), suggesting that its nucleolar location is dependent on the transcriptional state of the nucleolar chromatin. It is worth mentioning that, following γ-irradiation, poly (ADP-ribosyl)ation of authentic nucleolar proteins like numatrin/B23 and nucleolin/C23 (38, 39) does occur, indicating that PARP might directly or indirectly regulate ongoing ribosome biogenesis. The presence of PARP in the nucleolar compartment may be a part of a translation regulation process linked to the necessity to sustain or to resume cellular proliferation when the nucleolar chromatin has been damaged. A number of non-ribosomal proteins localize by a sequestration mechanism to the nucleolus and participate in important physiological processes, like p53 activation (40) and silencing or cell-cycle control (41). The absence of a putative nucleolar location signal in PARP protein sequences invites the speculation that PARP might be sequestered in the nucleolus by an unknown mechanism, which, however, still allows the possibility of shuttling between nucleolus and nucleoplasm when rDNA transcription is halted (37). Alternatively, the presence of PARP in the nucleolus may prevent undesired recombination within the rDNA-repeated sequences particularly during transcription.

The presence of PARP at the nuclear periphery (Plate 2A) is quite compatible with its association with the nuclear matrix (42–49), a proteinaceous structure that gives shape to the nucleus and remains after the removal of histones and all soluble proteins. Indeed, poly (ADP-ribosyl)ation of proteins associated with the nuclear matrix, such as lamins (44) and topoisomerase II (45, 47), has been reported, consistent with the preferential accumulation of PARP at the nuclear rim. The function of this possibly separate class of PARP molecules might rely on the ability of PARP to bind preferentially to supercoiled DNA (14, 50) and to form loops *in vitro* with linear duplexes (14), suggesting a possible role in anchoring DNA loops of interphase chromatin on to the nuclear envelope. With this postulation, the cleavage of PARP during the execution phase of apoptosis may facilitate nuclear shrinkage and chromatin condensation (see Chapter 4). Conversely, the non-cleavable PARP mutant (D214A) delays this key step of the cell-death process by almost 10 hours (51), pointing again to a structural role of PARP in organizing chromatin at the nuclear periphery. Interestingly, this particular distribution of PARP, both in the nucleolus and at the periphery of the nucleus, is shared by other enzymes important for cell survival and DNA metabolism such as Ku-proteins (52), DNA polα (53), pRB (54). In line with this observation it is important to note that PARP and Ku proteins form a complex that synergistically bind to matrix attachment sequences (55).

As shown in Plates 2C and D, anti-PARP or anti-poly ADP-ribose antibodies also stain the chromosomal rim in metaphase chromosomes. This distribution of PARP at the surface of chromosomes may provide markers to reorganize interphase chromatin structures; a similar distribution has been found for Ku-proteins (52).

2.3   PARP is a modular enzyme, highly conserved during evolution

The seminal contribution from Shizuta's laboratory established the multi-functional and modular nature of poly (ADP-ribose) polymerase (56–58). After limited proteolysis of PARP protein, three main functional domains were initially identified. First, there is a 46-kDa amino-terminal fragment including the DNA binding domain. A central 22-kDa fragment that contains the automodification sites follows. Third, there is a carboxy-terminal fragment of 54 kDa bearing the NAD$^+$ binding domain (Fig. 2.2). This organization of serial functional elements conserved during evolution is clearly reflected by the multiple alignment of the amino-acid sequences (Figs 2.3A, B) deduced from 10 cDNAs encoding the human (1, 2, 59, 60), mouse (61), rat (62), bovine (63), chicken (64), *Xenopus* (65), *Drosophila* (66), *Sarcophaga peregrina* (67), *Zea mays* (68, 69), and *Arabidopsis thaliana* (70) PARP proteins. Strikingly, functional domains of PARP, which were identified on the

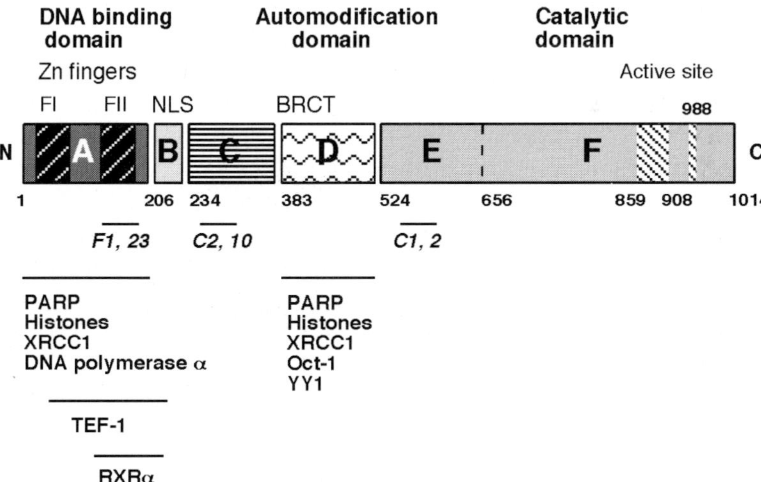

Fig. 2.2   Schematic representation of the domain structure of the human poly (ADP-ribose) polymerase (PARP). NLS, nuclear localization signal; BRCT, *BRCA1 C-t*erminus motif (131, 132). Indicated below the domain structure is the epitope map of various anti-human PARP antibodies (33). Shown below are the binding sites of various PARP partners, including PARP itself (81, 85), histones (85, 116), DNA polymerase-α (33), XRCC1 (113), Oct-1 (105), YY1 (106, 107), TEF-1 (108).

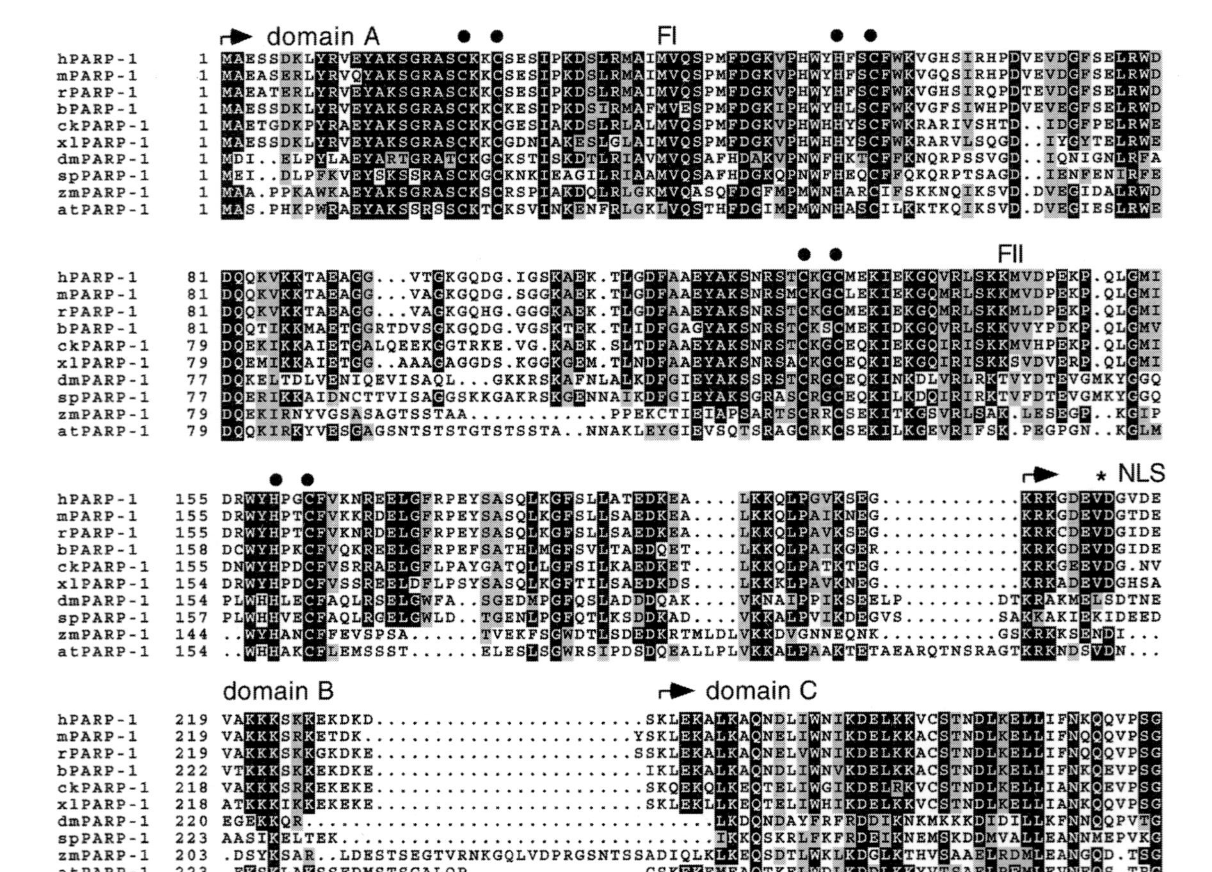

domain A   FI
hPARP-1   1 MAESSDKLYRVEYAKSGRASCKKCSESIPKDSLRMAIMVQSPMFDGKVPHWYHFSCFWKVGHSIRHPDVEVDGFSELRWD
mPARP-1   1 MAEASERLYRVQYAKSGRASCKKCSESIPKDSLRMAIMVQSPMFDGKVPHWYHFSCFWKVGQSIRHPDVEVDGFSELRWD
rPARP-1   1 MAEATERLYRVEYAKSGRASCKKCSESIPKDSLRMAIMVQSPMFDGKVPHWYHFSCFWKVGHSIRQPDTEVDGFSELRWD
bPARP-1   1 MAESSDKLYRVEYAKSGRASCKKCKESIPKDSIRMAFMVESPMFDGKIPHWYHLSCFWKVGFSIWHPDVEVEGFSELRWD
ckPARP-1   1 MAETGDKPYRAEYAKSGRASCKKCGESIAKDSLRLALMVQSPMFDGKVPHWHHYSCFWKRARIVSHTD..IDGFPELRWE
xlPARP-1   1 MAESSDKLYRVEYAKSGRASCKKCGDNIAKESLGLAIMVQSPMFDGKVPHWHHYSCFWKRARVLSQGD..IYGYTELRWE
dmPARP-1   1 MDI..ELPYLAEYARTGRATCKGCKSTISKDTLRIAVMVQSAFHDAKVPNWFHKTCFFKNQRPSSVGD..IQNIGNLRFA
spPARP-1   1 MEI..DLPFKVEYSKSSRASCKGCKNKIEAGILRIAAMVQSAFHDGKQPNWFHEQCFFQKQRPTSAGD..IENFENIRFE
zmPARP-1   1 MAA.PPKAWKAEYAKSGRASCKSCRSPIAKDQLRLGKMVQASQFDGFMPMWNHARCIFSKKNQIKSVD.DVEGIDALRWD
atPARP-1   1 MAS.PHKPWRAEYAKSGRSSCKTCKSVINKENFRLGKLVQSTHFDGIMPMWNHASCILKKTKQIKSVD.DVEGIESLRWE

                                        FII
hPARP-1   81 DQQKVKKTAEAGG...VTGKGQDG.IGSKAEK.TLGDFAAEYAKSNRSTCKGCMEKIEKGQVRLSKKMVDPEKP.QLGMI
mPARP-1   81 DQQKVKKTAEAGG...VAGKGQDG.SGGKAEK.TLGDFAAEYAKSNRSMCKGCLEKIEKGQMRLSKKMVDPEKP.QLGMI
rPARP-1   81 DQQKVKKTAEAGG...VAGKGQHG.GGGKAEK.TLGDFAAEYAKSNRSTCKGCMEKIEKGQMRLSKKMLDPEKP.QLGMI
bPARP-1   81 DQQTIKKMAETGGRTDVSGKGQDG.VGSKTEK.TLIDFGAGYAKSNRSTCKSCMEKIDKGQVRLSKKVVYPDKP.QLGMV
ckPARP-1   79 DQEKIKKAIETGALQEEKGGTRKE.VG.KAEK.SLTDFAAEYAKSNRSTCKGCEQKIEKGQIRISKKMVHPEKP.QLGMI
xlPARP-1   79 DQEMIKKAIETGG...AAAGAGGDS.KGGKGEM.TLNDFAAEYAKSNRSACKGCEQKIEKGQIRISKKSVDVERP.QLGMI
dmPARP-1   77 DQKELTDLVENIQEVISAQL...GKKRSKAFNLALKDFGIEYAKSSRSTCRGCEQKINKDLVRLRKTVYDTEVGMKYGGQ
spPARP-1   77 DQERIKKAIDNCTTVISAGCGSKKGAKRSKGENNAIKDFGIEYAKSGRASCGCEQKILKDQIRIRKTVFDTEVGMKYGGQ
zmPARP-1   79 DQEKIRNYVGSASAGTSSTAA...........PPEKCTIEIAPSARTSCRRCSEKITKGSVRLSAK.LESEGP.KGIP
atPARP-1   79 DQQKIRKYVESGAGSNTSTSTGTSTSSTA.NNAKLEYGIEVSQTSRAGCRKCSEKILKGEVRIFSK.PEGPGN..KGLM

                                                              domain C ➜   * NLS
hPARP-1   155 DRWYHPGCFVKNREELGFRPEYSASQLKGFSLLATEDKEA....LKKQLPGVKSEG...........KRKGDEVDGVDE
mPARP-1   155 DRWYHPTCFVKKRDELGFRPEYSASQLKGFSLLSAEDKEA....LKKQLPAIKNEG...........KRKGDEVDGTDE
rPARP-1   155 DRWYHPTCFVKNRDELGFRPEYSASQLKGFSLLSAEDKEA....LKKQLPAVKSEG...........KRKCDEVDGIDE
bPARP-1   158 DCWYHPKCFVQKREELGFRPEFSATHLMGFSVLTAEDQET....LKKQLPAIKGER...........KRKGDEVDGIDE
ckPARP-1 155 DNWYHPDCFVSRRAELGFLPAYGATQLLGFSILKAEDKET....LKKQLPATKTEG...........KRKGEEVDG.NV
xlPARP-1 154 DRWYHPDCFVSSREELDFLPSYSASQLKGFTILSAEDKDS....LKKKLPAVKNEG...........KRKADEVDGHSA
dmPARP-1 154 PLWHHLECFAQLRSELGWFA..SGEDMPGFQSIADDDQAK....VKNAIPPIKSEELP.......DTKRAKMELSDTNE
spPARP-1 157 PLWHHVECFAQLRGELGWLD..TGENLPGFQTLKSDDKAD....VKKALPVIKDEGVS........SAKKAKIEKIDEED
zmPARP-1 144 ..WYHANCFEVSPSA......TVEKFSGWDTLSDEDKRTMLDLVKKDVGNNEQNK......GSKRKKSENDI...
atPARP-1 154 ..WHHAKCFLEMSSST......ELESLSGWRSIPDSDQEALLPLVKKALPAAKTETAEARQTNSRAGTKRKNDSVDN...

              domain B   domain C ➜
hPARP-1   219 VAKKKSKKEKDKD...................SKLEKALKAQNDLIWNIKDELKKVCSTNDLKELLIFNKQQVPSG
mPARP-1   219 VAKKKSRKETDK....................YSKLEKALKAQNELIWNIKDELKKACSTNDLKELLIFNQQQVPSG
rPARP-1   219 VAKKKSKKGKDKE...................SSKLEKALKAQNELVWNIKDELKKACSTNDLKELLIFNQQQVPSG
bPARP-1   222 VTKKKSKKEKDKE...................IKLEKALKAQNDLIWNVKDELKKACSTNDLKELLIFNKQEVPSG
ckPARP-1 218 VAKKKSRKEKEKE...................SKQEKQLKEQTELIWGIKDELRKVCSTNDLKELLIANKQEVPSG
xlPARP-1 218 ATKKKIKKEKEKE...................SKLEKLLKEQTELIWHIKDELKKVCSTNDLKELLIANKQQVPSG
dmPARP-1 220 EGEKKQR.........................LKDQNDAYFRFRDDIKNKMKKKDIDILLKFNNQQPVTG
spPARP-1 223 AASIKELTEK......................IKKQSKRLFKFRDEIKNEMSKDDMVALLEANNMEPVKG
zmPARP-1 203 .DSYKSAR..LDESTSEGTVRNKGQLVDPRGSNTSSADIQLKLKEQSDTLWKLKDGLKTHVSAAELRDMLEANGQD.TSG
atPARP-1 223 .EKSKLAKSSFDMSTSGALQP...............CSKEKEMEAQTKELWDLKDDLKKYVTSAELREMLEVNEQS.TRG

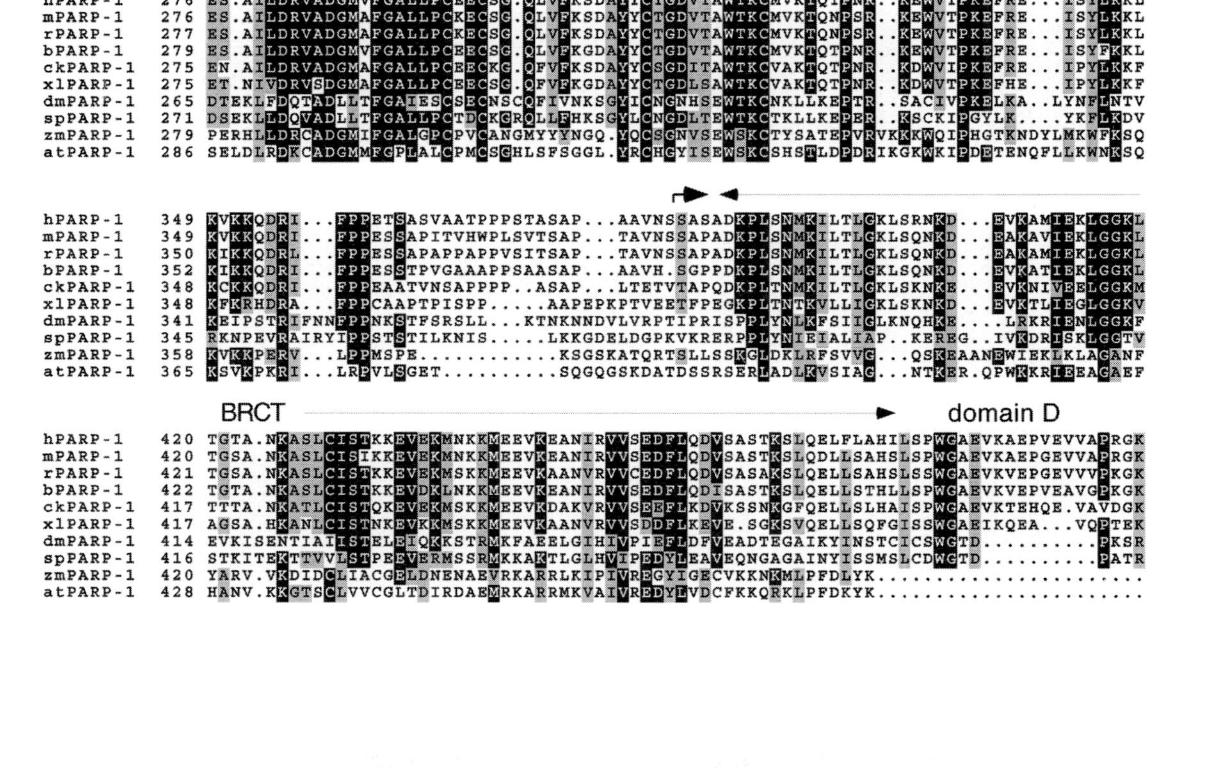

hPARP-1 276 ES.AILDRVADGMVFGALLPCEECSG.QLVFKSDAYYCTGDVTAWTKCMVKTQTPNR..KEWVTPKEFRE...ISYLKKL
mPARP-1 276 ES.AILDRVADGMAFGALLPCKECSG.QLVFKSDAYYCTGDVTAWTKCMVKTQNPSR..KEWVTPKEFRE...ISYLKKL
rPARP-1 277 ES.AILDRVADGMAFGALLPCKECSG.QLVFKSDAYYCTGDVTAWTKCMVKTQNPSR..KEWVTPKEFRE...ISYLKKL
bPARP-1 279 ES.AILDRVADGMVFGALLPCEECSG.QLVFKGDAYYCTGDVTAWTKCMVKTQTPNR..KEWVTPKEFRE...ISYFKKL
ckPARP-1 275 EN.AILDRVADGMAFGALLPCEECKG.QFVFKSDAYYCSGDITAWTKCVAKTQTPNR..KDWVIPKEFRE...IPYLKKF
xlPARP-1 275 ET.NIVDRVSDGMAFGALLPCEECSG.QFVFKGDAYYCTGDLSAWTKCVAKTQTPNR..KDWVTPKEFHE...IPYLKKF
dmPARP-1 265 DTEKLFDQTADLLTFGAIESCSECNSCQFIVNKSGYICNGNHSEWTKCNKLLKEPTR..SACIVPKELKA..LYNFLNTV
spPARP-1 271 DSEKLLDQVADLLTFGALLPCTDCKGRQLLFHKSGYLCNGDLTEWTKCTKLLKEPER..KSCKIPGYLK....YKFLKDV
zmPARP-1 279 PERHLLDRCADGMIFGALGPCPVCANGMYYYNGQ.YQCSGNVSEWSKCTYSATEPVRVKKKWQIPHGTKNDYLMKWFKSQ
atPARP-1 286 SELDLRDKCADGMMFGPLALCPMCSGHLSFSGGL.YRCHGYISEWSKCSHSTLDPDRIKGKWKIPDETENQFLLKWNKSQ

hPARP-1 349 KVKKQDRI...FPPETSASVAATPPPSTASAP...AAVNSSASADKPLSNMKILTLGKLSRNKD...EVKAMIEKLGGKL
mPARP-1 349 KVKKQDRI...FPPESSAPITVHWPLSVTSAP...TAVNSSAPADKPLSNMKILTLGKLSQNKD...EAKAVIEKLGGKL
rPARP-1 350 KIKKQDRL...FPPESSAPAPPAPPVSITSAP...TAVNSSAPADKPLSNMKILTLGKLSQNKD...EAKAMIEKLGGKL
bPARP-1 352 KIKKQDRI...FPPESSTPVGAAAPPSAASAP...AAVH.SGPPDKPLSNMKILTLGKLSQNKD...EVKATIEKLGGKL
ckPARP-1 348 KCKKQDRI...FPPEAATVNSAPPPP..ASAP...LTETVTAPQDKPLTNMKILTLGKLSKNKE...EVKNIVEELGGKM
xlPARP-1 348 KFKRHDRA...FPPCAAPTPISPP.....AAPEPKPTVEETFPEGKPLTNTKVLLIGKLSKNKD...EVKTLIEGLGGKV
dmPARP-1 341 KEIPSTRIFNNFPPNKSTFSRSLL...KTNKNNDVLVRPTIPRISPPLYNLKFSIIGLKNQHKE....LRKRIENLGGKF
spPARP-1 345 RKNPEVRAIRYIPPSTSTILKNIS.....LKKGDELDGPKVKRERPPLYNIEIALIAP..KEREG..IVKDRISKLGGTV
zmPARP-1 358 KVKKPERV...LPPMSPE...........KSGSKATQRTSLLSSKGLDKLRFSVVG...QSKEAANEWIEKLKLAGANF
atPARP-1 365 KSVKPKRI...LRPVLSGET.........SQGQGSKDATDSSRSERLADLKVSIAG...NTKER.QPWKKRIEEAGAEF

BRCT   domain D

hPARP-1 420 TGTA.NKASLCISTKKEVEKMNKKMEEVKEANIRVVSEDFLQDVSASTKSLQELFLAHILSPWGAEVKAEPVEVVAPRGK
mPARP-1 420 TGSA.NKASLCISIKKEVEKMNKKMEEVKEANIRVVSEDFLQDVSASTKSLQDLLSAHSLSPWGAEVKAEPGEVVAPRGK
rPARP-1 421 TGSA.NKASLCISTKKEVEKMSKKMEEVKAANVRVVCEDFLQDVSASAKSLQELLSAHSLSSWGAEVKVEPGEVVVPKGK
bPARP-1 422 TGTA.NKASLCISTKKEVDKLNKKMEEVKEANIRVVSEDFLQDISASTKSLQELLSTHLLSPWGAEVKVEPVEAVGPKGK
ckPARP-1 417 TTTA.NKATLCISTQKEVEKMSKKMEEVKDAKVRVVSEEFLKDVKSSNKGFQELLSHAISPWGAEVKTEHQE.VAVDGK
xlPARP-1 417 AGSA.HKANLCISTNKEVEKMSKKMEEVKAANVRVVSDDFLKEVE.SGKSVQELLSQFGISSWGAEIKQEA...VQPTEK
dmPARP-1 414 EVKISENTIAIISTELEIQKKSTRMKFAEELGIHIVPIEFLDFVEADTEGAIKYINSTCICSWGTD.........PKSR
spPARP-1 416 STKITEKTTVVLSTPEEVERMSSRMKKAKTLGLHVIPEDYLEAVEQNGAGAINYISSMSLCDWGTD.........PATR
zmPARP-1 420 YARV.VKDIDCLIACGELDNENAEVRKARRLKIPIVREGYIGECVKKNKMLPFDLYK..................
atPARP-1 428 HANV.KKGTSCLVVCGLTDIRDAEMRKARRMKVAIVREDYLVDCFKKQRKLPFDKYK..................

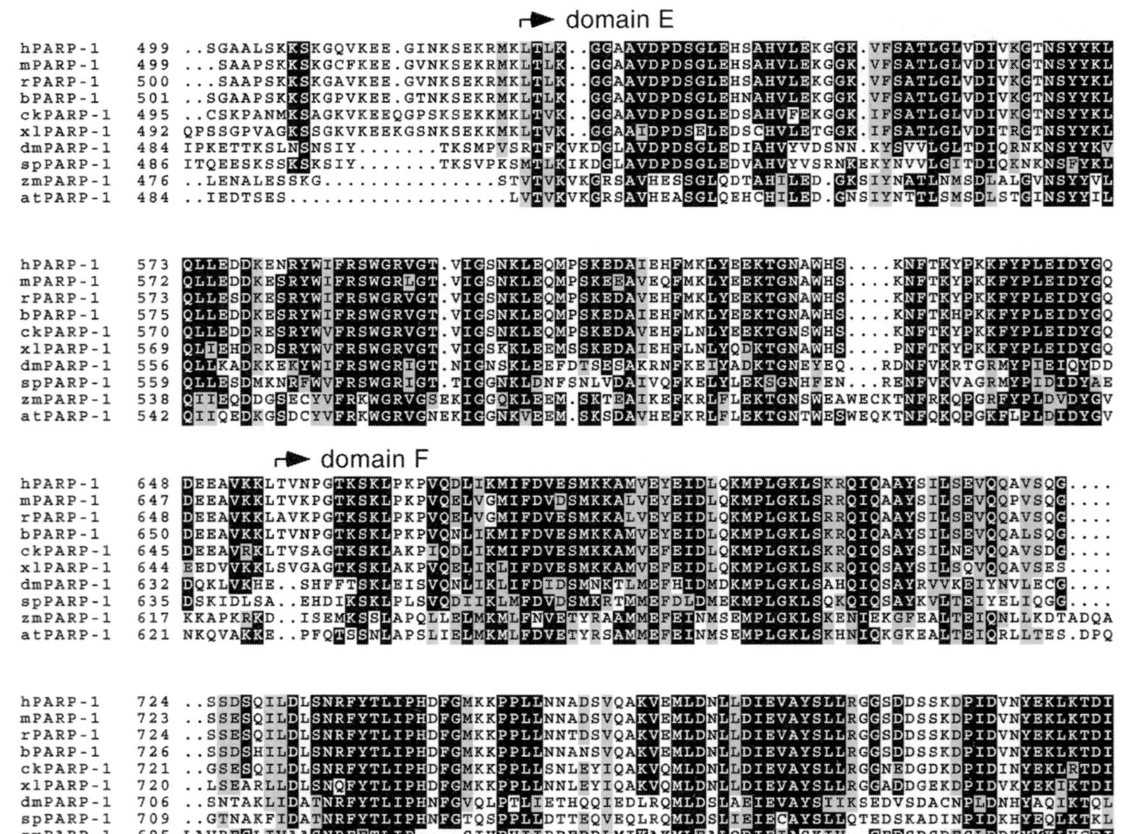

domain E

hPARP-1   499 ..SGAALSKKSKGQVKEE.GINKSEKRMKLTLK..GGAAVDPDSGLEHSAHVLEKGGK.VFSATLGLVDIVKGTNSYYKL
mPARP-1   499 ...SAAPSKKSKGCFKEE.GVNKSEKRMKLTLK..GGAAVDPDSGLEHSAHVLEKGGK.VFSATLGLVDIVKGTNSYYKL
rPARP-1   500 ..SAAPSKKSKGAVKEE.GVNKSEKRMKLTLK..GGAAVDPDSGLEHSAHVLEKGGK.VFSATLGLVDIVKGTNSYYKL
bPARP-1   501 .SGAAPSKKSKGPVKEE.GTNKSEKRMKLTLK..GGAAVDPDSGLEHNAHVLEKGGK.VFSATLGLVDIVKGTNSYYKL
ckPARP-1 495 ..CSKPANMKSAGKVKEEQGPSKSEKKMKLTVK..GGAAVDPDSGLEDSAHVFEKGGK.IFSATLGLVDIVKGTNSYYKL
xlPARP-1 492 QPSSGPVAGKSSGKVKEEKGSNKSEKKMKLTVK..GGAAIDPDSELEDSCHVLETGGK.IFSATLGLVDITRGTNSYYKL
dmPARP-1 484 IPKETTKSLNSNSIY.......TKSMPVSRTFKVKDGLAVDPDSGLEDIAHVYVDSNN.KYSVVLGLTDIQRNKNSYYKV
spPARP-1 486 ITQEESKSSKSKSIY.......TKSVPKSMTLKIKDGLAVDPDSGLEDVAHVYVSRNKEKYNVVLGITDIQKNKNSFYKL
zmPARP-1 476 ..LENALESSKG................STVTVKVKCRSAVHESSGLQDTAHILED.GKSIYNATLNMSDLALGVNSYYVL
atPARP-1 484 ..IEDTSES..................LVTVKVKCRSAVHEASGLQEHCHILED.GNSIYNTTLSMSDLSTGINSYYIL

hPARP-1   573 QLLEDDKENRYWIFRSWGRVGT.VIGSNKLEQMPSKEDAIEHFMKLYEEKTGNAWHS....KNFTKYPKKFYPLEIDYGQ
mPARP-1   572 QLLEDDKESRYWIFRSWGRLGT.VIGSNKLEQMPSKEEAVEQFMKLYEEKTGNAWHS....KNFTKYPKKFYPLEIDYGQ
rPARP-1   573 QLLESDKESRYWIFRSWGRVGT.VIGSNKLEQMPSKEDAVEHFMKLYEEKTGNAWHS....KNFTKYPKKFYPLEIDYGQ
bPARP-1   575 QLLEDDKESRYWIFRSWGRVGT.VIGSNKLEQMPSKEDAIEHFMKLYEEKTGNAWHS....KNFTKHPKKFYPLEIDYGQ
ckPARP-1 570 QLLEDDRESRYWVFRSWGRVGT.VIGSNKLEQMPSKEDAVEHFLNLYEEKTGNSWHS....KNFTKYPKKFYPLEIDYGQ
xlPARP-1 569 QLIEHDRDSRYWVFRSWGRVGT.VIGSKKLEEMSSKEDAIEHFLNLYQDKTGNAWHS....PNFTKYPKKFYPLEIDYGQ
dmPARP-1 556 QLLKADKKEKYIFRSWGRIGT.NIGNSKLEEFDTSESAKRNFKEIYADKTGNEYEQ...RDNFVKRTGRMYPIEIQYDD
spPARP-1 559 QLLESDMKNRFWVFRSWGRIGT.TIGGNKLDNFSNLVDAIVQFKELYLEKSGNHFEN...RENFVKVAGRMYPIDIDYAE
zmPARP-1 538 QITEQDDGSECYVFRKWGRVGSEKIGGQKLEEM.SKTEAIKEFKRLFLEKTGNSWEAWECKTNFRKQPGRFYPLDVDYGV
atPARP-1 542 QIIQEDKGSDCYVFRKWGRVGNEKIGGNKVEEM.SKSDAVHEFKRLFLEKTGNTWESWEQKTNFQKQPGKFLPLDIDYGV

domain F

hPARP-1   648 DEEAVKKLTVNPGTKSKLPKPVQDLIKMIFDVESMKKAMVEYEIDLQKMPLGKLSKRQIQAAYSILSEVQQAVSQG....
mPARP-1   647 DEEAVKKLTVKPGTKSKLPKPVQELVGMIFDVDSMKKALVEYEIDLQKMPLGKLSRRQIQAAYSILSEVQQPVSQG....
rPARP-1   648 DEEAVKKLAVKPGTKSKLPKPVQELVGMIFDVDSMKKALVEYEIDLQKMPLGKLSRRQIQAAYSILSEVQQAVSQG....
bPARP-1   650 DEEAVKKLTVNPGTKSKLPKPVQNLIKMIFDVESMKKAMVEYEIDLQKMPLGKLSKRQIQAAYSILSEVQQALSQG....
ckPARP-1 645 DEEAVRKLTVSAGTKSKLAKPIQDLIKMIFDVESMKKAMVEFEIDLQKMPLGKLSKRQIQSAYSILNEVQQAVSDG....
xlPARP-1 644 EEDVVKKLSVGAGTKSKLAKPVQELIKLIFDVESMKKAMVEFEIDLQKMPLGKLSKRQIQSAYSILSQVQQAVSES....
dmPARP-1 632 DQKLVKHE..SHFFTSKLEISVQNLIKLIFDIDSMNKTLMEFHIDMDKMPLGKLSAHQIQSAYRVVKEIYNVLECG....
spPARP-1 635 DSKIDLSA..EHDIKSKLPLSVQDIIKLMFDVDSMKRTMMEFDLDMEKMPLGKLSQKQIQSAYKVLTEIYELIQGG....
zmPARP-1 617 KKAPKRKD..ISEMKSSLAPQLLELMKMLFNVETYRAAMMEFEINSEMPLGKLSKENIEKGFEALTEIQNLLKDTADQA
atPARP-1 621 NKQVAKKE..PFQTSSNLAPSLIELMKMLFDVETYRSAMMEFEINSEMPLGKLSKHNIQKGKEALTEIQRLLTES.DPQ

hPARP-1   724 ..SSDSQILDLSNRFYTLIPHDFGMKKPPLLNNADSVQAKVEMLDNLLDIEVAYSLLRGGSDDSSKDPIDVNYEKLKTDI
mPARP-1   723 ..SSESQILDLSNRFYTLIPHDFGMKKPPLLNNADSVQAKVEMLDNLLDIEVAYSLLRGGSDDSSKDPIDVNYEKLKTDI
rPARP-1   724 ..SSESQILDLSNRFYTLIPHDFGMKKPPLLNNTDSVQAKVEMLDNLLDIEVAYSLLRGGSDDSSKDPIDVNYEKLKTDI
bPARP-1   726 ..SSDSHILDLSNRFYTLIPHDFGMKKPPLLNNANSVQAKVEMLDNLLDIEVAYSLLRGGSDDSSKDPIDVNYEKLKTDI
ckPARP-1 721 ..GSESQILDLSNRFYTLIPHDFGMKKPPLLSNLEYIQAKVQMLDNLLDIEVAYSLLRGGNEDGDKDPIDINYEKLRTDI
xlPARP-1 720 ..LSEARLLDLSNQFYTLIPHDFGMKKPPLLNNLEYIQAKVQMLDNLLDIEVAYSLLRGGADDGEKDPIDVKYEKIKTDI
dmPARP-1 706 ..SNTAKLIDATNRFYTLIPHNFGVQLPTLIETHQQIEDLRQMLDSLAEIEVAYSIIKSEDVSDACNPLDNHYAQIKTQL
spPARP-1 709 ..GTNAKFIDATNRFYTLIPHNFGTQSPPLLDTTEQVEQLRQMLDSLIEIECAYSLLQTEDSKADINPIDKHYEQLKTKL
zmPARP-1 695 LAVRESLIVAASNRFFTLIP....SIHPHIIRDEDDLMIKAKMLEALQDIEIASKIV..GFDSDSDESLDDKYMKLHCDI
atPARP-1 698 PIMKESLLVDASNRFFTMIP....IIHPHIIRDEDDFKSKVKMLEALQDIEIASRIV..GFDVDSTESLDDQNKKLHCDI

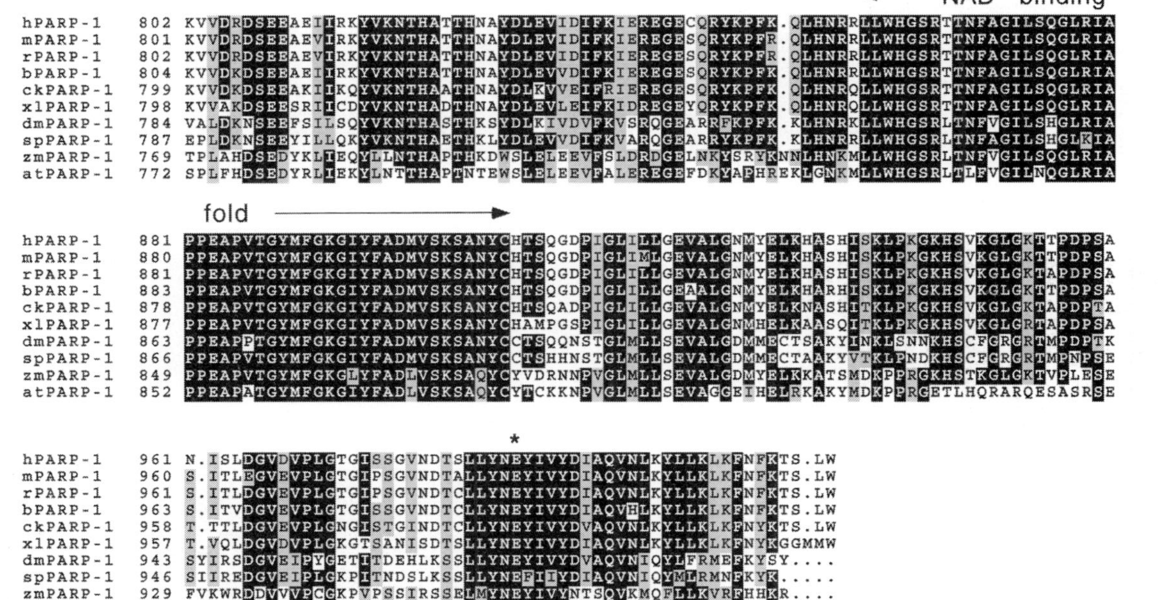

Fig. 2.3 Alignment of the deduced amino-acid sequences of the human poly (ADP-ribose) polymerase (hPARP-1, accession number P09874), mouse (mPARP-1, accession number P11103), rat (rPARP-1, accession number P27008), bovine (bPARP-1, accession number P18493), chicken (ckPARP-1 accession number P26446), *X. laevis* (xlPARP-1, accession number P31669), *D. melanogaster* (dmPARP-1, accession number P35875), *S. peregrina* (spPARP-1, accession number D16482), *Zea mays* (zmPARP-1 accession number AF093627; AJ222589), and *A. thaliana* (atPARP-1 accession number AJ131705). Identical amino-acid residues are boxed in black. Conserved substitutions are indicated in grey. All the functional domains start with arrows. Closed circles refer to amino acids involved in Zn^{2+} coordination. The asterisk (*) points to the caspases' cleavage after D214 at the DEVD sequence located within the bipartite nuclear localization signal (NLS) and to the glutamate E988 involved in polymer elongation and located in domain F.

basis of limited proteolysis (see above), correspond to blocks of conserved regions interspaced by highly variable sequences, probably organized in loops accessible to proteases. The multiple alignment in Figs 2.3A, B shows sequence-specific regions for plant PARPs (i.e. end of domain A; end of domain E), whereas other regions are specific to vertebrates (i.e. end of finger FII) or invertebrates (end of BRCT (*br*east *ca*ncer susceptibility protein, BRCA1, *C-t*erminus) motif).

The modular structure of the PARP molecule was fully confirmed by structure–function relationship analyses (71–85). This organization was dissected further into smaller regions (A to F in Fig. 2.2) sharing homology with previously identified functional modules present in other multifunctional enzymes and bearing specialized and autonomous functions.

2.4   The various DNA binding strategies of PARP

2.4.1   The zinc-finger domain (module A) is an autonomous recognition element of DNA breaks

The PARP DNA binding domain (DBD) is located in the amino-terminal region (Fig. 2.2). It contains a repeated sequence (amino acids 2–97 and 106–207). A zinc-finger motif of the form $Cx_2Cx_{28, 30}Hx_2C$ is strictly conserved from human to the plant *A. thaliana* within these sequences (Figs 2.3A and 2.4). This suggests that the PARP DBD might have arisen by sequence divergence of a duplicated primordial element, evolving to form two independently folded zinc-containing domains (FI, FII). Interestingly, a sequence of 97 residues corresponding to finger FI is also present in the N-terminal domain of the human DNA ligase III protein (29, 86, 87) (Fig. 2.4), which is an essential enzyme of the base excision repair (BER) pathway. In PARP, FI (residues 1–97) is encoded by two exons (4), whereas in DNA ligase III the homologous FI motif (residues 1–97) is encoded exactly by a single exon (A. Tomkinson, personal communication). The FI in both enzymes is recognized by a polyclonal antibody against the human PARP FI motif, but not by an anti-FII antibody (88). The N-terminal region of PARP by itself acts as a detector of single-strand breaks. It interacts with one and a half turns of the double helix, protecting seven nucleotides each side of the break, irrespective of the nucleotide sequence (89). Presumably, PARP exploits the flexibility created by a sugar–phosphate backbone interruption in DNA, to bend the nicked duplex by an angle of about 100 degrees. The characteristic V-shaped conformation of the PARP–DNA complex as visualized by dark-field electron microscopy (90), may favour the formation of an active PARP dimer ready to accept its substrate NAD^+ (91) (Fig. 2.5). A similar bent conformation of DNA has recently been observed in DNA polymerase-β, co-crystallized with a nicked duplex of DNA (92) or in interaction with nicked DNA

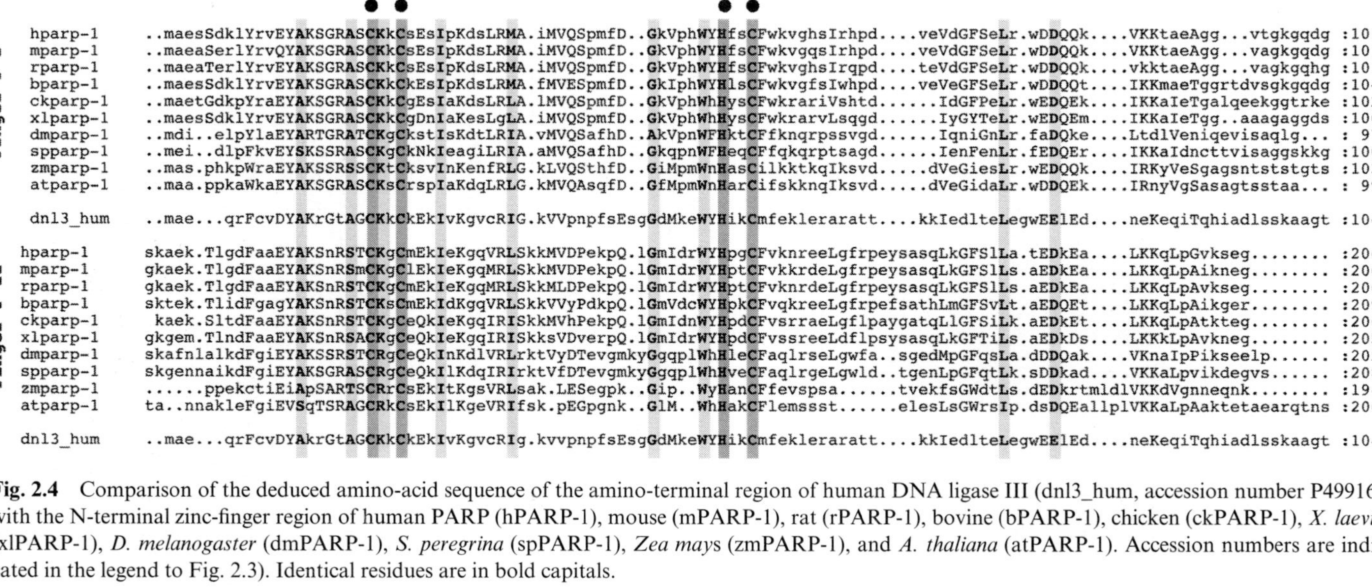

Fig. 2.4 Comparison of the deduced amino-acid sequence of the amino-terminal region of human DNA ligase III (dnl3_hum, accession number P49916) with the N-terminal zinc-finger region of human PARP (hPARP-1), mouse (mPARP-1), rat (rPARP-1), bovine (bPARP-1), chicken (ckPARP-1), *X. laevis* (xlPARP-1), *D. melanogaster* (dmPARP-1), *S. peregrina* (spPARP-1), *Zea mays* (zmPARP-1), and *A. thaliana* (atPARP-1). Accession numbers are indicated in the legend to Fig. 2.3). Identical residues are in bold capitals.

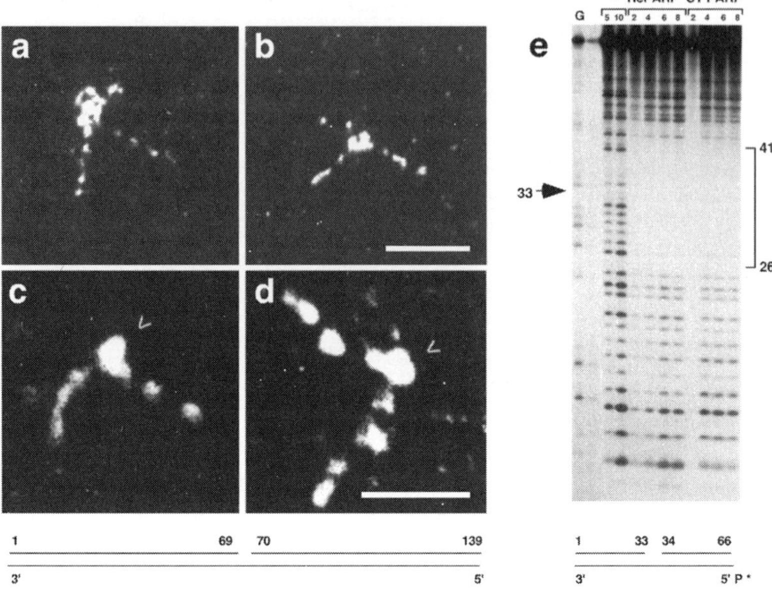

Fig. 2.5 Electron micrographs of PARP-nicked DNA complexes (a–d). PARP is located at the apex of a 139-bp DNA duplex containing a single-strand break at position 69. The bars represent 50 nm. DNase I footprinting of human (RePARP) and calf thymus PARP (CT PARP) bound to a 66-bp DNA duplex containing a single-strand break located at position 33. PARP protects the nucleotides 26–41, on the continuous strand.

and XRCC1 (93). Localized DNA flexibility, as well as an appropriate shape of the interacting proteins, may contribute *in vivo* to target site selection by DNA bending factors that may constitute road blocks for ongoing replication until repair is complete.

In prokaryotes, the HU protein, which is a ubiquitous low molecular weight protein, has been shown to bind to double-strand DNA containing a nick, a gap of 1 to 2 nucleotides (94, 95), and also to DNA with either cruciform or bulge structures (96). HU controls the architecture of genomic DNA in bacteria and participates in many cellular processes involving DNA including replication, transcription, and packaging DNA in chromosome-like structures. Although PARP and HU are not related in sequence, they may, however, represent two independent solutions to the same biological problem.

The precise location of the DNA binding site(s) in the N-terminal domain is still unknown. However, deletion experiments performed with the PARP full-length protein, indicate that the disruption of either of the two zinc fingers still leaves the capacity to recognize a single-strand DNA interruption (Gradwohl *et al.* unpublished results). Previous results derived from the over-expression of the N-terminal domain, suggested that the second zinc finger alone determined the specificity for single-strand breaks (73). However, this data needs therefore to be re-evaluated. In agreement with this view, the unique zinc finger in DNA ligase III produces the same DNA footprint, as does PARP (88). In both cases the footprint is independent of the nucleotide sequence, thus reinforcing the structural and functional similarities between the N-termini of PARP and DNA ligase III. These two enzymes may form a new class of zinc-finger proteins in which one or two zinc-finger modules serve primarily to target a repair function at a single-strand break in DNA. The crystal structure of the PARP zinc-finger domain should strongly improve our understanding of how this recognition element functions.

In a series of deletion and mutation experiments, the function of each zinc finger has been evaluated through the DNA binding capacity and the consequent enzymatic activation of PARP mutants (74) (Gradwohl *et al.*, unpublished results). The results clearly show that FI is a determinant structure for PARP activation whichever DNA end is used, whereas an FII deletion mutant can still be activated to a certain extent by double-strand DNA, suggesting that the position of FI may be critical with respect to the active site. The glucocorticoid receptor (GR) also contains two zinc fingers that are involved in the specific recognition of GR-responsive elements. When the PARP zinc-finger domain was replaced by the DNA binding domain of the GR, the chimeric enzyme was targeted to GRE sites but failed, not surprisingly, to stimulate PARP activity in a DNA-dependent manner (82). Finally, it is also important to keep in mind that PARP activation depends not just on residues located in the zinc fingers. Two random mutations (K249E and G313E) in domain C were recently found that abolish PARP activity without affecting its DNA binding capacity (80).

2.4.2　PARP binds to various bent DNA structures

Although PARP is exclusively activated by DNA ends (97, 98), *in vitro* PARP binds to other DNA structures such as cruciform (13), curved (99, 100) (Gradwohl, unpublished results), or supercoiled DNA (14, 50) with a high affinity and in a cooperative manner. The common feature between all these structures and DNA fragments containing a nick or a gap, is the possible occurrence of a sharp angle which could meet the capacity of PARP to bind to two helices simultaneously to form loops with any type of DNA (Fig. 2.6). This, as yet, unexplored property of PARP is shared by eukaryotic topoisomerases II and I (101, 102) and by yeast chromatin remodelling complexes, SWI/SNF (103). This property could be relevant to its function at the nuclear matrix (see above). Alternatively, this could explain the frequently detected association between PARP and a number of transcription factors or co-activators that have been reported recently, including AP-2 (104), Oct-1 (105), YY1 (106, 107), TEF-1 (108), and RXR and (109) NF-κB (110, 111).

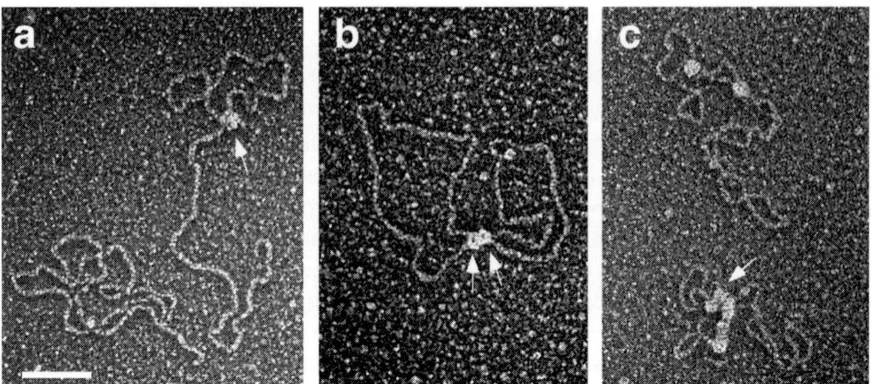

Fig. 2.6   Visualization of PARP–DNA complexes. The formation of DNA loops are particularly visible in PARP–pBR322 (form III) complexes (panel a) or in PARP–pBR322 (form II) (panel b). The recognition of DNA crossovers present in pBR322 (form I) by PARP is shown in panel c. The bar represents 100 nm.

Stereospecific protein–nucleic acid complexes may be formed that are in turn recognized by other factors and/or by the basal transcription machinery for possible synergy. The DNA binding site(s) responsible for this specific function reside mainly in the central part of the PARP molecule (domains C–D, 36 kDa, Fig. 2.2) that binds to curved DNA (99, 112) (Gradwohl *et al.* unpublished results).

2.4.3   The N-terminal domain also binds PARP partners

It has recently been shown that the N-terminal region of PARP not only binds to DNA but is also an interface for protein–protein interaction with several partners, including PARP, histones (81), DNA polymerase-α (53), XRCC1 (113), and the transcription factors TEF-1 (108) and RxRα (109) (Fig. 2.2). Interestingly, the integrity of the zinc fingers is required not only for DNA binding, but also for protein binding (53). However, in the case of the XRCC1–PARP complex the DNA binding function of PARP is maintained (113). Clearly, this dual function of the PARP zinc-finger domain, also demonstrated in several other zinc-finger proteins (114, 115), needs to be further investigated to be fully understood.

2.4.4   The activation and the regulatory mechanisms

Little is known about the mechanism by which the basal PARP activity is stimulated 500-fold in response to binding to a DNA break. The multiple enzymatic activities catalysed by PARP (initiation, elongation, branching, self-modification) are the subject of a complex regulatory mechanism, which may involve allosteric changes associated with the formation of a catalytic dimer at a DNA break and with torsional constraints imposed on the activator DNA, upon DNA binding (90, 116). Most of the physical evidence supporting the formation of a catalytic dimer of PARP comes from kinetic experiments, which convincingly demonstrate that optimal activity occurs at a strict stoichiometry of two polymerase molecules per DNA end (91, 117). Deviation from this ratio results in decreased polymer formation, thus suggesting that strict automodification of PARP molecules, cycling on and off DNA strand breaks, seems unlikely. Instead, a heteromodification mechanism (modification of another PARP molecule by the 'activated' one) has been put forward by Panzeter and Althaus (117). This model excludes both molecules functioning simultaneously as catalyst and acceptor.

The formation of a catalytic dimer at a DNA end is compatible with the symmetry of the DNase I footprint of PARP on a DNA break (89) and with chemical cross-linking experiments performed on purified PARP (118). The strong dependence of PARP activity on PARP concentration can be explained by the local cooperative binding of PARP to DNA as followed by electron microscopy (EM) (Gradwohl *et al.* unpublished results) or gel retardation (Molinete *et al.* unpublished results). Both techniques clearly demonstrate a local colligative property of PARP, at an increasing ratio of enzyme to damaged or undamaged DNA. The heteromodification model, based on a ratio of one catalyst to one acceptor molecule, is in agreement with the distal-end addition of ADP-ribose residues on the growing polymer chain (119–121) and with the geometry of the donor and the acceptor sites in the catalytic domain (see Section 2.7). However, it is not clear how a multi-branched ADP-ribose polymer can be synthesized. More work will be necessary to understand at the molecular level how the NAD^+ concentration determines the average ADP-ribose polymer size (polymerization reaction), and how the frequency of DNA strand breaks determines the total number of polymer chains (122). It is now essential to identify the amino-acid residues involved in the formation of the catalytic dimer.

2.5　The bipartite nuclear localization signal (NLS) contains a cleavage site for Caspase-3

An autonomous module of 38 amino acids (domain B, Fig. 2.2) containing a bipartite motif of the consensus sequence 2K/R–X 10–12–3K/R constitutes the PARP nuclear homing sequence (77). Extensive site-directed mutagenesis and deletion experiments performed on this short region encoded by exon 5 (5), revealed that each of the two basic clusters (D1, D2) was essential but not sufficient on its own for this function. Crucial residues for nuclear targeting (K207,

R208, and K222, human PARP numbering) are strictly conserved from humans to *Xenopus* (Fig. 2.3). The precise distance between the two motifs, D1 and D2, is tolerant to change, but the correct positioning of a lysine at the surface of D2 is crucial to fit perhaps within the long surface groove of the nuclear import factor that contains an asparagine array (123). In invertebrates and plants (Fig. 2.3), the nuclear targeting motif resembles a monopartite NLS.

Domain B is not only recognized by the nuclear transport machinery, but it also contains the major proteolytic cleavage sites (D214, in the sequence 211DEVD214) of several apoptotic proteases (caspases) (see Chapter 4) (124), of plasmin (125), and granzyme B (J. Tschopp, personal communication). This confirms that it is an exposed regulatory region of PARP even when it is complexed to nicked DNA (126). Although this crucial cleavage motif is lost in lower species, PARP is processed during apoptosis induced by DNA damage in *Drosophila* S2 cells (127). Finally, it is important to note that in PARP, as well as in DNA ligase III, the N-terminal region bears two close targeting motifs: one for nuclear localization and a second for DNA breaks.

2.6   The automodification domain (module D) contains a BRCT motif: an interface for protein–protein contacts

PARP appears to be the main acceptor of ADP-ribose polymers after DNA damage both *in vitro* (121) and *in vivo* (128). The auto poly (ADP-ribosyl)ation of PARP which takes place mainly at glutamate residues interferes with DNA binding due to charge repulsion between covalently bound polymers and nicked DNA (129, 130) (Figs 2.7 and 2.8). This unique property is thought to play a role in the downregulation of the enzymatic activity.

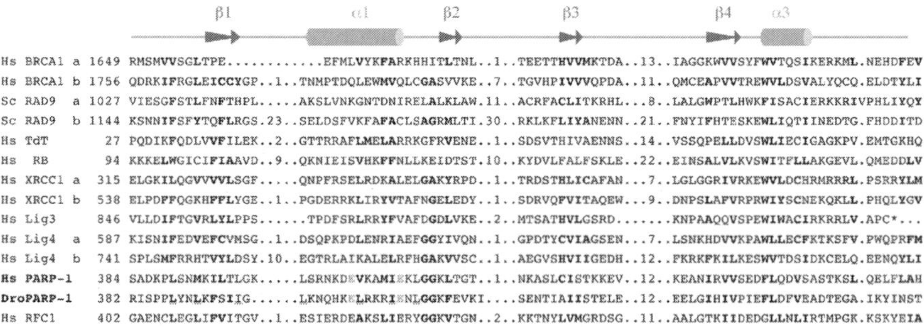

Fig. 2.7   Multiple alignment of some BRCT family members (131, 132). The secondary structure of the second XRCC1 BRCT domain is shown on the top according to Zhang *et al.* (135). The relative locations of the BRCT domains in XRCC1 (Hs XRCC1), PARP (Hs PARP), and DNA ligase III (Hs Lig3) are indicated in Fig. 3.6 (see Chapter 3). The residues that conform with the BRCT consensus motifs are indicated in bold. Amino acids involved in the putative leucine zipper of *Drosophila* PARP (Dro PARP) (66) are underlined. Two conserved potential automodification sites (glutamate residues) are shown in grey in both PARP sequences.

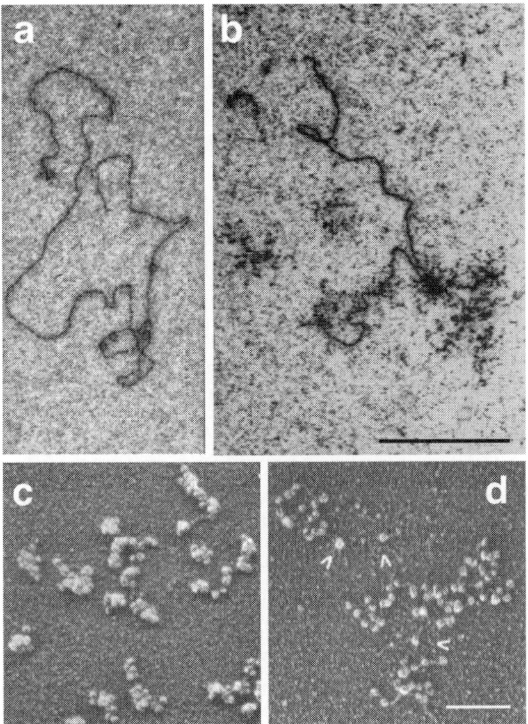

Fig. 2.8 Visualization by dark-field electron microscopy of PARP–DNA complexes (a) before and (b) after auto-poly (ADP-ribosylation) *in vitro*. Automodified unbound PARP and ADP-ribose polymers are clearly visible on the carbon film (179). Relaxation of chromatin superstructure (30-nm fibre) induced by poly (ADP-ribosylation) of calf thymus polynucleosomes (180,181). (c) Native chromatin; (d) poly (ADP-ribosylated) chromatin. The arrow heads point to automodified PARP. In (c) and (d) the samples were rotatory shadowed with carbon/platinum at an angle of 7°. The bars represent 100 nm.

Interestingly, a leucine zipper has been found in the automodification domain of the *Drosophila* PARP (Fig. 2.7) (66), suggesting that this region might be responsible for protein–protein interactions in homo- and/or heteromodification reactions. However, in vertebrates this leucine zipper is poorly conserved. Instead, this region contains a motif termed BRCT (*br*east *ca*ncer susceptibility protein, BRCA1, *C-t*erminus) that was recently identified in a number of proteins that respond to DNA damage or which function in DNA repair or in cell-cycle checkpoints (131, 132). This module of about 100 residues appears to be typical of specific protein–protein contacts in several pairs of DNA repair proteins: XRCC1/DNA ligase III (87) and XRCC4/DNA ligase IV (133). In PARP, the BRCT module (aa 372–476) has been shown to be involved in the interaction with several partners such as XRCC1 (113), hUbc9 (134), histones (81, 85), and transcription factors such as Oct-1 (105) and YY1 (107) (Fig. 2.2). The crystal structure of the second BRCT domain of XRCC1 (motif b, Fig. 2.7) has recently been solved (135). It identifies an α-helix (α1, Fig. 2.7) that plays an essential role in protein–protein interactions. Strikingly, in PARP the homologous region contains two highly conserved glutamate residues (E407 and E413), located on the same face of the corresponding α1 helix, that are totally accessible to the solvent (P. Freemont, personal communication). Hence, we speculate that in PARP the protein interface function mediated by the BRCT domain might be negatively regulated by poly (ADP-ribosyl)ation.

In *Drosophila*, in addition to the major 3.2-kb band, a minor transcript of 2.6 kb has recently been isolated by Miwa's group (136). It encodes a 92.3-kDa enzyme named PARPII (which should not confused with PARP-2, see Section 2.8), a polypeptide of 804 residues, spliced at the automodification domain which is entirely comprised in exon 5 in *Drosophila*. Ectopic expression of PARPII in Rat-1 cells affects cellular adhesion and causes growth retardation. Because it is a catalytically inactive enzyme, PARPII may act as a dominant-negative mutant in these conditions (137).

A region of about 300 amino-acid residues encompassing domains C and D (Fig. 2.2) has been shown to display DNA binding affinity for bent DNA (99, 112). This result is in accord with the fact that PARP forms nucleosome-like structures in which a 140-bp DNA wraps around the enzyme (100). The existence of a second DNA binding site in PARP, independent of the zinc finger (99, 112), may explain its ability to form loops with DNA and to bind preferentially to supercoiled DNA (14) (see Section 2.4.2.). The central region of the PARP molecule therefore plays a dual role: it binds protein partners as well as DNA. It is not known whether the same sequence or motif is involved in both functions. It is worth mentioning that the BRCT module of the replication factor RFC p140 (138) and BRCA1 (139) also bind DNA ends.

2.7   The catalytic domain and the connection with ADP-ribosylating toxins

The carboxy-terminal region of PARP (Fig. 2.2, domain F) bears all the different catalytic activities associated with the full-length enzyme: NAD⁺ hydrolysis, initiation, elongation, branching, and termination of ADP-ribose polymers (75, 76). This basal activity of the enzyme, located in this fragment is independent of the presence of DNA breaks. The K_M of this domain alone for NAD⁺ is quite comparable to that of the full-length PARP (50 μM). However, its specific activity is about 500-fold lower than that of the entire enzyme that is fully activated by DNA strand breaks (75, 76)

This part of the molecule is the region most conserved during evolution

(Fig. 2.2). It includes residues 654 to 1014 of PARP called PARP-CF (PARP catalytic fragment). PARP-CF contains a block of 50 amino acids (aa 859–908) representing 'the PARP signature' that is virtually unchanged from humans to plants (68–70), forming, in fact, the catalytic site of PARP.

The chicken PARP-CF has been overexpressed in an insect cell/baculovirus expression system (140). This protein was purified by affinity chromatography and crystallized (141). Its X-ray structure was determined both in the absence and in the presence of various inhibitors, at 2.8 Å and 2.2 Å resolution, respectively (142). The binding geometry of the ADP-ribose donor substrate NAD^+ could be deduced from the X-ray structure of complexes of PARP-CF with nicotinamide analogues (143). The crystal structure of a complex with a substrate analogue yielded the binding mode of the ADP-ribose acceptor substrate. Based on the binding geometry of the ADP-ribose donor and acceptor substrates, a model was proposed for the enzymatic mechanism in the elongation and the branching reaction of poly (ADP-ribose) polymerase (144).

2.7.1   The native structure of PARP-CF

The sequence of PARP-CF from chicken has 87% identity with the cognate human sequence. The structures described here are for the chicken protein, but all the sequence numbers correspond to the numbering in the human PARP sequence, in order to facilitate comparison with the abundant work on human PARP. The N- and C-terminal residues of PARP-CF (654 to 661 and 1013 to 1014) are disordered in the crystal and could not be located in the electron density maps.

PARP-CF is monomeric and is composed of two domains: a purely α-helical N-terminal domain containing residues 662–784 (helices A–F in Plate 3) and a C-terminal domain of residues 785–1012 that bears the catalytic site (Plate 3). The N-terminal helical domain of PARP-CF has no significant structural similarity to any other protein domain found in the Protein Data Bank as judged with the program DALI (145). Interestingly, the C-terminal domain of PARP-CF has a chain fold similar to the NAD^+ binding domains of bacterial toxins that are ADP-ribosyl transferases. These include diphtheria toxin (146), *Pseudomonas* exotoxin A (147), heat-labile enterotoxin from *Escherichia coli* (148), and cholera toxin (149). This structural homology indicates that PARP belongs to the ADP-ribosyl-transferase (ART) protein family showing a common NAD^+ binding fold (142). PARP is a special member of the ART protein family because it not only ADP-ribosylates a protein, but also elongates the bound ADP-ribose, thus synthesizing a linear or branched poly (ADP-ribose) polymer. The core of the PARP-CF ART domain is formed by a five-stranded antiparallel sheet and a four-stranded mixed β-sheet. Surrounding this core are eight α-helices (G–N in Plate 3) and a three-and two-stranded β-sheet.

PARP-CF (domain F, residues 654 to 1014) is the smallest PARP fragment retaining catalytic activity (75). The ART domain alone, comprising residues 785–1014, is not catalytically active, despite the homology with the ADP-ribosylating domains of the bacterial toxins, which are catalytically active on their own. This can be explained by the crystal structure of PARP-CF. Removing the N-terminal helix (residues 662–676) of PARP-CF (helix A in Plate 3) will destabilize the NAD binding domain of PARP, because numerous hydrophobic side chains will then be exposed to the solvent (143).

2.7.2 PARP–inhibitor complexes

PARP inhibitors are of pharmaceutical interest for several reasons. PARP inhibitors potentiate the cytotoxic effects of radiation and of monofunctional alkylating agents used in cancer therapy (reviewed in ref. 150) (see Chapter 6). PARP is also a potential drug target in conditions caused by cell death via the so-called nitric oxide pathway (151). Such conditions include ischaemia–reperfusion injuries, neurodegenerative disorders, or endotoxic shock (see Chapter 5). The structures of inhibitor complexes of PARP-CF are useful for the rational design of new PARP inhibitors.

The structures of PARP-CF in complexes with four different nicotinamide-analogue PARP inhibitors (IC_{50} in the range from 0.16 μM to 10 μM) have been determined at 2.4- to 2.8-Å resolution (143). There is no induced fit upon binding of these four inhibitors. There are no significant movements of the polypeptide chain, with the exception of a tighter packing of one surface loop (residues 823–827) that is not directly involved in inhibitor binding. The root mean square deviation of all C? atoms is only 0.24 to 0.27 Å between ligated and unligated structures. But the residues involved in inhibitor binding do become more ordered upon binding, as indicated by a reduction of their crystallographic temperature factor (143). So far, all inhibitors do bind to the nicotinamide site of the NAD substrate-binding pocket that is close to the active site Glu988. The residues His862, Gly863, Tyr896, Phe897, Ala898, Lys903, Ser904, Tyr907, and Glu988 (Plate 4) line this binding pocket. All the inhibitors form hydrogen bonds with their amide group, to the peptide atoms of Gly863, and to the hydroxyl group of Ser904. The inhibitor 4-amino-1, 8-naphthalimide forms an additional hydrogen bond with Glu988 and 8-hydroxy-2-methyl-3-hydro-quinazolinone forms a weak one with Tyr907. These additional hydrogen bonds give these two inhibitors no additional affinity to PARP, as judged from the IC_{50} values that are comparable to the other inhibitors.

2.7.3 NAD⁺ binding to PARP

The structure of the complex between PARP-CF and NAD$^+$ could not be obtained directly by co-crystallization experiments because of the auto poly (ADP-ribosylation) of PARP-CF (75). That reaction renders heterogeneously modified PARP-CF unsuitable for crystallization. Soaking PARP-CF crystals with NAD$^+$ yielded only electron density for the nicotinamide moiety of NAD$^+$, even for crystals of the inactive mutant E988K (unpublished results). The binding of NAD$^+$ to PARP could only be deduced by homology modelling, using the structure of the homologous complex of diphtheria toxin and NAD$^+$ (Fig. 2.9). The nicotinamide binding position that was determined from the binding coordinates of the nicotinamide-analogue inhibitors, was also used in this homology modelling approach (143). The resulting model of the binding of NAD$^+$ is consistent with mutants found in the random mutagenesis experiments on human PARP. Rolli *et al.* found that five mutants with reduced PARP activity affected residues directly involved in putative NAD$^+$ binding and four mutants were at nearby residues (152). Site-directed mutagenesis of residues that were thought to bind NAD$^+$, confirmed the model. The mutants G863A, Y896N, and Y907N show an increased K_M for NAD$^+$ and reduced polymerase activity (144).

In our model of NAD$^+$ binding to PARP-CF, the nicotinamide binding is in agreement with the observed PARP inhibitor binding and is very similar to the binding of the nicotinamide moiety in the diphtheria toxin–NAD$^+$ complex. The C1$'_\mathrm{N}$ atom is exposed to the attacking nucleophile (Plate 5). In diphtheria toxin the NAD$^+$ is very exposed to the solvent, whereas in the

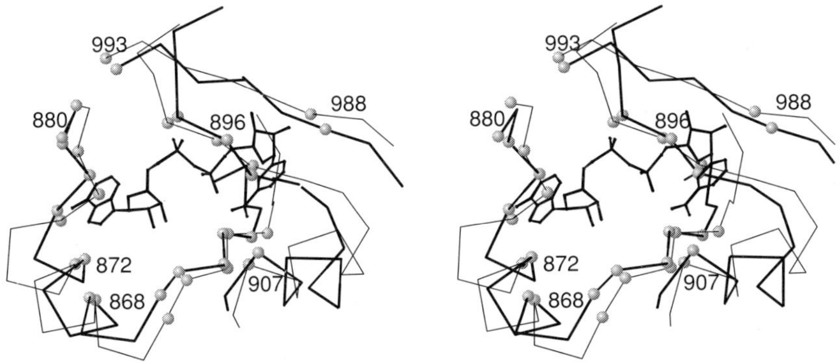

Fig. 2.9 NAD$^+$-binding as derived from homology-modelling. PARP (thin lines) and diphtheria toxin/NAD$^+$ complex (thick lines) in a superposition based on the 17 residues (marked red) which are in contact with NAD$^+$ in the diphtheria toxin/NAD$^+$ complex. This alignment was used as the start for the PARP-CF/NAD$^+$ homology modelling. For the final NAD$^+$ model the binding coordinates of the nicotinamide-analogue PARP inhibitors were also taken into account (143, 144).

PARP-CF complex 90% of the surface of NAD$^+$ is covered. The adenine moiety is buried by helix F of the N-terminal domain. This helix covers part of the NAD$^+$ binding pocket (Plate 3). The adenine of NAD$^+$ is fixed by specific hydrogen bonds to residues Gly876, Asp770, and Arg878 and by hydrophobic interactions to Ile872 and Leu877. The interactions between the NAD$^+$ riboses and PARP-CF are very similar to those observed in diphtheria toxin. In the complex of PARP-CF and NAD$^+$, the phosphates in the NAD$^+$ are hydrogen-bonded both to a water molecule and to Gln763 and Asp766 (Plate 5). The modelling resulted in a very compact NAD$^+$ conformation, similar to that observed previously in other ADP-ribosyl transferases (153, 154), but different to the elongated conformations usually found in dehydrogenases and related enzymes (155). The internal strain at the adenosine that results from this compact conformation is believed to help in the ADP-ribosyl transfer reaction, because it is relieved on ADP-ribose transfer (153).

In diphtheria toxin and in *Pseudomonas* exotoxin A, the so-called active site loop (residues 881–888 in PARP-CF) becomes mobile upon binding of NAD$^+$ (153, 154). In the PARP-CF crystal structure this loop is in contact with the N-terminal domain and it has a low mobility as judged from the atomic temperature factors. This correlates with the finding that mutant L713F in the N-terminal domain of PARP-CF increases PARP-activity nine-fold (156). This mutation is located near this loop and is likely to disturb its conformation. It was therefore suggested that the N-terminal domain participates in PARP activity modulation (142, 143).

2.7.4   Binding of the ADP-ribosyl acceptor

The ADP-ribosyl acceptor binding site was detected by chance in a co-crystallization experiment with carba-NAD, that was originally designed to establish NAD$^+$ binding (144). Carba-NAD is an NAD-analogue in which a methylene group replaces the ring oxygen of the nicotinamide ribose. It lacks the scissile glycosidic bond to the nicotinamide and can therefore not be used as an ADP-ribosyl donor. But it is still a competitive inhibitor of the ADP-ribosyl transferase cholera toxin (157, 158). In the co-crystal carba-NAD did not bind in the NAD$^+$ binding pocket but in a different pocket nearby. The nicotinamide- and carba-ribose moiety of carba-NAD were disordered. Only the rigid part of the bound carba-NAD, that was visible in the electron density maps, is included in the X-ray structure. It was proposed that this bound carba-NAD marks the binding site of the poly (ADP-ribose) acceptor, because mutagenesis of residues forming this site modulated PARP chain elongation and branching activity (144, 152).

This acceptor-binding site is formed by the loops following helices *H* and *K*, by helix *L*, by the loop following the β–strand *f*, and by strands *l* and *m*.

This acceptor site is unique to PARP and is absent in the structure of other ADP-ribosyl transferases (144). The loosely bound adenine group of carba-NAD has hydrophobic interactions with the side-chain of Met890 but makes no hydrogen bonds to the protein (Plate 6). Its solvent accessibility is twice as high (40%) as the average of protein-bound adenine (159). The ribose forms hydrogen bonds with Tyr907, Glu988, and a water molecule. This water is at the position of the attacked $C1'_N$ atom of the modelled donor NAD^+. The pyrophosphate is the part most tightly bound in the observed ligand. It is hydrogen-bonded to the N-atom of His826 and to the ε-amino group of Lys903 and to the backbone amides 985-N and 986-N (Plate 6).

The region around the acceptor binding site, especially the loop around His826, which is very mobile in the unligated structure, solidifies upon ligand binding (144). During catalysis, when both the donor and acceptor substrates bind simultaneously, further stiffening of the protein can be expected, because binding of nicotinamide-analogue inhibitors alone diminished the B-factors at the active centre as well (143).

Mutagenesis confirmed that the observed binding site of the ADP-moiety of carba-NAD, is indeed the poly (ADP-ribose) acceptor site. The site-directed mutants K903Q, K903E, and K903E/E988K, which are residues involved in phosphate and ribose binding of the observed ADP moiety, lost poly (ADP-ribose) elongation activity (144). In random mutagenesis experiments the strongest effect on PARP activity (200-fold reduction) was found with the rather conservative mutation M890V, which would displace the accepting ribose from its proper position in the transfer reaction (152).

2.7.5   The reaction mechanism

The reaction catalysed by PARP is the transfer of the ADP-ribose unit of the donor substrate NAD^+ to a nucleophilic acceptor. This can be the carboxy-late of the glutamate side-chain in a protein (initiation), or the 2'-hydroxyl of an adenine ribose (elongation), or a nicotinamide ribose (branching) in a growing poly (ADP-ribose) chain. In this reaction the β-glycosidic bond between the C1' atom and the nicotinamide in NAD^+ is broken and a new α-glycosidic bond to the respective nucleophilic acceptor is formed. The ratio of branching to elongation is about 2% (160).

PARP differs from other ART enzymes because, in general, they only mono (ADP-ribosylate) a specific residue in specific acceptor proteins. By contrast, PARP mainly ADP-ribosylates growing ADP-ribose chains, as well as proteins. ART enzymes have a common binding mode for the donor NAD^+. The NAD^+ binding conformation is unique to this protein family and is a consequence of their chain fold. All ART enzymes have the catalytic glutamate in common, and presumably they all catalyse by a similar mechanism via a transition state with a substantial oxocarbenium ion character, similar to the reaction mechanism of inverting glycosidases (143, 150).

The role of NAD+ conformation in catalysis

The internal strain of the very compact *syn*-conformation of NAD^+ bound to ART helps to drive the reaction, because the strain is relieved upon ADP-ribosyl transfer (153). In the known structures where NAD^+ is complexed with ART enzymes, NAD^+ is always bound with the nicotinamide ribose in the 3'-*endo* conformation (153, 154). The 3'-*endo* conformation helps catalysis in two respects. In the 3'-*endo* conformation the ribose atoms C2', C1', O4', and C4' are planar, which is close to the geometry of the expected oxocarbenium transition state. Lysozyme binds the reacting pyranose similarly in a suitably deformed conformation (143). The 3'-*endo* conformation leaves the C1' atom very accessible to nucleophilic attack, whereas it is shielded in the 2'-*endo* conformation.

The role of the catalytic glutamate

Glu988 is strongly conserved in all ART enzymes. Kinetic analysis of pertussis toxin showed that the ADP-ribosyl transfer reaction has S_N2-character with a transition through an oxocarbenium ribose (161). In PARP-CF Glu988 binds the acceptor nucleophile as well as the donor NAD^+ by forming hydrogen bonds to their 2'-hydroxyls. Therefore Glu 988 is able to polarize both the acceptor nucleophile and the donor ribose, and can stabilize the oxocarbenium in the donor ribose and simultaneously increase the nucleophilicity of the attacking acceptor (144, 162).

The elongation reaction

In the elongation reaction ADP-ribose is transferred from the donor substrate NAD^+ to the 2'-hydroxyl group of the adenine ribose of a growing ADP-ribose chain. The binding geometry of the two reactants can be directly deduced from the NAD^+ donor binding site in our model and from the observed ADP moiety of carba-NAD, that most likely represents the acceptor, a poly (ADP-ribose) chain end in this case (144) (Plates 7 and 8). The attacking 2'-hydroxyl of the acceptor ribose is hydrogen-bonded to Glu988, which is the only residue in the region that is able to act as the deprotonating catalytic base. The acceptor ribose is preoriented for reaction by two additional hydrogen bonds from its 3'-hydroxyl to Tyr907 and Glu988. This is in agreement with the finding that 3'-deoxy NAD is a poor substrate for chain elongation (163). It is not known how many units of the polymer are bound by PARP.

The branching reaction

The active site cleft of the PARP-CF is open on both sides, giving enough space to bind the acceptor polymer in two opposite directions (Plate 8C). The phosphates are the most strongly bound part of the acceptor substrate, stronger than the ribose and far more specific and strongly bound than the adenine (144). From this, and with the observed reaction geometry, the substrate binding in the branching reaction can be proposed. In the branching reaction the polymer is bound in an orientation reversed to that observed in elongation. In this orientation the phosphates are again bound to the phosphate sites. The nicotinamide ribose now occupies the acceptor site previously occupied by the adenine ribose in the elongation reaction. In the region occupied by the non-specifically bound adenine during elongation there is enough space for the next unit of ADP-ribose during branching (Fig. 2.10). In this branching geometry with a reversed polymer binding, subsequent ADP-ribose transfer will produce a fork (that is, branching) in poly (ADP-ribose). There is only one PARP mutant that increases the branching/elongation ratio—the random mutant Y986H (152). This mutant is consistent with the proposed

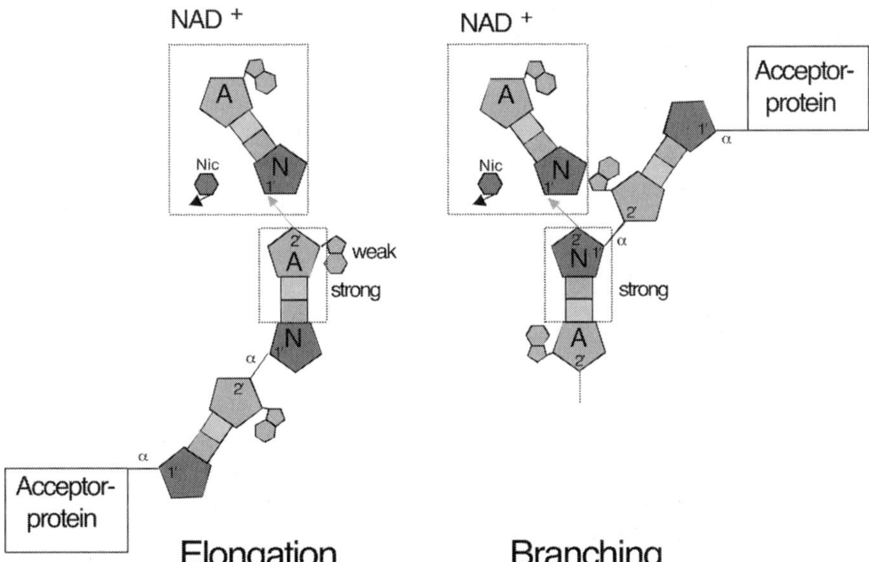

Fig. 2.10 Proposed mechanism of the PARP branching reaction. One ADP-ribose unit of the polymer consists of A (adenine–ribose), P (phosphate), N (nicotinamide–ribose), and Ad (adenine). At the acceptor site (marked by dotted rectangles) the adenine is only weakly bound and there is enough space for the polymer to bind in two opposite directions. Because of the fixed donor position, the two reversed binding modes of the polymer give rise to the two different reactions: elongation and branching of poly ADP-ribose) (144).

branching model. The mutant His986 can bind the pyrophosphates more with an additional hydrogen bond from one of its side-chain nitrogens (Plate 7). This increases the likelihood of binding in the branching orientation.

2.8　Plant PARPs and mammalian PARP homologues

2.8.1　Plant PARPs

PARP activity in plant cells was first demonstrated by following the incorporation of ^{3}H from radioactive NAD$^+$ into the nuclei of root-tip cells (164). The enzymatic activity was purified from maize seedlings and found to be associated with a protein of an apparent molecular mass of 113 kDa (165). These authors also demonstrated the presence of a product 'ladder'. The limited amount of information currently available on the biological function of plant PARP comes from experiments involving PARP inhibitors, and suggests a possible *in vivo* role in the prevention of homologous recombination at sites of DNA damage (166). With the discovery of a gene coding for a PARP-related polypeptide APP (*A. thaliana* homologue of *PARP*, APP) with a calculated molecular mass of 72 kDa (Fig. 2.11) and having a different DNA binding domain (167), it became evident that two structurally different PARP proteins, both possessing DNA-dependent poly (ADP-ribose) activities, were present both in *A. thaliana* and in maize (68–70) (see Table 2.1).

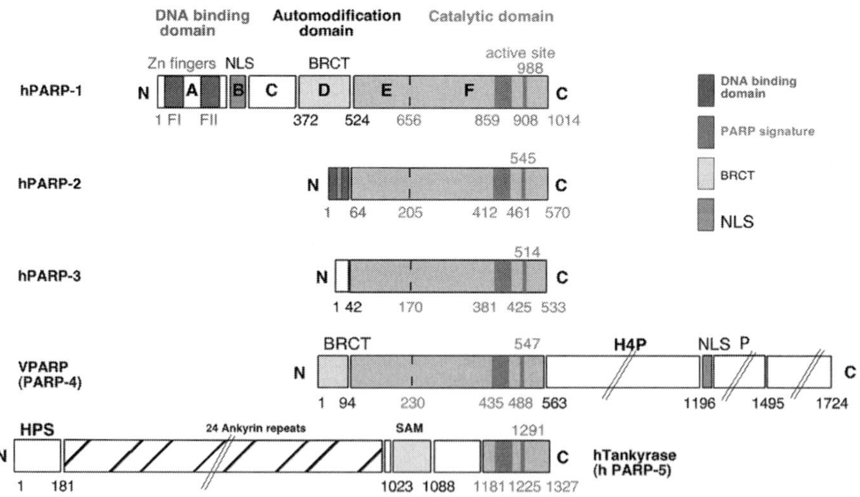

Fig. 2.11 Schematic representation of the domain structure of the members of the PARP family. See Table 2.1 for references. NLS, nuclear location signal; BRCT, homologous to the BRCA1 C-terminus; H4P, homologous to inter-alpha-inhibitor family; P, proline-rich region; HPS, region containing homopolymeric runs of His, Pro, and Ser; SAM, sterile alpha module.

Table 2.1 Characteristics of the various plant homologues

Name	Origin	Chr. mapping	aa no.	Mr (kDa)	Subcellular location	DNA binding	ADP-ribose	Other modules	GenBank	Ref.
atPARP-1	*A. thaliana*	II	983	111.2	nucleus	+	poly	zinc-fingers NLS, BRCT	AJ131705	70
zPARP-1	*Z. mais*	ND	969	109.1	nucleus	+	poly	*idem*	AF093627 AJ222589	69 68
APP	*A. thaliana*	IV	637	72.1	nucleus	+	poly	DBD	Z48243	167
NAP	*Z. mais*	ND	653	73	nucleus	ND	ND	*idem*	AJ222588	68
PM38	Glycine max	V (*A. thaliana*)	798	89.8	ND	ND	ND	ND	AF169023	

Interestingly, in *A. thaliana* the promoter of both enzymes (PARP-1 and APP) is strongly upregulated following a sublethal dose of gamma rays (70). Similarly, the *app* gene promoter is transcriptionally activated in DNA ligase-I deficient cells (68). Contrary to mammalian cells, both PARP homologues in plants are therefore regulated at the transcriptional level in response to DNA breaks, pointing, however, to a similar role in DNA repair and cell survival. The possibility of rapidly developing *A. thaliana* transgenic plants expressing antisense RNA or dominant-negative mutants by T-DNA insertion mutation should illuminate the function of both PARP enzymes. Given the apparently universal structure–function conservation, the outcome may well be important not only for plants but also for mammalian cells.

2.8.2 Mammalian PARP homologues

In mammals too, additional members of the PARP family have been identified this last year (see Table 2.2). In fact, the discovery of two *bona fide* plant PARP enzymes, APP and NAP (68), together with the presence of residual PARP activity induced by DNA damage in embryonic fibroblasts derived from PARP knockout mice (168, 169) have suggested the existence of previously unknown PARP homologs. On the basis of these premises a cDNA encoding a novel PARP homologue of 62 kDa (named PARP-2) was recently cloned by screening EST databanks with the APP sequence (169). PARP-2 is a 62-kDa protein that shares considerable homology with the catalytic domain of the classical PARP (113 kDa), which should be renamed PARP-1, and which is responsive to DNA damage (Fig. 2.12). In spite of the absence of the characteristic zinc-finger module that acts as a nick sensor in PARP-1, the most intriguing feature of PARP-2 is its DNA binding property and its activation by DNA that has been treated with DNase I. The PARP-2 mRNA (2 kb) is not transcriptionally induced following damage. Moreover, no compensation for the PARP-1 deficiency is observed in PARP-1$^{-/-}$ cells by upregulation of PARP-2 gene expression (169).

While the function of PARP-2 is still largely unknown, a third member of the PARP family, called tankyrase, was identified in a yeast two-hybrid screen with the telomeric protein TRF1 as bait (170). Tankyrase has been localized to human telomeres (170).Tankyrase is a 142-kDa protein containing 26 ankyrin repeats and a PARP catalytic fragment of limited size that mostly corresponds to the β-domain defined in the PARP crystal structure (Fig. 2.12). Apparently, this enzyme synthesizes poly (ADP-ribose) on itself and on TRF1, independently of the presence of DNA. TRF1 is a regulator of telomere maintenance (171, 172) that can be poly (ADP-ribosylated) *in vitro* by tankyrase. The modified TRF1 subsequently loses its affinity for telomeric DNA, raising the interesting possibility that the function of human telomeres could be regulated by poly (ADP-ribosyl)ation.

Table 2.2 Characteristics of the various mammalian homologues

Name	Origin	Chr. mapping	aa no.	M_r (kDa)	Subcellular location	DNA binding	ADP-ribose	Other modules	GenBank	Ref.
PARP-1	human	1q41–q42	1014	113.2	nucleus	+	poly	zinc fingers, NLS, BRCT	J03473	1–3
PARP-1	mouse	1H5	1013	112.9	nucleus	+	poly	zinc fingers, NLS, BRCT	X14206	61
PARP-2	human	14q11.2	570	64.8	nucleus	+	poly	NLS	AJ236912	169
PARP-2	mouse	14C1	559	63.4	nucleus	+	poly	NLS	AJ007780	169
PARP-3	human	3p21	533	60.1	ND	ND	ND		NM-005485	173
VPARP	human	13q11	1724	192.7	Vaults particles cytoplasmic and nuclear, mitotic spindle	ND	poly	BRCT, NLS, IαI related H5, proline-rich domain	AF158255 AF057160	174 175 176
Tankyrase (PARP-5)	human	8p22–p23	1327	142	telomeres, nuclear pores, pericentriolar matrix	—	poly	HPS-rich 24 ankyrin repeats (TRF1 binding) SAM motif	AF0825556	170 184
Tankyrase (PARP-5)	mouse	8							AA415426	178

hPARP-1   352 KQDRIFPPETSASVAATPPPSTASAPAAVNSSASADKPLSNMKILTLGKLSRNKDEVKAMIEKLGGKLTGTANKASLCISTKKEVEKMNKKMEEVKEANI
hPARP-2   1 --MAARRR
mPARP-2   1 --MAPRRQ
NAP   1 MSARLRVADVRAELQRRGLDVSGTKPALVRRLDAAICEAEKAVVAAAPTSVANGYDVAVDGKRNCGNNK-RKRSGDGGEEGNGDTCTDVTKLEGMSYREL
APP   1 MANKLKVDELRLKLAERGLSTTGVKAVLVERLEEAIAEDTKKEESKSKR----------------------------KRNSSNDTYESNKLIAIGEFRGMIVKEL
hPARP-3   1 --
VPARP   1 --MVMGIFANCIFCLKVKYLPQQ
TANKYRASE 1011 --

hPARP-1   452 RVVSEDFLQDVSASTKSLQELFLAHILSPWGAEVKAEPVEVVAPRGKSGAALSKKSKGQVKEEGINKSEKRMKLTLKGGAAVDPDSGLEHS---AHVLEK
hPARP-2   7 RSTGGGRARALNESKRVNNGNTAPEDSSP-AKKTRRCQRQESKKMPVAGGKANKDRTEDKQDE------SVKALLLKGKAPVDPECTAKVG--KAHVYCE
mPARP-2   7 RSGSG--RRVLNEAKKVDNGNKATEDDSPPGKKMRTCQR----KGPMAGGKD-ADRTKDNRD------SVKTLLLKGKAPVDPECAAKLG--KAHVYCE
NAP   100 QGLAKARGVAANGGKKDVIQRLLSATAGPAAVADGGPLG---AKEVIKGD----EEVEVKKE------KMVTATKKGAAVLDQHIPDHIKV-NYHVLQV
APP   78 REEAIKRGLDTTGTKKDLLERLCNDANNVSNAPVKSSNG---TDEAEDDNNG---FEEEKKEE------KIVTATKKGAAVLDQWIPDEIKS-QYHVLQR
hPARP-3   1 ----------------MAPKPKPWVQTEGPEKK----KGRQAGREE-DPFRSTAEA------LKAIPAEKRIIRVDPTCPLSSN-----PGTQ
VPARP   22 QKKKLQTDIKENGGKFSFSLNPQCTHIILDNADVLSQYQLNSIQKNHVHIANPDFIWKSIREK-----RLLDVKNYDPYKPLDITPPPDQKASSSEVKTE
TANKYRASE 1011 --

hPARP-1   549 GGKVFSATLGLVDIVKGTNSYYKLQLLEDDKENRYWIFRSWGRVGTVIGSNKLE--QMPSKEDAIEHFMKLYEEKTGNAWHS-KNFTKYPKKFYPLEIDY
hPARP-2   98 GNDVYDVMLNQTNLQFNNNKYYLIQLLEDDAQRNFSVWMRWGRVGKMGQHSLVA--CSGNLNKAKEIFQKKFLDKTKNNWEDREKFEKVPGKYDMLQMDY
mPARP-2   91 GDDVYDVMLNQTNLQFNNNKYYLIQLLEDDAQRNFSVWMRWGRVGKTGQHSLVT--CSGDLNKAKEIFQKKFLDKTKNNWEDRENFEKVPGKYDMLQMDY
NAP   186 GDEIYDATLNQTNVGDNNNKFYIIQVLESDAGGSFMVYNRWGRVGVRGQDKLHG--PSPTRDQAIYEFEGKFHNKTNNHWSDRKNFKCYAKKYTWLEMDY
APP   165 GDDVYDAILNQTNVRDNNNKFFVLQVLESDSKKTYMVYTRWGRVGVKGQSKLDG--PYDSWDRAIEIFTNKFNDKTKNYWSDRKEFIPHPKSYTWLEMDY
hPARP-3   62 VYEDYNCTLNQTNIENNNKKFYIIQLLQ-DSNRFFTCWNRWGRVGEVGQSKIN---HFTRLEDAKKDFEKKFREKTKNNWAERDHFVSHPGKYTLIEVQA
VPARP   117 GLCPDSATEEEDTVELTEFGMQNVEIPHLPQDFEVAKYNTLEKVGMEGGQEAVVVELQCSRDSRDCPFLISSHFLLDDGMETRRQFAIKKTSEDASEYFE
TANKYRASE 1011 --

                                                  A   B   C   D   E
hPARP-1   646 G---QDE--EAVKKLTVNPG----TKSKLPKPVQDLIKMIFDVESMKKAMVEYEIDLQKMPLGKLSKRQIQAAYSILSEVQQAVSQGSSD-SQILDLSNR
hPARP-2   196 ATNTQD---EEETKKEESLKSPLKPESQLDLRVQELIKLICNVQAMEEMMMEMKYNTKPAPLGKLTVAQIKAGYQSLKKIEDCIRACQHG-RALMEACNE
mPARP-2   189 AASTQD---ESKTKEEETLK----PESQLDLRVQELLKLICNVQTMEEMMIEMKYDTKRAPLGKLTVAQIKAGYQSLKKIEDCIRAGQHG-RALVEACNE
NAP   284 G-------ETEKEIEKGSITDQIKETKLETRIAQFISLICNISMMKQRMVEIGYNAEKLPLGKLRKATILKGYHVLKRISDVISKADR--RHLEQLTGE
APP   263 GKEENDS--PVNNDIPSSSEVKPEQSKLDTRVAKFISLICNVSMMAQHMMEIGYNANKLPLGKISKSTISKGYEVLKRISEVIDRYDR--TRLEELSGE
hPARP-3   158 EDEAQEAV-VKVDRAPVRTVTKRVQPCSLDPATQKLITNIFSKEMFKNTMALMDLDVKKMPLGKLSKQQIARGFEALEALEEALKGPTDGGQSLEELSSH
VPARP   217 NYIEELKKQGFLLREHFTPEATQLASEQLQALLEEVMNSSTLSQEVSDLVEMIWAEALGHLEHMLLKPVNRISLNDVSKAEGILLLVKA-ALKNGETAE
TANKYRASE 1011 --------------------------------GDGAAGTERKEGEVAGLDMNISQFLKSLGLEHLRDIFETEQITLD

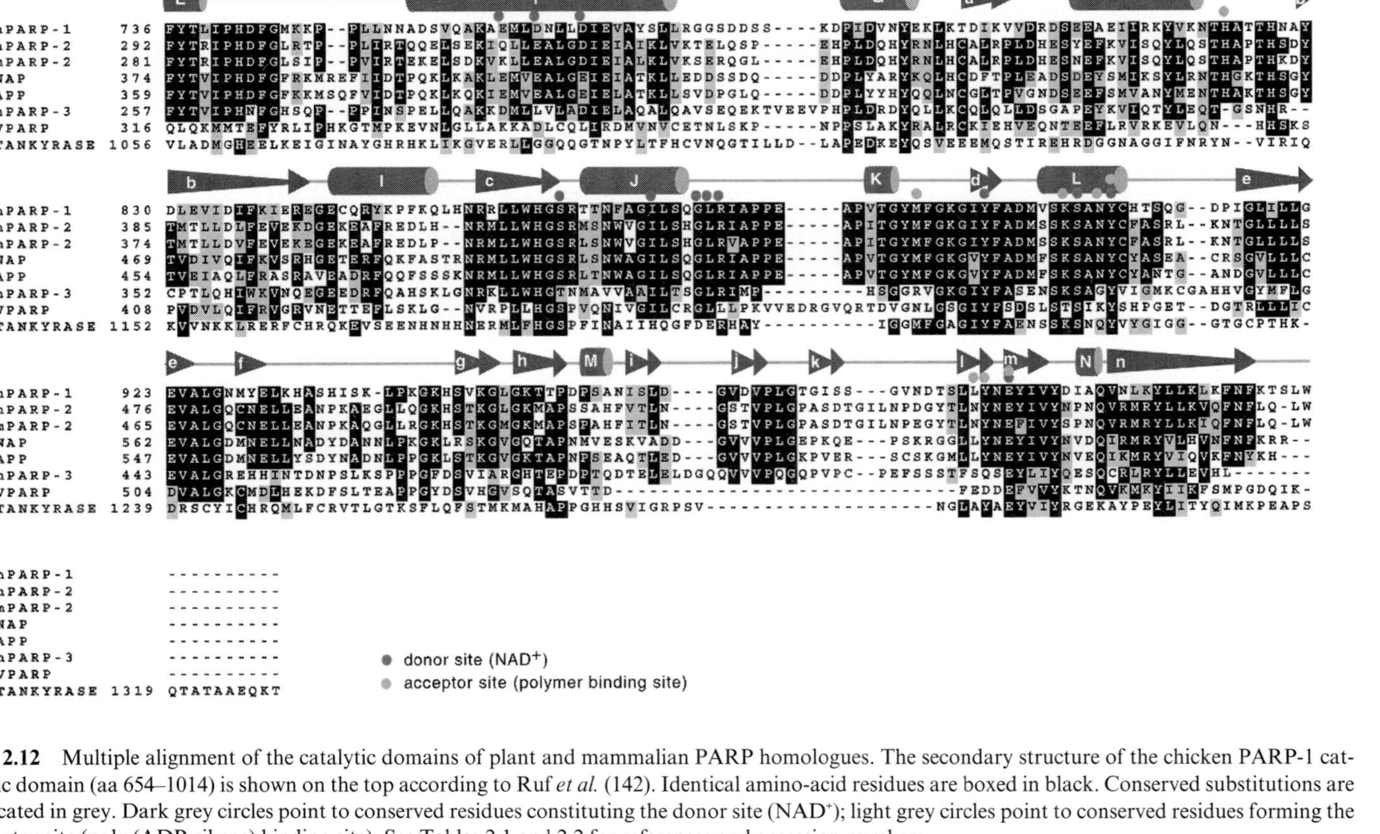

Fig. 2.12 Multiple alignment of the catalytic domains of plant and mammalian PARP homologues. The secondary structure of the chicken PARP-1 catalytic domain (aa 654–1014) is shown on the top according to Ruf *et al.* (142). Identical amino-acid residues are boxed in black. Conserved substitutions are indicated in grey. Dark grey circles point to conserved residues constituting the donor site (NAD+); light grey circles point to conserved residues forming the acceptor site (poly (ADP-ribose) binding site). See Tables 2.1 and 2.2 for references and accession numbers.

Furthermore, two other PARP homologues have been identified recently (see Table 2.2): (1) PARP-3, a 60-kDa protein of 533 amino acids (173) whose function is totally unknown; and (2) PH5P (termed PARP-4 in Table 2.2), a multifunctional enzyme of about 192 kDa which belongs to the inter-α-inhibitor family (174). The PARP-4 cDNA was re-isolated as a one of the three proteins present in Vault-particles in a two-hybrid screen (175). It encodes an enzyme named VPARP (for *V*ault PARP) containing a domain homologous (28% identity) to the catalytic domain of PARP-1, a BRCT domain, an inter-alpha trypsin inhibitor domain, and a bipartite NLS (Fig. 2.12).Vault particles are large ribonucleoprotein complexes of 13 MDa, located primarily in cytoplasm and ubiquitously conserved throughout eukaryotes. Although the cellular role of the Vault particle has remained elusive, several findings support the notion that vaults may have a transport function (175, 176). VPARP contains a domain that interacts with MVP, the major protein component of vaults. MVP is a substrate for a DNA-independent VPARP modification *in vitro*. Surprisingly, VPARP, but not the two other components of vaults, is associated with the mitotic spindle microtubules in HeLa cells, suggesting a role for this enzyme at the end of cellular division. Even more intriguing is the fact that the third protein component of vaults is a protein (TEP-1) that is also a component of the telomerase complex (177), again suggesting a possible link with the maintenance of genomic integrity.

Considering the sequence of the catalytic domain of all the members of the PARP family, PARP-1 sequences are highly conserved from human to plants (Fig. 2.13), whereas APP and NAP form a group which includes the mammalian PARP-2. PARP-3, VPARP, and tankyrase are clearly the proteins most distantly related to PARP-1.

The unsuspected existence of a family of PARP proteins—recently discovered, in part, in PARP-1 knockout cells—raises a number of important questions with regard to their specific cellular function(s) and their specific targets, which might be determined by their variable N-terminal modules. The global PARP activity measured in a cell after DNA damage may represent, in fact, the addition of at least five (and perhaps more) distinct enzymatic activities in which PARP-1 seems to account for 85–90% of the total. Perhaps the disruption of each *PARP* gene will be necessary to elucidate the physiological role of each member of the family and may reveal possible functional redundancy.

Acknowledgements

We are indebted to our colleagues of the PARP group: J. C. Amé, J. Oliver, F. Dantzer, C. Niedergang, B. Rinaldi, G. de la Rubia, A. Huber, and V. Schreiber for their contribution to the work summarized in this review. We

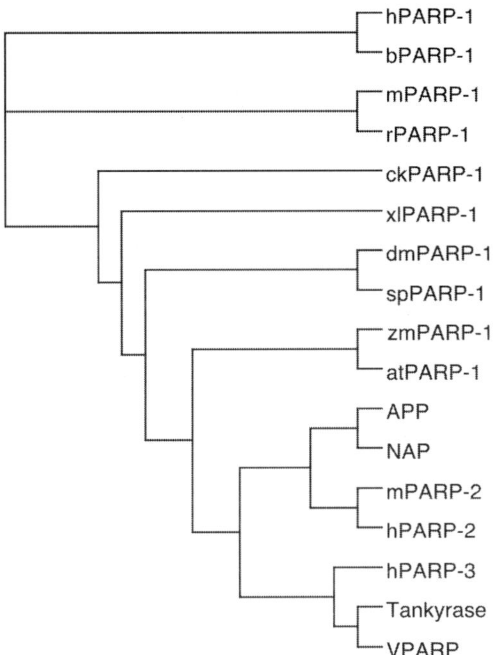

Fig. 2.13 Phylogenetic tree (Clustalw) based on sequence comparison of catalytic domain from various PARP-1 species and PARP homologues from mammals and plants. (See legend Fig. 2.3).

thank O. Poch for his fine contribution to the sequence alignments and Eric LeCam and Etienne Delain for their help with electron microscopy. This work was supported by CNRS, Association pour la Recherche contre le Cancer, la Ligue Contre le Cancer, Electricité de France, le Commissariat à l'Energie Atomique, and la Fondation pour la Recherche Médicale.

References

1. Cherney, B. W., McBride, O. W., Chen, D., Alkhatib, H., Bhatia, K., Hensley, P., and Smulson, M. E. (1987). cDNA sequence, protein structure, and chromosomal location of the human gene for poly(ADP-ribose) polymerase. *Proc. Natl Acad. Sci. USA*, **84**, 8370.
2. Herzog, H., Zabel, B. U., Schneider, R., Auer, B., Hirsch-Kauffmann, M., and Schweiger, M. (1989). Human nuclear NAD+ ADP-ribosyltransferase: localization of the gene on chromosome 1q41–q42 and expression of an active human enzyme in *Escherichia coli. Proc. Natl Acad. Sci. USA*, **86**, 3514.
3. Baumgartner, M., Schneider, R., Auer, B., Herzog, H., Schweiger, M., and Hirsch-Kauffmann, M. (1992). Fluorescence *in situ* mapping of the human nuclear NAD+ ADP-ribosyltransferase gene (ADPRT) and two secondary sites

to human chromosomal bands 1q42, 13q34, and 14q24. *Cytogenet. Cell. Genet.*, **61**, 172.

4. Auer, B., Nagl, U., Herzog, H., Schneider, R., and Schweiger, M. (1989). Human nuclear NAD+ ADP-ribosyltransferase(polymerizing): organization of the gene. *DNA*, **8**, 575.

5. Berghammer, H., Schweiger, M., and Auer, B. (1992). In ADP-ribosylation reactions (G. Poirier, P. Moreau Eds.), pp. 69. Springer-Verlag, Heidelberg, Germany. Structure and organisation of the mouse pADPRT gene.

6. Schweiger, M., Auer, B., Herzog, H., Hirsch-Kauffmann, M., Kaiser, P., Flick, K., Nagl, U. and Schneider, R. (1992). Molecular Biology of Nuclear NADt ADP-ribosyl transferase. In *ADP-ribosylation reactions* (G. G. Poirier, P. Moreau Editors), pp 20. Springer-Verlag, Heidelberg, Germany.

7. Hanai, S., Uchida, M., Kobayashi, S., Miwa, M., and Uchida, K. (1998). Genomic organization of Drosophila poly(ADP-ribose) polymerase and distribution of its mRNA during development. *J. Biol. Chem.*, **273**, 11881.

8. Ogura, T., Nyunoya, H., Takahashi-Masutani, M., Miwa, M., Sugimura, T., and Esumi, H. (1990). Characterization of a putative promoter region of the human poly(ADP-ribose) polymerase gene: structural similarity to that of the DNA polymerase beta gene. *Biochem. Biophys. Res. Commun.*, **167**, 701.

9. Yokoyama, Y., Kawamoto, T., Mitsuuchi, Y., Kurosaki, T., Toda, K., Ushiro, H., Terashima, M., Sumimoto, H., Kuribayashi, I., Yamamoto, Y., *et al.* (1990). Human poly(ADP-ribose) polymerase gene. Cloning of the promoter region. *Eur. J. Biochem.*, **194**, 521.

10. Potvin, F., Roy, R. J., Poirier, G. G., and Guerin, S. L. (1993). The US-1 element from the gene encoding rat poly(ADP-ribose) polymerase binds the transcription factor Sp1. *Eur. J. Biochem.*, **215**, 73.

11. Potvin, F., Thibodeau, J., Kirkland, J. B., Dandenault, B., Duchaine, C., and Poirier, G. G. (1992). Structural analysis of the putative regulatory region of the rat gene encoding poly(ADP-ribose) polymerase. *FEBS Lett.*, **302**, 269.

12. Schweiger, M., Oei, S. L., Herzog, H., Menardi, C., Schneider, R., Auer, B., and Hirsch-Kauffmann, M. (1995). Regulation of the human poly(ADP-ribosyl) transferase promoter via alternative DNA racket structures. *Biochimie*, **77**, 480.

13. Sastry, S. S. and Kun, E. (1990). The interaction of adenosine diphosphoribosyl transferase (ADPRT) with a cruciform DNA. *Biochem. Biophys. Res. Commun.*, **167**, 842.

14. Gradwohl, G., Mazen, A., and de Murcia, G. (1987). Poly(ADP-ribose) polymerase forms loops with DNA. *Biochem. Biophys. Res. Commun.*, **148**, 913.

15. Menegazzi, M., Gerosa, F., Tommasi, M., Uchida, K., Miwa, M., Sugimura, T., Suzuki, H., and Gelosa, F. (1988). Induction of poly(ADP-ribose) polymerase gene expression in lectin-stimulated human T lymphocytes is dependent on protein synthesis. *Biochem. Biophys. Res. Commun.*, **156**, 995.

16. McNerney, R., Tavasolli, M., Shall, S., Brazinski, A., and Johnstone, A. (1989). Changes in mRNA levels of poly(ADP-ribose) polymerase during activation of human lymphocytes. *Biochim. Biophys. Acta*, **1009**, 185.

17. Menegazzi, M., De Prati, A. C., Ledda-Columbano, G. M., Columbano, A., Uchida, K., Miwa, M., and Suzuki, H. (1990). Regulation of poly(ADP-ribose) polymerase mRNA levels during compensatory and mitogen-induced growth of rat liver. *Arch. Biochem. Biophys.*, **279**, 232.

18. Cesarone, C. F., Scarabelli, L., Scovassi, A. I., Izzo, R., Menegazzi, M., Carcereri De Prati, A., Orunesu, M., and Bertazzoni, U. (1990). Changes in activity and mRNA levels of poly(ADP-ribose) polymerase during rat liver regeneration. *Biochim. Biophys. Acta*, **1087**, 241.

19. Ogura, T., Takenouchi, N., Yamaguchi, M., Matsukage, A., Sugimura, T., and Esumi, H. (1990). Striking similarity of the distribution patterns of the poly(ADP-ribose) polymerase and DNA polymerase beta among various mouse organs. *Biochem. Biophys. Res. Commun.*, **172**, 377.

20. Menegazzi, M., Grassi-Zucconi, G., Carcerero De Prati, A., Ogura, T., Poltronieri, P., Nyunoya, H., Shiratori-Nyunoya, Y., Miwa, M., and Suzuki, H. (1991). Differential expression of poly(ADP-ribose) polymerase and DNA polymerase beta in rat tissues. *Exp. Cell Res.*, **197**, 66.

21. Cosi, C., Suzuki, H., Milani, D., Facci, L., Menegazzi, M., Vantini, G., Kanai, Y., and Skaper, S. D. (1994). Poly(ADP-ribose) polymerase: early involvement in glutamate-induced neurotoxicity in cultured cerebellar granule cells. *J. Neurosci. Res.*, **39**, 38.

22. Concha, I. I., Figueroa, J., Concha, M. I., Ueda, K., and Burzio, L. O. (1989). Intracellular distribution of poly(ADP-ribose) synthetase in rat spermatogenic cells. *Exp. Cell Res.*, **180**, 353.

23. Zucconi, G. G., Carcereri de Prati, A., Menegazzi, M., Cosi, C., and Suzuki, H. (1992). DNA repair enzymes in the brain. DNA polymerase beta and poly(ADP-ribose) polymerase. *Ann. N. Y. Acad. Sci.*, **663**, 432.

24. Alcivar, A. A., Hake, L. E., and Hecht, N. B. (1992). DNA polymerase-beta and poly(ADP)ribose polymerase mRNAs are differentially expressed during the development of male germinal cells. *Biol. Reprod.*, **46**, 201.

25. Dantzer, F., de la Rubia, G., Ménissier-de Murcia, J., Hostomsky, Z. de Murcia, G. and Schreiber, V. (2000). Base Excision Repair is impaired in mammalian cells lacking poly(ADP-ribose) polymerase-1. *Biochemistry*, **39**, 7559.

26. Wang, Z. Q., Auer, B., Stingl, L., Berghammer, H., Haidacher, D., Schweiger, M., and Wagner, E. F. (1995). Mice lacking ADPRT and poly(ADP-ribosyl)ation develop normally but are susceptible to skin disease. *Genes Dev.*, **9**, 509.

27. Jacobs, H., Fukita, Y., van der Horst, G. T., de Boer, J., Weeda, G., Essers, J., de Wind, N., Engelward, B. P., Samson, L., Verbeek, S., Ménissier de Murcia, J., de Murcia, G., te Riele, H., and Rajewsky, K. (1998). Hypermutation of immunoglobulin genes in memory B cells of DNA repair-deficient mice. *J. Exp. Med.*, **187**, 1735.

28. Zhou, Z. Q. and Walter, C. A. (1995). Expression of the DNA repair gene XRCC1 in baboon tissues. *Mutat. Res.*, **348**, 111.

29. Mackey, Z. B., Ramos, W., Levin, D. S., Walter, C. A., McCarrey, J. R., and Tomkinson, A. E. (1997). An alternative splicing event which occurs in mouse pachytene spermatocytes generates a form of DNA ligase III with distinct biochemical properties that may function in meiotic recombination. *Mol. Cell. Biol.*, **17**, 989.

30. Almon, E., Goldfinger, N., Kapon, A., Schwartz, D., Levine, A. J., and Rotter, V. (1993). Testicular tissue-specific expression of the p53 suppressor gene. *Dev. Biol.*, **156**, 107.

31. Pecker, I., Avraham, K. B., Gilbert, D. J., Savitsky, K., Rotman, G., Harnik, R., Fukao, T., Schrock, E., Hirotsune, S., Tagle, D. A., Collins, F. S., Wynshaw-

Boris, A., Ried, T., Copeland, N. G., Jenkins, N. A., Shiloh, Y., and Ziv, Y. (1996). Identification and chromosomal localization of Atm, the mouse homolog of the ataxia-telangiectasia gene. *Genomics*, **35**, 39.

32. Chambon, P., Weil, J. D., and Mandel, P. (1963). Nicotinamide mononucleotide activation of a new DNA-dependent polyadenylic acid synthesizing nuclear enzyme. *Biochem. Biophys. Res. Commun.*, **11**, 39.

33. Lamarre, D., Talbot, B., de Murcia, G., Laplante, C., Leduc, Y., Mazen, A., and Poirier, G. G. (1988). Structural and functional analysis of poly(ADP ribose) polymerase: an immunological study. *Biochim. Biophys. Acta*, **950**, 147.

34. Dantzer, F., Nasheuer, H. P., Vonesch, J. L., de Murcia, G., and Menissier-de Murcia, J. (1998). Functional association of poly(ADP-ribose) polymerase with DNA polymerase alpha-primase complex: a link between DNA strand break detection and DNA replication. *Nucleic Acids Res.*, **26**, 1891.

35. Fakan, S., Leduc, Y., Lamarre, D., Brunet, G., and Poirier, G. G. (1988). Immunoelectron microscopical distribution of poly(ADP-ribose)polymerase in the mammalian cell nucleus. *Exp. Cell Res.*, **179**, 517.

36. Mosgoeller, W., Steiner, M., Hozak, P., Penner, E., and Wesierska-Gadek, J. (1996). Nuclear architecture and ultrastructural distribution of poly (ADP-ribosyl)transferase, a multifunctional enzyme. *J. Cell Sci.*, **109**, 409.

37. Desnoyers, S., Kirkland, J. B., and Poirier, G. G. (1996). Association of poly(ADP-ribose) polymerase with nuclear subfractions catalyzed with sodium tetrathionate and hydrogen peroxide crosslinks. *Mol. Cell. Biochem.*, **159**, 155.

38. Leitinger, N. and Wesierska-Gadek, J. (1993). ADP-ribosylation of nucleolar proteins in HeLa tumor cells. *J. Cell. Biochem.*, **52**, 153.

39. Ramsamooj, P., Notario, V., and Dritschilo, A. (1995). Modification of nucleolar protein B23 after exposure to ionizing radiation. *Radiat. Res.*, **143**, 158.

40. Weber, J., Taylor, L., Rousell, M., Sherr, C., and Bar-Sagi, D. (1999). Nucleolar Arf sequesters Mdm2 and activates p53. *Nature Cell. Biol.*, **1**, 20.

41. Garcia, S. N. and Pillus, L. (1999). Net results of nucleolar dynamics. *Cell*, **97**, 825.

42. Cardenas-Corona, M. E., Jacobson, E. L., and Jacobson, M. K. (1987). Endogenous polymers of ADP-ribose are associated with the nuclear matrix. *J. Biol. Chem.*, **262**, 14863.

43. Alvarez-Gonzalez, R. and Ringer, D. P. (1988). Nuclear matrix associated poly(ADP-ribose) metabolism in regenerating rat liver. *FEBS Lett.*, **236**, 362.

44. Pedraza-Reyes, M. and Alvarez-Gonzalez, R. (1990). Oligo(3'-deoxy ADP-ribosyl)ation of the nuclear matrix lamins from rat liver utilizing 3'-deoxyNAD as a substrate. *FEBS Lett.*, **277**, 88.

45. Scovassi, A. I., Mariani, C., Negroni, M., Negri, C., and Bertazzoni, U. (1993). ADP-ribosylation of nonhistone proteins in HeLa cells: modification of DNA topoisomerase II. *Exp. Cell Res.*, **206**, 177.

46. Quesada, P., d'Erme, M., Parise, G., Faraone-Mennella, M. R., Caiafa, P., and Farina, B. (1994). Nuclear matrix-associated poly (ADPribosyl)ation system in rat testis chromatin. *Exp. Cell Res.*, **214**, 351.

47. Kaufmann, S. H., Brunet, G., Talbot, B., Lamarre, D., Dumas, C., Shaper, J. H., and Poirier, G. (1991). Association of poly(ADP-ribose) polymerase with the nuclear matrix: the role of intermolecular disulfide bond formation, RNA retention, and cell type. *Exp. Cell Res.*, **192**, 524.

48. Wesierska-Gadek, J. and Sauermann, G. (1985). Modification of nuclear matrix

proteins by ADP-ribosylation. Association of nuclear ADP-ribosyltransferase with the nuclear matrix. *Eur. J. Biochem.*, **153**, 421.
49. D'Erme, M., Malanga, M., Quesada, P., Faraone-Mennella, M. R., Farina, B., and Caiafa, P. (1990). Tightly-bound form of poly(ADP-ribose)polymerase in the higher order of chromatin organization. *Biochem. Int.*, **20**, 887.
50. Zahradka, P. and Ebisuzaki, K. (1984). Poly(ADP-ribose) polymerase is a zinc metalloenzyme. *Eur. J. Biochem.*, **142**, 503.
51. Oliver, F. J., de la Rubia, G., Rolli, V., Ruiz-Ruiz, M. C., de Murcia, G., and Ménissier-de Murcia, J. (1998). Importance of poly(ADP-ribose) polymerase and its cleavage in apoptosis. Lesson from an uncleavable mutant. *J. Biol. Chem.*, **273**, 33533.
52. Higashiura, M., Shimizu, Y., Tanimoto, M., Morita, T., and Yagura, T. (1992). Immunolocalization of Ku-proteins (p80/p70): localization of p70 to nucleoli and periphery of both interphase nuclei and metaphase chromosomes. *Exp. Cell Res.*, **201**, 444.
53. Dantzer, F., Nasheuer, H. P., de Murcia, G., and Ménissier-de Murcia, J. (1998). Poly(ADP-ribose)polymerase links the DNA damage surveillance network to the DNA replication machinery. *Nucleic Acids Res.*, **26**, 1891.
54. Mancini, M. A., Shan, B., Nickerson, J. A., Penman, S., and Lee, W. H. (1994). The retinoblastoma gene product is a cell cycle-dependent, nuclear matrix-associated protein. *Proc. Natl Acad. Sci. USA*, **91**, 418.
55. Galande, S. and Kohwi-Shigematsu, T. (1999). Poly(ADP-ribose) polymerase and Ku autoantigen form a complex and synergistically bind to matrix attachment sequences. *J. Biol. Chem.*, **274**, 20521.
56. Shizuta, Y., Kameshita, I., Ushiro, H., Matsuda, M., Suzuki, S., Mitsuuchi, Y., Yokoyama, Y., and Kurosaki, T. (1986). The domain structure and the function of poly(ADP-ribose) synthetase. *Adv. Enzyme Regul.*, **25**, 377.
57. Nishikimi, M., Ogasawara, K., Kameshita, I., Taniguchi, T., and Shizuta, Y. (1982). Poly(ADP-ribose) synthetase. The DNA binding domain and the automodification domain. *J. Biol. Chem.*, **257**, 6102.
58. Kameshita, I., Matsuda, Z., Taniguchi, T., and Shizuta, Y. (1984). Poly (ADP-ribose) synthetase. Separation and identification of three proteolytic fragments as the substrate-binding domain, the DNA-binding domain, and the automodification domain. *J. Biol. Chem.*, **259**, 4770.
59. Kurosaki, T., Ushiro, H., Mitsuuchi, Y., Suzuki, S., Matsuda, M., Matsuda, Y., Katunuma, N., Kangawa, K., Matsuo, H., Hirose, T., *et al.* (1987). Primary structure of human poly(ADP-ribose) synthetase as deduced from cDNA sequence. *J. Biol. Chem.*, **262**, 15990.
60. Uchida, K., Morita, T., Sato, T., Ogura, T., Yamashita, R., Noguchi, S., Suzuki, H., Nyunoya, H., Miwa, M., and Sugimura, T. (1987). Nucleotide sequence of a full-length cDNA for human fibroblast poly(ADP-ribose) polymerase. *Biochem. Biophys. Res. Commun.*, **148**, 617.
61. Huppi, K., Bhatia, K., Siwarski, D., Klinman, D., Cherney, B., and Smulson, M. (1989). Sequence and organization of the mouse poly (ADP-ribose) polymerase gene. *Nucleic Acids Res.*, **17**, 3387.
62. Beneke, S., Meyer, R., and Burkle, A. (1997). Isolation of cDNA encoding full-length rat (*Rattus norvegicus*) poly (ADP-ribose) polymerase. *Biochem. Mol. Biol. Int.*, **43**, 755.

63. Saito, I., Hatakeyama, K., Kido, T., Ohkubo, H., Nakanishi, S., and Ueda, K. (1990). Cloning of a full-length cDNA encoding bovine thymus poly (ADP-ribose) synthetase: evolutionarily conserved segments and their potential functions. *Gene*, **90**, 249.
64. Ittel, M. E., Garnier, J. M., Jeltsch, J. M., and Niedergang, C. P. (1991). Chicken poly(ADP-ribose) synthetase: complete deduced amino acid sequence and comparison with mammalian enzyme sequences. *Gene*, **102**, 157.
65. Uchida, K., Uchida, M., Hanai, S., Ozawa, Y., Ami, Y., Kushida, S., and Miwa, M. (1993). Isolation of the poly(ADP-ribose) polymerase-encoding cDNA from *Xenopus laevis*: phylogenetic conservation of the functional domains. *Gene*, **137**, 293.
66. Uchida, K., Hanai, S., Ishikawa, K., Ozawa, Y., Uchida, M., Sugimura, T., and Miwa, M. (1993). Cloning of cDNA encoding *Drosophila* poly(ADP-ribose) polymerase: leucine zipper in the auto-modification domain. *Proc. Natl Acad. Sci. USA*, **90**, 3481.
67. Masutani, M., Nozaki, T., Hitomi, Y., Ikejima, M., Nagasaki, K., de Prati, A. C., Kurata, S., Natori, S., Sugimura, T., and Esumi, H. (1994). Cloning and functional expression of poly(ADP-ribose) polymerase cDNA from *Sarcophaga peregrina. Eur. J. Biochem.*, **220**, 607.
68. Babiychuk, E., Cottrill, P. B., Storozhenko, S., Fuangthong, M., Chen, Y., O'Farrell, M. K., Van Montagu, M., Inze, D., and Kushnir, S. (1998). Higher plants possess two structurally different poly(ADP-ribose) polymerases. *Plant J.*, **15**, 635.
69. Mahajan, P. B. and Zuo, Z. (1998). Purification and cDNA cloning of maize Poly(ADP)-ribose polymerase. *Plant Physiol.*, **118**, 895.
70. Doucet-Chabeaud, G., Godon, C., Brutesco, C., de Murcia, G. and Kazmaier, M. (2000). Ionising radiation induces expression of PARP-1 and PARP-2 genes in Arabidopsis. *Molecular-Gen.* (in press).
71. Mazen, A., Menissier-de Murcia, J., Molinete, M., Simonin, F., Gradwohl, G., Poirier, G., and de Murcia, G. (1989). Poly(ADP-ribose)polymerase: a novel finger protein. *Nucleic Acids Res.*, **17**, 4689.
72. Gradwohl, G., Menissier de Murcia, J., Molinete, M., Simonin, F., and de Murcia, G. (1989). Expression of functional zinc finger domain of human poly(ADP-ribose)polymerase in *E. coli. Nucleic Acids Res.*, **17**, 7112.
73. Gradwohl, G., Menissier-de Murcia, J., Molinete, M., Simonin, F., Koken, M., Hoeijmakers, J. H., and de Murcia, G. (1990). The second zinc-finger domain of poly(ADP-ribose) polymerase determines specificity for single-stranded breaks in DNA. *Proc. Natl Acad. Sci. USA*, **87**, 2990.
74. Ikejima, M., Noguchi, S., Yamashita, R., Ogura, T., Sugimura, T., Gill, D. M., and Miwa, M. (1990). The zinc fingers of human poly(ADP-ribose) polymerase are differentially required for the recognition of DNA breaks and nicks and the consequent enzyme activation. Other structures recognize intact DNA. *J. Biol. Chem.*, **265**, 21907.
75. Simonin, F., Ménissier de Murcia, J., Poch, O., Muller, S., Gradwohl, G., Molinete, M., Penning, C., Keith, G., and de Murcia, G. (1990). Expression and site-directed mutagenesis of the catalytic domain of human poly(ADP-ribose)polymerase in *Escherichia coli*. Lysine 893 is critical for activity. *J. Biol. Chem.*, **265**, 19249.

76. Simonin, F., Hofferer, L., Panzeter, P. L., Muller, S., de Murcia, G., and Althaus, F. R. (1993). The carboxyl-terminal domain of human poly(ADP-ribose) polymerase. Overproduction in *Escherichia coli*, large scale purification, and characterization. *J. Biol. Chem.*, **268**, 13454.

77. Schreiber, V., Molinete, M., Boeuf, H., de Murcia, G., and Menissier-de Murcia, J. (1992). The human poly(ADP-ribose) polymerase nuclear localization signal is a bipartite element functionally separate from DNA binding and catalytic activity. *EMBO J.*, **11**, 3263.

78. Molinete, M., Vermeulen, W., Burkle, A., Menissier-de Murcia, J., Kupper, J. -H., Hoeijmakers, J. H. J., and de Murcia, G. (1993). Overproduction of the poly(ADP-ribose) polymerase DNA-binding domain blocks alkylation-induced DNA repair synthesis in mammalian cells. *EMBO J.*, **12**, 2109.

79. Avila, M. A., Velasco, J. A., Smulson, M. E., Dritschilo, A., Castro, R., and Notario, V. (1994). Functional expression of human poly(ADP-ribose) polymerase in *Schizosaccharomyces pombe* results in mitotic delay at G1, increased mutation rate, and sensitization to radiation. *Yeast*, **10**, 1003.

80. Trucco, C., Flatter, E., Fribourg, S., de Murcia, G., and Menissier-de Murcia, J. (1996). Mutations in the amino-terminal domain of the human poly(ADP-ribose) polymerase that affect its catalytic activity but not its DNA binding capacity. *FEBS Lett.*, **399**, 313.

81. Griesenbeck, J., Oei, S. L., Mayer-Kuckuk, P., Ziegler, M., Buchlow, G., and Schweiger, M. (1997). Protein–protein interaction of the human poly(ADP-ribosyl)transferase depends on the functional state of the enzyme. *Biochemistry*, **36**, 7297.

82. Rosenthal, D., Hong, T., Cherney, B., Zhang, S., Shima, T., Danielsen, M., and Smulson, M. (1994). Expression and characterization of a fusion protein between the catalytic domain of poly(ADP-ribose) polymerase and the DNA binding domain of the glucocorticoid receptor. *Biochem. Biophys. Res. Commun.*, **202**, 880.

83. Smulson, M., Istock, N., Ding, R., and Cherney, B. (1994). Deletion mutants of poly(ADP-ribose) polymerase support a model of cyclic association and dissociation of enzyme from DNA ends during DNA repair. *Biochemistry*, **33**, 6186.

84. Thibodeau, J., Simonin, F., Favazza, M., Gradwohl, G., Poirier, G., and de Murcia, G. (1990). Expression in *E. coli* of the catalytic domain of rat poly(ADP-ribose)polymerase. *FEBS Lett.*, **264**, 81.

85. Buki, K. G., Bauer, P. I., Hakam, A., and Kun, E. (1995). Identification of domains of poly(ADP-ribose) polymerase for protein binding and self-association. *J. Biol. Chem.*, **270**, 3370.

86. Wei, Y. F., Robins, P., Carter, K., Caldecott, K., Pappin, D. J., Yu, G. L., Wang, R. P., Shell, B. K., Nash, R. A., Schar, P., *et al.* (1995). Molecular cloning and expression of human cDNAs encoding a novel DNA ligase IV and DNA ligase III, an enzyme active in DNA repair and recombination. *Mol. Cell. Biol.*, **15**, 3206.

87. Nash, R. A., Caldecott, K. W., Barnes, D. E., and Lindahl, T. (1997). XRCC1 protein interacts with one of two distinct forms of DNA ligase III. *Biochemistry*, **36**, 5207.

88. Mackey, Z. B., Niedergang, C., Ménissier-de Murcia, J., Leppard, J., Au, K., Chen, J., de Murcia, G., and Tomkinson, A. E. (1999). DNA ligase III is

recruited to DNA strand breaks by a zinc finger motif homologous to that of Poly(ADP-ribose) polymerase. Identification of two functionally distinct DNA binding regions within DNA ligase III. *J. Biol. Chem.*, **274**, 21679.

89. Menissier-de Murcia, J., Molinete, M., Gradwohl, G., Simonin, F., and de Murcia, G. (1989). Zinc-binding domain of poly(ADP-ribose)polymerase participates in the recognition of single strand breaks on DNA. *J. Mol. Biol.*, **210**, 229.

90. Le Cam, E., Fack, F., Ménissier de Murcia, J., Cognet, J. A., Barbin, A., Sarantoglou, V., Revet, B., Delain, E., and de Murcia, G. (1994). Conformational analysis of a 139 base-pair DNA fragment containing a single-stranded break and its interaction with human poly(ADP-ribose) polymerase. *J. Mol. Biol.*, **235**, 1062.

91. Mendoza-Alvarez, H. and Alvarez-Gonzalez, R. (1993). Poly(ADP-ribose) polymerase is a catalytic dimer and the automodification reaction is intermolecular. *J. Biol. Chem.*, **268**, 22575.

92. Sawaya, M. R., Prasad, R., Wilson, S. H., Kraut, J., and Pelletier, H. (1997). Crystal structures of human DNA polymerase β complexed with gapped and nicked DNA: evidence for an induced fit mechanism. *Biochemistry*, **36**, 11205.

93. Marintchev, A., Mullen, M. A., Maciejewski, M. W., Pan, B., Gryk, M. R., and Mullen, G. P. (1999). Solution structure of the single-strand break repair protein XRCC1 N-terminal domain. *Nature Struct. Biol.*, **6**, 884.

94. Castaing, B., Zelwer, C., Laval, J., and Boiteux, S. (1995). HU protein of *Escherichia coli* binds specifically to DNA that contains single-strand breaks or gaps. *J. Biol. Chem.*, **270**, 10291.

95. Kamashev, D., Balandina, A., and Rouviere-Yaniv, J. (1999). The binding motif recognized by HU on both nicked and cruciform DNA. *EMBO J.*, **18**, 5434.

96. Pinson, V., Takahashi, M., and Rouviere-Yaniv, J. (1999). Differential binding of the *Escherichia coli* HU, homodimeric forms and heterodimeric form to linear, gapped and cruciform DNA. *J. Mol. Biol.*, **287**, 485.

97. Benjamin, R. C. and Gill, D. M. (1980). Poly(ADP-ribose) synthesis *in vitro* programmed by damaged DNA. A comparison of DNA molecules containing different types of strand breaks. *J. Biol. Chem.*, **255**, 10502.

98. Benjamin, R. C. and Gill, D. M. (1980). ADP-ribosylation in mammalian cell ghosts. Dependence of poly(ADP-ribose) synthesis on strand breakage in DNA. *J. Biol. Chem.*, **255**, 10493.

99. Sastry, S. S., Buki, K. G., and Kun, E. (1989). Binding of adenosine diphosphoribosyltransferase to the termini and internal regions of linear DNAs. *Biochemistry*, **28**, 5670.

100. Ittel, M. E., Jongstra-Bilen, J., Niedergang, C., Mandel, P., and Delain, E. (1985). DNA–poly(ADP-ribose) polymerase complex: isolation of the DNA wrapping the enzyme molecule. In *ADP-ribosylation of proteins* (ed. F. R. Althaus, H. Hilz, and S. Shall), p. 000. Springer-Verlag, Berlin, Germany.

101. Zechiedrich, E. L. and Osheroff, N. (1990). Eukaryotic topoisomerases recognize nucleic acid topology by preferentially interacting with DNA crossovers. *EMBO J.*, **9**, 4555.

102. West, K. L. and Austin, C. A. (1999). Human DNA topoisomerase II beta binds and cleaves four-way junction DNA *in vitro*. *Nucleic Acids Res.*, **27**, 984.

103. Bazett-Jones, D. P., Cote, J., Landel, C. C., Peterson, C. L., and Workman, J. L.

(1999). The SWI/SNF complex creates loop domains in DNA and polynucleosome arrays and can disrupt DNA–histone contacts within these domains. *Mol. Cell. Biol.*, **19**, 1470.

104. Kannan, P., Yu, Y., Wankhade, S., and Tainsky, M. A. (1999). PolyADP-ribose polymerase is a coactivator for AP-2-mediated transcriptional activation. *Nucleic Acids Res.*, **27**, 866.

105. Nie, J., Sakamoto, S., Song, D., Qu, Z., Ota, K., and Taniguchi, T. (1998). Interaction of Oct-1 and automodification domain of poly(ADP-ribose) synthetase. *FEBS Lett.*, **424**, 27.

106. Oei, S. L., Griesenbeck, J., Schweiger, M., Babich, V., Kropotov, A., and Tomilin, N. (1997). Interaction of the transcription factor YY1 with human poly(ADP-ribosyl) transferase. *Biochem. Biophys. Res. Commun.*, **240**, 108.

107. Griesenbeck, J., Ziegler, M., Tomilin, N., Schweiger, M., and Oei, S. L. (1999). Stimulation of the catalytic activity of poly(ADP-ribosyl) transferase by transcription factor Yin Yang 1. *FEBS Lett.*, **443**, 20.

108. Butler, A. J. and Ordahl, C. P. (1999). Poly(ADP-ribose) polymerase binds with transcription enhancer factor 1 to MCAT1 elements to regulate muscle-specific transcription. *Mol. Cell. Biol.*, **19**, 296.

109. Miyamoto, T., Kakizawa, T., and Hashizume, K. (1999). Inhibition of nuclear receptor signalling by poly(ADP-ribose) polymerase. *Mol. Cell. Biol.*, **19**, 2644.

110. Hassa, P. O. and Hottiger, M. O. (1999). A role of poly (ADP-ribose) polymerase in NF-kappaB transcriptional activation. *Biol. Chem.*, **380**, 953.

111. Oliver, F. J., Ménissier-de Murcia, J., Nacci, C., Decker, P., Andriantsitohaina, R., Muller, S., de la Rubia, G., Stoclet, J. C., and de Murcia, G. (1999). Resistance to endotoxic shock as a consequence of defective NF-kappaB activation in poly (ADP-ribose) polymerase-1 deficient mice. *EMBO J.*, **18**, 4446.

112. Thibodeau, J., Potvin, F., Kirkland, J. B., and Poirier, G. (1993). Expression in *Escherichia coli* of the 36 kDa domain of poly(ADP-ribose) polymerase and investigation of its DNA binding properties. *Biochim. Biophys. Acta*, **1163**, 49.

113. Masson, M., Niedergang, C., Schreiber, V., Muller, S., Menissier-de Murcia, J., and de Murcia, G. (1998). XRCC1 is specifically associated with poly(ADP-ribose) polymerase and negatively regulates its activity following DNA damage. *Mol. Cell. Biol.*, **18**, 3563.

114. Rhodes, D. and Klug, A. (1993). Zinc fingers. *Sci. Am.*, **268**, 56.

115. Schwabe, J. W. and Klug, A. (1994). Zinc mining for protein domains. *Nature Struct. Biol.*, **1**, 345.

116. Sastry, S. S. and Kun, E. (1988). Molecular interactions between DNA, poly(ADP-ribose) polymerase, and histones. *J. Biol. Chem.*, **263**, 1505.

117. Panzeter, P. L. and Althaus, F. R. (1994). DNA strand break-mediated partitioning of poly(ADP-ribose) polymerase function. *Biochemistry*, **33**, 9600.

118. Bauer, P. I., Buki, K. G., Hakam, A., and Kun, E. (1990). Macromolecular association of ADP-ribosyltransferase and its correlation with enzymic activity. *Biochem. J.*, **270**, 17.

119. Taniguchi, T. (1987). Reaction mechanism for automodification of poly(ADP-ribose) synthetase. *Biochem. Biophys. Res. Commun.*, **147**, 1008.

120. Panzeter, P. L., Zweifel, B., and Althaus, F. R. (1992). Synthesis of poly(ADP-ribose)–agarose beads: an affinity resin for studying (ADP-ribose)n–protein interactions. *Anal. Biochem.*, **207**, 157.

121. Kawaichi, M., Ueda, K., and Hayaishi, O. (1981). Multiple autopoly(ADP-ribosyl)ation of rat liver poly(ADP-ribose) synthetase. Mode of modification and properties of automodified synthetase. *J. Biol. Chem.*, **256**, 9483.

122. Alvarez-Gonzalez, R., Watkins, T. A., Gill, P. K., Reed, J. L., and Mendoza-Alvarez, H. (1999). Regulatory mechanisms of poly(ADP-ribose) polymerase. *Mol. Cell. Biochem.*, **193**, 19.

123. Conti, E., Uy, M., Leighton, L., Blobel, G., and Kuriyan, J. (1998). Crystallographic analysis of the recognition of a nuclear localization signal by the nuclear import factor karyopherin alpha. *Cell*, **94**, 193.

124. Kaufmann, S. H., Desnoyers, S., Ottaviano, Y., Davidson, N. E., and Poirier, G. G. (1993). Specific proteolytic cleavage of poly(ADP-ribose) polymerase: an early marker of chemotherapy-induced apoptosis. *Cancer Res.*, **53**, 3976.

125. Buki, K. G. and Kun, E. (1988). Polypeptide domains of ADP-ribosyltransferase obtained by digestion with plasmin. *Biochemistry*, **27**, 5990.

126. D'Amours, D., Germain, M., Orth, K., Dixit, V. M., and Poirier, G. G. (1998). Proteolysis of poly(ADP-ribose) polymerase by caspase 3: kinetics of cleavage of mono(ADP-ribosyl)ated and DNA-bound substrates. *Radiat. Res.*, **150**, 3.

127. Poltronieri, P., Yokota, T., Koyama, Y., Hanai, S., Uchida, K., and Miwa, M. (1997). PARP cleavage in the apoptotic pathway in S2 cells from *Drosophila melanogaster*. *Biochem. Cell. Biol.*, **75**, 445.

128. Kreimeyer, A., Wielckens, K., Adamietz, P., and Hilz, H. (1984). DNA repair-associated ADP-ribosylation *in vivo*. Modification of histone H1 differs from that of the principal acceptor proteins. *J. Biol. Chem.*, **259**, 890.

129. Ferro, A. M. and Olivera, B. M. (1982). Poly(ADP-ribosylation) *in vitro*. Reaction parameters and enzyme mechanism. *J. Biol. Chem.*, **257**, 7808.

130. Zahradka, P. and Ebisuzaki, K. (1982). A shuttle mechanism for DNA–protein interactions. The regulation of poly(ADP-ribose) polymerase. *Eur. J. Biochem.*, **127**, 579.

131. Bork, P., Hofmann, K., Bucher, P., Neuwald, A. F., Altschul, S. F., and Koonin, E. V. (1997). A superfamily of conserved domains in DNA damage-responsive cell cycle checkpoint proteins. *FASEB J.*, **11**, 68.

132. Callebaut, I. and Mornon, J. P. (1997). From BRCA1 to RAP1: a widespread BRCT module closely associated with DNA repair. *FEBS Lett.*, **400**, 25.

133. Critchlow, S. E., Bowater, R. P., and Jackson, S. P. (1997). Mammalian DNA double-strand break repair protein XRCC4 interacts with DNA ligase IV. *Curr. Biol.*, **7**, 588.

134. Masson, M., Menissier-de Murcia, J., Mattei, M. G., de Murcia, G., and Niedergang, C. P. (1997). Poly(ADP-ribose) polymerase interacts with a novel human ubiquitin conjugating enzyme: hUbc9. *Gene*, **190**, 287.

135. Zhang, X., Morera, S., Bates, P. A., Whitehead, P. C., Coffer, A. I., Hainbucher, K., Nash, R. A., Sternberg, M. J., Lindahl, T., and Freemont, P. S. (1998). Structure of an XRCC1 BRCT domain: a new protein–protein interaction module. *EMBO J.*, **17**, 6404.

136. Kawamura, T., Hanai, S., Yokota, T., Hayashi, T., Poltronieri, P., Miwa, M., and Uchida, K. (1998). An alternative form of poly(ADP-ribose) polymerase in *Drosophila melanogaster* and its ectopic expression in rat-1 cells. *Biochem. Biophys. Res. Commun.*, **251**, 35.

137. Miwa, M., Hanai, S., Poltronieri, P., Uchida, M., and Uchida, K. (1999). Functional analysis of poly(ADP-ribose) polymerase in *Drosophila melanogaster*. *Mol. Cell. Biochem.*, **193**, 103.

138. Allen, B. L., Uhlmann, F., Gaur, L. K., Mulder, B. A., Posey, K. L., Jones, L. B., and Hardin, S. H. (1998). DNA recognition properties of the N-terminal DNA binding domain within the large subunit of replication factor C. *Nucleic Acids Res.*, **26**, 3877.

139. Yamane, K. and Tsuruo, T. (1999). Conserved BRCT regions of TopBP1 and of the tumor suppressor BRCA1 bind strand breaks and termini of DNA. *Oncogene*, **18**, 5194.

140. Giner, H., Simonin, F., de Murcia, G., and Menissier-de Murcia, J. (1992). Overproduction and large-scale purification of the human poly(ADP-ribose) polymerase using a baculovirus expression system. *Gene*, **114**, 279.

141. Jung, S., Miranda, E. A., de Murcia, J. M., Niedergang, C., Delarue, M., Schulz, G. E., and de Murcia, G. M. (1994). Crystallization and X-ray crystallographic analysis of recombinant chicken poly(ADP-ribose) polymerase catalytic domain produced in Sf9 insect cells. *J. Mol. Biol.*, **244**, 114.

142. Ruf, A., Menissier-de Murcia, J., de Murcia, G., and Schulz, G. E. (1996). Structure of the catalytic fragment of poly(ADP-ribose) polymerase from chicken. *Proc. Natl Acad. Sci. USA*, **93**, 7481.

143. Ruf, A., de Murcia, G., and Schulz, G. E. (1998). Inhibitor and NAD+ binding to poly(ADP-ribose) polymerase as derived from crystal structures and homology modeling. *Biochemistry*, **37**, 3893.

144. Ruf, A., Rolli, V., de Murcia, G., and Schulz, G. E. (1998). The mechanism of the elongation and branching reaction of poly(ADP-ribose) polymerase as derived from crystal structures and mutagenesis. *J. Mol. Biol.*, **278**, 57.

145. Holm, L. and Sander, C. (1993). Protein structure comparison by alignment of distance matrices. *J. Mol. Biol.*, **233**, 123.

146. Bennett, M. J., Choe, S., and Eisenberg, D. (1994). Refined structure of dimeric diphtheria toxin at 2. 0 A resolution. *Protein Sci.*, **3**, 1444.

147. Li, M., Dyda, F., Benhar, I., Pastan, I., and Davies, D. R. (1995). The crystal structure of *Pseudomonas aeruginosa* exotoxin domain III with nicotinamide and AMP: conformational differences with the intact exotoxin. *Proc. Natl Acad. Sci. USA*, **92**, 9308.

148. Sixma, T. K., Kalk, K. H., van Zanten, B. A., Dauter, Z., Kingma, J., Witholt, B., and Hol, W. G. (1993). Refined structure of *Escherichia coli* heat-labile enterotoxin, a close relative of cholera toxin. *J. Mol. Biol.*, **230**, 890.

149. Zhang, R. G., Scott, D. L., Westbrook, M. L., Nance, S., Spangler, B. D., Shipley, G. G., and Westbrook, E. M. (1995). The three-dimensional crystal structure of cholera toxin. *J. Mol. Biol.*, **251**, 563.

150. Griffin, R. J., Curtin, N. J., Newell, D. R., Golding, B. T., Durkacz, B. W., and Calvert, A. H. (1995). The role of inhibitors of poly(ADP-ribose) polymerase as resistance-modifying agents in cancer therapy. *Biochimie*, **77**, 408.

151. Szabo, C. and Dawson, V. L. (1998). Role of poly(ADP-ribose) synthetase in inflammation and ischaemia–reperfusion. *Trends Pharmacol Sci.*, **19**, 287.

152. Rolli, V., O'Farrell, M., Menissier-de Murcia, J., and de Murcia, G. (1997). Random mutagenesis of the poly(ADP-ribose) polymerase catalytic domain reveals amino acids involved in polymer branching. *Biochemistry*, **36**, 12147.

153. Bell, C. E. and Eisenberg, D. (1996). Crystal structure of diphtheria toxin bound to nicotinamide adenine dinucleotide. *Biochemistry*, **35**, 1137.
154. Li, M., Dyda, F., Benhar, I., Pastan, I., and Davies, D. R. (1996). Crystal structure of the catalytic domain of Pseudomonas exotoxin A complexed with a nicotinamide adenine dinucleotide analog: implications for the activation process and for ADP ribosylation. *Proc. Natl Acad. Sci. USA*, **93**, 6902.
155. Moodie, S. L. and Thornton, J. M. (1993). A study into the effects of protein binding on nucleotide conformation. *Nucleic Acids Res.*, **21**, 1369.
156. Miranda, E. A., Dantzer, F., O'Farrell, M., de Murcia, G. and Ménissier-de Murcia, J. M. (1995). Characterisation of a gain-of-function mutant of poly(ADP-ribose) polymerase. *Biochem. Biophys. Res. Commun.*, **212**, 317.
157. Slama, J. T. and Simmons, A. M. (1988). Carbanicotinamide adenine dinucleotide: synthesis and enzymological properties of a carbocyclic analogue of oxidized nicotinamide adenine dinucleotide. *Biochemistry*, **27**, 183.
158. Slama, J. T. and Simmons, A. M. (1989). Inhibition of NAD glycohydrolase and ADP-ribosyl transferases by carbocyclic analogues of oxidized nicotinamide adenine dinucleotide. *Biochemistry*, **28**, 7688.
159. Moodie, S. L., Mitchell, J. B., and Thornton, J. M. (1996). Protein recognition of adenylate: an example of a fuzzy recognition template. *J. Mol. Biol.*, **263**, 486.
160. Keith, G., Desgres, J., and de Murcia, G. (1990). Use of two-dimensional thin-layer chromatography for the components study of poly(adenosine diphosphate ribose). *Anal. Biochem.*, **191**, 309.
161. Scheuring, J. and Schramm, V. L. (1997). Pertussis toxin: transition state analysis for ADP-ribosylation of G-protein peptide alphai3C20. *Biochemistry*, **36**, 8215.
162. Marsischky, G. T., Wilson, B. A., and Collier, R. J. (1995). Role of glutamic acid 988 of human poly-ADP-ribose polymerase in polymer formation. Evidence for active site similarities to the ADP-ribosylating toxins. *J. Biol. Chem.*, **270**, 3247.
163. Alvarez-Gonzalez, R. (1994). DeoxyNAD and deoxyADP-ribosylation of proteins. *Mol. Cell. Biochem.*, **138**, 213.
164. Payne, J. G. and Bol, A. K. (1976). Cytological detection of poly (ADP-ribose) polymerase. *Exp. Cell Res.*, **99**, 428.
165. Chen, Y. M., Shall, S., and O'Farrell, M. (1994). Poly(ADP-ribose) polymerase in plant nuclei. *Eur. J. Biochem.*, **224**, 135.
166. Puchta, H., Swoboda, P., and Hohn, B. (1995). Induction of intrachromosomal recombination in whole plants. *Plant J.*, **7**, 203.
167. Lepiniec, L., Babiychuk, E., Kushnir, S., Van Montagu, M., and Inze, D. (1995). Characterization of an *Arabidopsis thaliana* cDNA homologue to animal poly(ADP-ribose) polymerase. *FEBS Lett.*, **364**, 103.
168. Shieh, W. M., Amé, J.-C., Wilson, M. V., Wang, Z.-Q., Koh, D. W., Jacobson, M. K., and Jacobson, E. K. (1998). Poly(ADP-ribose) polymerase null mouse cells synthesize poly(ADP-ribose) polymers. *J. Biol. Chem.*, **273**, 30069.
169. Amé, J.-C., Rolli, V., Schreiber, V., Niedergang, C., Apiou, F., Decker, P., Muller, S., Höger, T., Ménissier-de Murcia, J., and de Murcia, G. (1999). PARP-2, a novel mammalian DNA-damage dependent poly(ADP-ribose) polymerase. *J. Biol. Chem.*, **274**, 17860.
170. Smith, S., Giriat, I., Schmitt, A., and de Lange, T. (1998). Tankyrase, a poly(ADP-ribose) polymerase at human telomeres. *Science*, **282**, 1484.

171. de Lange, T. (1998). Length control of human telomeres. *Cancer J. Sci. Am.*, **4** (Suppl. 1), S22.
172. van Steensel, B. and de Lange, T. (1997). Control of telomere length by the human telomeric protein TRF1. *Nature*, **385**, 740.
173. Johansson, M. (1999). A human poly(ADP-ribose) polymerase gene family (ADPRTL): cDNA cloning of two novel poly(ADP-ribose) polymerase homologues. *Genomics*, **57**, 442.
174. Jean, L., Risler, J. L., Nagase, T., Coulouarn, C., Nomura, N., and Salier, J. P. (1999). The nuclear protein PH5P of the inter-alpha-inhibitor superfamily: a missing link between poly(ADP-ribose)polymerase and the inter-alpha-inhibitor family and a novel actor of DNA repair? *FEBS Lett.*, **446**, 6.
175. Kickhoefer, V. A., Siva, A. C., Kedersha, N. L., Inman, E. M., Ruland, C., Streuli, M., and Rome, L. H. (1999). The 193-kD vault protein, VPARP, is a novel Poly(ADP-ribose) polymerase. *J. Cell. Biol.*, **146**, 917.
176. Kickhoefer, V. A., Rajavel, K. S., Scheffer, G. L., Dalton, W. S., Scheper, R. J., and Rome, L. H. (1998). Vaults are up-regulated in multidrug-resistant cancer cell lines. *J. Biol. Chem.*, **273**, 8971.
177. Harrington, L., McPhail, T., Mar, V., Zhou, W., Oulton, R., Bass, M. B., Arruda, I., and Robinson, M. O. (1997). A mammalian telomerase-associated protein. *Science*, **275**, 973.
178. Zhu, L., Smith, S., de Lange, T. and Seldin, M. F. (1999). Chromosomal mapping of the tankyrase gene, in human and mouse. *Genomics*, **57**, 320.
179. de Murcia, G., Jongstra-Bilen, J., Ittel, M. E., Mandel, P., and Delain, E. (1983). Poly(ADP-ribose) polymerase auto-modification and interaction with DNA: electron microscopic visualization. *EMBO J.*, **2**, 543.
180. Poirier, G., de Murcia, G., Jongstra-Bilen, J., Niedergang, C., and Mandel, P. (1982). Poly (ADP-ribosyl)ation of polynucleosomes causes relaxation of chromatin structure. *Proc. Natl Acad. Sci. USA*, **79**, 3423.
181. de Murcia, G., Huletsky, A., Lamarre, D., Gaudreau, A., Pouyet, J., Daune, M., and Poirier, G. (1986). Modulation of chromatin superstructure induced by poly(ADP-ribose) synthesis and degradation. *J. Biol. Chem.*, **261**, 7011.
182. Kraulis P. J. (1991). MOLSCRIPT: A program to produce both detailed and schematic plots of protein structures. *J. Appl. Crystallogr.*, **24**, 946.
183. Nicholls, A., Sharp, K. A., and Honig, B. (1991). Protein folding and association: insights from the interfacial and thermodynamic properties of hydrocarbons. *Proteins: Struct. Funct. Genet.*, **11**, 282.
184. Smith, S. and de Lange. (1999). Cell cycle localization of the telomeric PARP, to nuclear pore complexes and centrosomes. *J. Cell Sci.*, **112**, 3649.

---------- **3** ----------

Biological significance of poly (ADP-ribosyl)ation reactions: molecular and genetic approaches

Alexander Bürkle, Valerie Schreiber, Françoise Dantzer, F. Javier Oliver, Claude Niedergang, Gilbert de Murcia, and Josiane Ménissier-de Murcia

3.1   Introduction

Inhibition of PARP activity with NAD^+ analogues (such as nicotinamide, benzamide, and derivatives thereof) has constituted the major approach to elucidating the physiological role of poly (ADP-ribosyl)ation) reactions. The non-specific inhibitor effects that were repeatedly noted (1–3)—as well as the existence of three classes of ADP-ribosyl transferases—with varying sensitivity to ADP-ribosylation inhibitors (4), made it difficult to draw definitive conclusions based on inhibitor experiments alone.

During the past decade, alternative approaches were developed to abrogate PARP activity in living cells and organisms, with both high selectivity and efficiency. One strategy was to select for cell clones that display reduced levels of PARP protein and consequently PARP activity, either spontaneously or after mutagen treatment of cultures. Different molecular and genetic approaches have been chosen to reduce the expression of PARP in living cells. These include the use of antisense RNA or PARP antisense oligonucleotides, the overexpression of a dominant-negative version of PARP, and more recently the disruption of the *PARP* gene by homologous recombination. Conversely, overexpression of the cDNA encoding full-length PARP has been set up in various cell types. In this chapter, a particular emphasis has been put on these approaches which have considerably increased our knowledge of the physiological role of PARP.

3.2   PARP-deficient cell lines established by random mutagenesis

A series of papers describing the isolation and characterization of Chinese hamster V79 cell clones deficient in poly (ADP-ribose) synthesis have been published by Berger and co-workers. Two different protocols were developed (5, 6). In the first, cell lines deficient in PARP enzyme activity were established. The selection strategy for obtaining such cells was based on the ability of this enzyme to mediate acute cell killing by depleting cellular NAD^+ and ATP in response to high levels of DNA damage induced by the alkylating agent 1-methyl-3-nitro-1-nitroso-guanidine (MNNG). After the first round of selection, cell clones were obtained which displayed 37–82% PARP activity compared to the parental cells under DNA damage conditions. It has been established that the apparent decrease in PARP activity was not due to increases in NAD^+ glycohydrolase, poly (ADP-ribose) glycohydrolase (PARG), or phosphodiesterase activities. By performing repeated rounds of mutagenesis and selection on one of the first-generation clones, subclones ADPRT 54 and ADPRT 351 were isolated which displayed 5–11% enzyme activity compared to the parental V79 cells. The second approach was to develop cell lines capable of growing at NAD^+ levels sufficiently low to limit substrate availability for poly (ADP-ribose) synthesis. The group isolated spontaneous mutants from V79 cells that could grow stably in culture medium free of nicotinamide or any of its analogues. These cell lines (N2, N3, and N4) maintained NAD^+ levels between 1.5 and 3% of those found in parental V79 cells grown in complete medium. Further characterization of the mutant lines showed that under conditions that restricted poly (ADP-ribose) synthesis, all had prolonged doubling times and increased sister-chromatid exchange frequencies.

The mutant cell lines proved to be resistant to the cytotoxic effects of the topoisomerase II inhibitor etoposide (7). Resistance was stable in the ADPRT 54 and ADPRT 351 cell lines, whereas in the N2, N3, and N4 cell lines resistance was dependent on cellular NAD^+ deficiency, as induced by growth in nicotinamide-free medium, and could be reverted by growth in nicotinamide-containing complete medium. All the cell lines responded to etoposide treatment by the formation of protein-cross-linked DNA strand breaks. Upon drug removal all the cell lines reversed the DNA strand breaks at similar rates, thus revealing a dissociation between the induction of DNA strand breaks by etoposide and cytotoxicity. The poly (ADP-ribosylation)-deficient cell lines were also significantly resistant to the topoisomerase II-targeted DNA intercalators adriamycin and *m*-AMSA (8, 9). Furthermore, the mutant cell lines showed constitutively increased levels of the glucose-regulated stress protein GRP78 (10). Similarly, in N2, N3, and N4 cells GRP78 was induced under nicotinamide-deficient conditions. The induction of GRP78 was associated with elevated levels of GRP78 mRNA and appeared to be regulated at the transcriptional level. This work thus revealed an association between a deficiency of the NAD^+-poly (ADP-ribose) synthesis system, induction of GRP78 synthesis, and resistance to etoposide.

Chatterjee *et al.* (11) reported on the testing of several inhibitors of NAD^+-PARP metabolism (including 3-aminobenzamide, PD128763, and 6-aminonicotinamide) for their ability to induce GRP78 expression and etoposide resistance. Interestingly, 6-aminonicotinamide treatment, which results in a lowering of cellular NAD^+ levels, was indeed highly effective both in the induction of GRP78 and the subsequent development of resistance to etoposide. In contrast, treatment with 3-aminobenzamide or PD128763 did not induce GRP78 nor did it result in etoposide resistance. Therefore, it is the low cellular level of NAD^+ rather than a deficiency of PARP that seems to play a role in GRP78 expression and etoposide resistance.

In contrast to the phenotype of resistance to topoisomerase II inhibitors, the same mutant V79 Chinese hamster cell lines were hypersensitive to camptothecin, a topoisomerase I inhibitor (12). Cell lines ADPRT 54 and ADPRT 351 were also found to be hypersensitive to several alkylating agents (8, 9) including alkylsulfonates, alkylnitrosoureas, nitrosoguanidine, melphalan, 1,3-bis(2-chloroethyl)-1-nitrosourea (BCNU; Carmustine), and mitomycin C as well as to bleomycin, UV-irradiation, the UV-mimetic agent 4-nitroquinoline-1-oxide, and X-irradiation, suggesting some DNA repair deficiency. Ray *et al.* (13) studied DNA repair in transcribed and non-transcribed nuclear sequences in ADPRT 351 cells by using quantitative Southern blot analysis. Cells were exposed to methylnitrosourea for 1 h and allowed to repair for 8 or 24 h. It was shown that ADPRT 351 cells (showing a 95% reduction in PARP activity) were equally as efficient in the repair of *N*-methylpurines in the transcribed sequence containing the dihydrofolate reductase gene as the parental V79 cell line. However, the same cells did show a deficiency in the repair of such lesions in the non-transcribed sequence containing the IgE gene. Furthermore, nucleoid sedimentation assays showed that these cells were deficient in repair across the entire genome when compared to the parental V79 cells. These studies indicated that PARP activity is not required for the repair of *N*-methylpurines in transcribed nuclear DNA sequences, but is necessary for the repair of such lesions in non-transcribed nuclear DNA sequences as well as across the entire genome. It should be noted that this finding is in apparent contradiction to that reported in the paper by Stevnsner *et al.* (14) who used a PARP antisense system and reported an involvement of PARP in the repair of the actively transcribed *DHFR* gene.

Whitacre *et al.* (15) reported on a possible link between this phenomenon and the p53 tumour suppressor protein: the NAD^+/PARP-deficient cell lines exhibited a significant reduction in both baseline p53 expression and its activity compared to the parental V79 cells. Furthermore, etoposide, a drug known to cause an increase in p53 expression and subsequent apoptosis in V79 cells, failed to produce any significant increase in p53 expression or apoptotic DNA fragmentation in NAD^+/PARP-deficient cell lines. Thus, it was suggested that the NAD^+/poly (ADP-ribose) system might be involved in the regulation of p53 and its dependent pathways.

The latter interpretation was recently challenged, however, by the observation of another group (16) on a relationship between *mono* (ADP-ribosylation), rather than *poly* (ADP-ribosylation), and apoptosis. This group examined the role of intracellular NAD^+ in tumour necrosis factor (TNF-α)- and UV-induced activation of a 24-kDa apoptotic protease (AP24) leading to internucleosomal DNA fragmentation and death. The results showed that nutritional depletion of NAD^+ to undetectable levels in the two human leukaemia lines U937 and HL-60 rendered cells completely resistant to apoptosis. This was attributed to a block in the activation of AP24 and subsequent DNA cleavage. Normal cells showed an elevation of an ADP-ribosyl transferase activity in both the cytosol and nucleus after exposure to TNF-α, before DNA fragmentation. Both ADP-ribosyl transferase activity and cell death were suppressed by an inhibitor of mono (ADP-ribosyl) transfer reactions. Nuclei from NAD^+-depleted cells were still sensitive to DNA fragmentation induced by exogenous AP24, indicating a selective function for NAD^+ upstream of AP24 activation in the apoptotic pathway. The requirement for intracellular NAD^+ and the activation of mono (ADP-ribosyl) transferase activity during apoptosis was confirmed in three other leukaemia cell lines. This mechanism, however, does not seem to be universal, since BJAB and Jurkat leukaemia cells underwent apoptosis normally even in the absence of detectable intracellular NAD^+. Furthermore, the status of p53 under the experimental conditions chosen was not reported. Nevertheless, this work suggested that after the exposure of some cell lines to TNF-α or UV light NAD^+-dependent mono (ADP-ribosyl) transferase activities may be critically involved in the signal transduction cascade, which ultimately leads to activation of AP24, subsequent DNA fragmentation, and cell death.

Other groups have also isolated PARP-deficient cell lines. MacLaren *et al.* (17) described a method for isolating Chinese hamster ovary (CHO) cells deficient in PARP activity by directly screening colonies for enzyme activity. By using this method, a series of CHO cell clones were isolated which had 50% or less PARP activity. These mutants displayed normal generation times and were 20% more sensitive to the effects of DNA methylating or ethylating agents than the parental cell, but showed normal sensitivity to γ-rays. In subsequent experiments a CHO mutant (PADR-9) was isolated which displayed approximately 17% of the wild-type level of PARP activity (18). Biochemical analysis of this mutant, using activity blot, as well as Western and Northern blot techniques, indicated that, relative to its parent cell, the mutant's enzyme activity, antibody recognition, and mRNA levels had been reduced to approximately the same extent. These findings are consistent with a mutation in the PADR-9 cell clone, which has resulted in a reduction in enzyme synthesis due to reduced mRNA synthesis and/or stability. Relative to wild-type CHO cells, the PADR-9 mutant displayed increased sensitivity to killing by DNA-alkylating agents while its sensitivity to γ-irradiation remained normal.

By applying a sequential mutation and selection protocol, Yoshihara and colleagues (19) succeeded in isolating a PARP-defective mutant clone (Cl-3527) from a mouse L1210 cell clone (Cl-3). The enzyme activity per cell in Cl-3527 cells was only 8% of that in wild-type L1210 cells, due to the absence of wild-type PARP and expression of a functionally altered PARP protein present at low levels only. A prolonged doubling time and increased temperature sensitivity characterized the phenotype of the mutant cells. Importantly, this group has addressed the main drawback of the approach of selecting mutant cell clones. Selected cell clones which, at least in some cases, had to survive a high-dose mutagen exposure are likely to carry additional mutations in loci other than *PARP*, thus making it difficult to unambiguously assign any resulting phenotype to the deficiency in poly (ADP-ribose) formation. Yoshihara *et al.* (19) could show that PARP activity and normal thermotolerance could be partially restored by introducing a wild-type *PARP* gene into Cl-3527 cells.

3.3   PARP antisense strategies

A common molecular genetic strategy to interfere selectively with the expression of a given gene is the expression of antisense RNA in transfected cells or, alternatively, the supplementation of cell-culture medium with synthetic single-stranded antisense DNA oligonucleotides that may be taken up by cells as intact molecules. In general, antisense approaches are particularly useful if the protein target is turned over rapidly, as is the case for many regulatory proteins. In contrast, the turnover of PARP protein in cell cultures is known to be rather slow. Nevertheless, several groups have succeeded in establishing PARP antisense systems.

Smulson and co-workers pursued the antisense strategy in a variety of cell systems. They have described the isolation and characterization of HeLa transfectants with hormone-inducible expression of PARP antisense RNA (14, 20, 21). An expression plasmid was constructed with the mouse mammary tumour virus (MMTV) promoter upstream of the full-length, antisense-oriented human PARP cDNA. Consistent with the relatively long half-life of PARP protein in cultured cells, 48–72 h were required after the induction of antisense RNA expression by dexamethasone for the large amount of PARP normally present in HeLa cells to be reduced by more than 80%. The PARP depletion achieved was reflected in an 80% decrease in enzyme activity and a more than 90% reduction in the cellular content of PARP protein. The following biological consequences of antisense RNA expression were observed:

1. The chromatin of PARP-depleted cells had an altered structure as assessed by nuclease susceptibility and other parameters.
2. Cell morphology was altered, with multinucleated cells being evident 72 h after the induction of antisense RNA expression.
3. The cells had lost their ability to efficiently rejoin DNA strand breaks at early time points after the infliction of DNA damage to the cells.

4. Cells were deficient in the gene-specific repair of lesions induced by the alkylating agent nitrogen mustard, as assessed in the essential dihydrofolate reductase gene, but not in the repair of UV-induced pyrimidine dimers in the same gene.

Thus, PARP appears to participate in the gene-specific repair of alkylation damage, but not in the repair of UV-induced pyrimidine dimers. It should be noted that the finding of an involvement of PARP in the gene-specific repair of active genes is in apparent contradiction to that reported in the paper by Ray *et al.* (13) on PARP-deficient mutant cell clones, which presented data on an involvement of PARP in the repair of non-transcribed DNA only.

6. The survival of cells exposed to the alkylating agent methyl methanesulfonate or nitrogen mustard was significantly reduced relative to controls.
7. The incidence of CAD (carbamoyl phosphate synthetase/aspartate transcarbamylase/dihydro-orotase) gene amplification was increased.

The same group used the PARP antisense strategy in a cellular differentiation system (22, 23), in that cultured 3T3-L1 preadipocytes were induced to differentiate into adipocytes. Differentiation induction was accompanied by increases in PARP protein and activity. Induction of PARP antisense RNA in stably transfected 3T3-L1 cells prevented the increase in PARP expression; the cells neither differentiated nor did they undergo the DNA replication that is associated with the initiation of the differentiation process, due to a block at the G_0/G_1 transition of the cell cycle. Co-immunoprecipitation of PARP and DNA polymerase-α was observed in differentiating control cells, whereas no such complex formation was detected in the induced antisense cells, nor in uninduced control cells.

A third PARP antisense cell system (24) was designed by Smulson and colleagues. Human keratinocyte lines were stably transfected with a PARP antisense construct under the control of the dexamethasone-inducible MMTV promoter. Induction of antisense RNA in cultured cells was found to selectively decrease the levels of PARP mRNA, protein, and enzyme activity. When keratinocyte clones containing the antisense construct or empty vector alone were grafted on to nude mice, this resulted in the formation of histologically normal human skin. The PARP antisense construct also proved to be inducible *in vivo*, by topical application of dexamethasone to the reconstituted epidermis. In addition, poly (ADP-ribose) could be induced and detected *in vivo* following the topical application of a DNA alkylating agent to the grafted transfected skin layers.

Taniguchi and colleagues employed PARP antisense RNA expression in their studies on the influence of PARP on the expression of MHC class II molecules. This group had previously shown that *PARP* gene expression is decreased during interferon-γ (IFN-γ)-induced activation of murine macrophage P388D1 tumour cells, which is accompanied by an increased expression of MHC class II molecules (25). In subsequent work, the group asked whether there might be a causal relationship at the level of gene expression. To address this question, they studied the effect of PARP on the IFN-γ-inducible expression of class II molecules by expressing PARP antisense RNA (26, 27). One of the expression plasmids carried a 0.7-kbp fragment of the 5'-coding region (pAS-5'), the other carried the full-length PARP cDNA (pAS-FL) in the antisense orientation under the control of the metallothionein I promoter. Human leukaemia THP-1 cells were transfected, and stable transformants isolated. A metal-inducible decrease in PARP activity, along with the expected metal-dependent antisense RNA expression, was observed in a small proportion of the neo-resistant pAS-5'-transfected clones examined. By contrast, in clones expressing antisense-orientated, full-length PARP cDNA (pAS-FL), there was a metal-independent constitutive lowering of PARP activity. In the pAS-5' clones, PARP activity had decreased by over 90% three days after the addition of metal ions. The expression of MHC class II molecules was amplified, as revealed by Northern blot analysis and flow cytometry, when these clones were incubated in the presence of metal ions for 3 days and then treated with IFN-γ. The same group also performed the reverse experiment, that is the overexpression of full-length PARP cDNA in sense orientation, in several different mouse and human macrophage cell systems (28–30). Consistent with the above observations, the IFN-γ- or phorbol ester-induced expression of both MHC class II molecules and interleukin-1 (IL-1) were considerably reduced in these transformants. Viewed together, the results indicate that the level of PARP critically regulates the expression of MHC class II molecules induced by IFN-γ, with PARP apparently acting as a suppressor of MHC class II gene expression. To gain insight into the underlying mechanism, they searched for DNase I-hypersensitive sites in the mouse MHC class II, I-A beta gene (27). Two sites were found: one in the promoter region and the other in the first intron. The hypersensitive site in the first intron proved to be less sensitive to DNase I in those transformants in which PARP was synthesized in large quantity. They could show that the suppressive effect of PARP overexpression on the mouse MHC class II, I-A beta gene was indeed dependent on the presence of sequences located in the first intron of this gene. This was achieved by transfecting lacZ reporter constructs comprising either the promoter region only or sequences spanning the promoter region up to a part of the second exon of the MHC class II gene.

Farzaneh and co-workers reported on the strategy of PARP antisense DNA oligonucleotides that are taken up intact by cells, and which can thus exert antisense effects if added to the culture medium of cells (31). A 74% decrease in PARP expression could be demonstrated in L1210 cells after 72 h of incubation with a partially sulfurized antisense oligonucleotide targeting the PARP mRNA translation initiation site. The specificity of this effect was indicated by the failure of sense-oriented or 'scrambled' oligonucleotides to lower cellular PARP activity. The biological endpoint of interest was the integration of the proviral DNA of retroviruses into the host-cell genome, since this process involves coordinate DNA strand-break formation and rejoining reactions. While the role of the retrovirus-encoded integrase protein (IN) had been studied in great detail, the cellular components acting in concert with the viral integrase had not been well characterized. The working hypothesis was posited that PARP could be one of the cellular factors participating in retroviral integration. The data revealed that abrogation of PARP activity by antisense oligonucleotides blocks the infection of mammalian cells by recombinant retroviral vectors. The same effect was observed when two different strategies of interfering with PARP activity were employed—either low molecular-weight PARP inhibitors or *trans*-dominant inhibition by overexpression of the PARP DNA-binding domain (see Chapter 3.3 for details). The requirement for PARP activity appeared to be restricted to processes involved in the integration of provirus into the host-cell DNA. PARP inhibition apparently did not affect viral entry into the host cell, reverse transcription of the viral RNA genome, postintegration synthesis of viral gene products, synthesis of the viral RNA genome, or the generation of infective virions. The authors concluded that it is the efficient retroviral infection of mammalian cells which is blocked by inhibition or PARP activity. It should be noted that in contrast to these results, another group studying foreign DNA integration in a hepadna-viral system found an increase in the integration frequency as a consequence of PARP inhibition by 3-aminobenzamide (32). The reasons for this discrepancy are not known, but may be related to biological differences between the two viral systems studied. The two viruses use quite different integration mechanisms.

3.4   Expression of truncated and full-length PARP in various systems

3.4.1   *Trans* dominant-negative PARP mutants

The expression of mutant proteins conferring a so-called dominant-negative phenotype in the presence of the wild-type protein is another commonly used molecular genetic approach to selectively interfere with the function of a specific protein in living cells. Since PARP consists of different functional domains, it was conceivable to create a dominant-negative mutant either by overexpressing a full-length enzyme mutated at the catalytic site, or a truncated version selectively coding for a regulatory function that would interfere with the activity of the wild-type protein. In fact, the two characteristic properties of the N-terminal domain of PARP, targeting to the cell nucleus and DNA strand-break detection, were exploited to generate such a mutant.

Küpper *et al.* (33) were the first to show that overexpression of the PARP DNA binding domain (DBD) (domains A to C, see Fig. 2.2) in transiently transfected cell cultures did indeed lead to *trans*-dominant inhibition of poly (ADP-ribosylation) due, in part, to the competition for DNA breaks.

Molinete *et al.* (34) have employed this feature of the DBD to characterize the DNA break-binding ability of PARP enzyme in the context of living cells, and to study the consequences of *trans*-dominant PARP inhibition on DNA repair. The DBD of human PARP, and several derivatives thereof, mutated in its first or second zinc finger, were overproduced in the following systems:

(1) in *Escherichia coli*, to study DNA break-binding *in vitro*;
(2) in transiently transfected CV-1 monkey cells, to monitor the *trans*-dominant inhibition of poly (ADP-ribosylation) triggered by MNNG treatment; and
(3) in human fibroblasts to study DNA repair.

A positive correlation was found between the *in vitro* DNA-binding capacity of the recombinant DBD polypeptides and their inhibitory effect on endogenous cellular PARP activity after MNNG treatment. Furthermore, overproduced wild-type DBD blocked unscheduled DNA synthesis (DNA repair synthesis) induced in living cells by MNNG treatment, but not that induced by UV-C irradiation. These results demonstrated that an intact second zinc finger of PARP is required for successful binding to DNA single-stranded breaks. They also underscored the importance of PARP to act as a critical regulatory component in the repair of DNA damage induced by alkylating agents.

Subsequently, more detailed analyses of the biological consequences of inhibiting poly (ADP-ribosylation) by DBD overexpression were carried out in stably transfected cell lines. Schreiber *et al.* (35) established stable HeLa transfectants that constitutively produce the PARP DBD, thus leading to *trans*-dominant inhibition of resident PARP activity. As a control, a cell line was constructed producing a point-mutated version of the DNA binding domain, which had no affinity for DNA *in vitro*. Expression of the PARP DBD had only a slight effect on undamaged cells, but had drastic consequences for cells treated with genotoxic agents. Exposure of cell lines expressing wild-type or mutated PARP DBD to MNNG at low concentrations resulted in an increase in cell doubling time, a G_2/M accumulation, and a marked reduction in cell survival. As expected, UV-C irradiation had no effect on cell growth or on the viability of cell lines overexpressing the PARP DBD. Cells treated with MNNG showed the characteristic nucleosomal DNA ladder, one of the hallmarks of cell death by apoptosis. These cultures also exhibited chromosomal instability as demonstrated by higher frequencies of both spontaneous and MNNG-induced, sister-chromatid exchanges. Surprisingly, the cell line producing the mutated DBD showed the same behaviour as those producing the wild-type DBD, suggesting that the mechanism of action of the dominant-negative mutant involved more than its DNA binding function. These results suggest that PARP is involved in the maintenance of genomic integrity in mammalian cells.

Independently, Küpper *et al.* (36) established stably transfected Chinese hamster cell lines, which drastically overexpress the PARP DBD under the control of the glucocorticoid-inducible MMTV promoter. In one cell line analysed in great detail (COM3), induction of the DBD led to about a 90% reduction of poly (ADP-ribose) accumulation after γ-irradiation and sensitized the cells to the cytotoxic effect of γ-irradiation as well as MNNG, in agreement with the data obtained by Schreiber *et al.* (35). Induction of DBD expression did not affect normal cellular proliferation or the replication of a super-transfected polyomavirus replicon. It was concluded that *trans*-dominant inhibition of the poly (ADP-ribose) synthesis, which would normally occur after γ-irradiation or MNNG treatment, is specifically associated with a disturbance of cellular recovery from the inflicted damage. In the same cell system, *trans*-dominant inhibition of poly (ADP-ribosylation) was also shown to potentiate carcinogen-induced amplification of chromosomally integrated SV40 DNA sequences (37), in agreement with previous data obtained using PARP-inhibitory drugs (38, 39). Recently, when focusing on mutagenesis, which represents another biological endpoint contributing to the phenomenon of genomic instability, Tatsumi-Miyajima *et al.* (40) demonstrated that *trans*-dominant inhibition of poly (ADP-ribosylation) potentiated methyl nitrosourea-induced shuttle-vector mutagenesis in the same cell system. These results provided further evidence for a role of poly (ADP-ribosylation) in limiting the extent of genomic instability caused by DNA damage.

Another interesting and highly relevant manifestation of genomic instability is the integration of foreign DNA sequences into the chromosomes of the host cell. By applying—among other strategies—the principle of *trans*-dominant PARP inhibition on a cell-culture system of retrovirus integration, Gäken *et al.* (31) could show that efficient retroviral infection of mammalian cells is blocked by overexpressing PARP DBD. The same results were obtained using ADP-ribosylation inhibitors or PARP antisense oligonucleotides (see Section 3.3), further extending the parallel between the *trans*-dominant inhibition approach and the use of PARP inhibitors.

3.4.2   Overexpression of wild-type, full-length PARP

The full-length human *PARP* gene has been expressed in various homologous and heterologous systems in a search for novel phenotypes. The two yeasts, *Saccharomyces cerevisiae* and *Schizosaccharomyces pombe* are the only nucleated organisms that are devoid of endogenous *PARP* genes and

PARP activity. Kaiser *et al.* (41) were the first to develop a system of heterologous PARP expression in *S. cerevisiae*. Constitutive expression of the human enzyme was possible only with the simultaneous inhibition of enzyme activity by 3-methoxybenzamide. The inducible expression of human PARP was accompanied by inhibition of cell division and morphological changes reminiscent of cell-cycle mutants. The presence of the amino terminus of PARP comprising the zinc-finger region was necessary for the appearance of this growth-inhibition phenotype. Nuclei of transformed yeasts revealed a large accumulation of poly (ADP-ribose), probably due to the lack of adequate polymer-degrading enzyme activities. The growth inhibition could be correlated with the presence of poly (ADP-ribose) rather than with metabolic disturbances such as NAD^+ depletion.

Avila *et al.* (42) chose *Schizosaccharomyces pombe* as another simple eukaryotic system in which to study PARP function. They first looked for an endogenous yeast PARP by DNA and RNA hybridization to mammalian probes under low-stringency conditions and by PARP activity assays. The data indicated that fission yeasts, like budding yeasts, are naturally devoid of PARP. Furthermore, the group isolated *S. pombe* strains expressing PARP by transformation with a human full-length PARP cDNA under the control of the SV40 early promoter. The PARP construct was transcribed and translated in *S. pombe*, generating a 113-kDa polypeptide, identical in size to the native human PARP protein. Recombinant PARP expressed in yeast was enzymatically active and was recognized by antibodies to human PARP. *S. pombe* cells expressing PARP showed a stable phenotype that was characterized by:

(1) cell-cycle retardation as a result of a specific delay in the G_1 phase;
(2) decreased cell viability in stationary cultures;
(3) enhanced rates of spontaneous and radiation-induced *ade6–ade7* mutations; and
(4) increased sensitivity to radiation.

Fritz *et al.* (43) reported on PARP overexpression in mammalian cells. Chinese hamster CHO-9 cells and a mutagen-hypersensitive derivative thereof were stably transfected with human PARP. The transfectants studied displayed reduced sensitivity towards the alkylating agent methyl methanesulfonate as compared with the respective parental cell line.

By using a similar approach, van Gool *et al.* (44) isolated stable transfectants of the Chinese hamster cell line CO60 that constitutively overexpress human PARP (COCF clones). First, the status of poly (ADP-ribose) accumulation in these transfectants was assessed both by immunofluorescence and by biochemical analyses. When cells were γ-irradiated to trigger poly (ADP-ribose) formation, the fraction of cells positive for poly (ADP-ribose)-specific immunofluorescence was consistently higher in the COCF clones than in the control clones (COCFN) which failed to express human PARP. Quantitative determinations of *in vivo* levels of poly (ADP-ribose) confirmed this result, revealing significantly higher polymer levels in the COCF clones. In contrast to the data obtained by Fritz *et al.* (43), cell-survival studies showed that increased poly (ADP-ribosylation) in the COCF clones is associated with sensitization to γ-irradiation and MNNG. Apparently, this could be explained neither by the depletion of cellular NAD^+ nor of ATP pools. Viewed together with the well-known cellular sensitization induced by poly (ADP-ribosyl)ation inhibition, the data suggested that an optimum level of cellular poly (ADP-ribose) accumulation exists for cellular survival from DNA damage. The molecular mechanisms underlying this unexpected behaviour of the cells with respect to survival is currently under investigation. Recently, Bernges *et al.* (45) characterized several stable transfectants of a rat cell line overexpressing human PARP cDNA. In line with the data obtained by van Gool *et al.* (44) in the hamster cell system, none of the rat cell clones displaying increased poly (ADP-ribose) accumulation upon treatment with γ-irradiation or the alkylating agent MNNG was protected with respect to cell survival. A similar result was obtained with HeLa cell lines constitutively overexpressing the full-length human PARP (F. Dantzer, unpublished results).

3.5   Knockout mice: the survival role of PARP in mammals under genotoxic stress

The *PARP* gene was inactivated in the mouse germline, and PARP-deficient (KO mice) were generated to evaluate the biological contribution of the enzyme *in vivo* during mouse development, postnatal life, and following genotoxic stress. Exon 1 (46, 47), exon 2 (48), or exon 4 (49), respectively, have been interrupted by homologous recombination. In each case, the knockout was complete: no fragment or functional domain of PARP could be detected in the corresponding mutant mouse. PARP-deficient mice are viable and fertile, indicating that PARP is dispensable for mouse development and tissue differentiation. They do not display any overt abnormal phenotype, although Wang *et al.* (48) have observed a spontaneous development of skin hyperplasia in 6-month-old KO mice and obesity in female mice older than 15 months. These findings were not observed in the two other animals.

Historically, the first paper on PARP-deficient mice (50) was published in May 1995 by Kolb's group, in collaboration with Wagner's laboratory. In a way, this article, dealing as it does with the protection of pancreatic islet cells of PARP-deficient mice treated with reactive oxygen intermediates (ROI) and nitric oxide (NO), pointed to the ability of their unexpected phenotype to resist inflammatory injury. This subject is currently a very active field of investigation (see Chapter 5 for details). The generation of PARP-deficient mice by several laboratories has in fact considerably enlarged our view of the physiological role of PARP. The following sections summarize the reported results, dealing mostly with their DNA repair capacity and maintenance of genomic integrity in response to genotoxic stress (Table 3.1) in the three animal models (see also refs 51–53 for recent reviews on these aspects).

3.5.1   PARP-deficient mice are sensitive to genotoxic stress

The previous chapters have shown that both monofunctional alkylating agents and γ-rays trigger the immediate synthesis of poly (ADP-ribose). To evaluate the role of PARP under stress conditions, PARP-deficient mice were therefore challenged with genotoxic agents known to stimulate PARP activity. PARP-deficient mice given an intraperitoneal (i.p.) injection of the alkylating agent *N*-methyl-*N*-nitrosourea (MNU) died by 2–4 weeks after administration, while 60% of the wild-type mice recovered and were still alive 8 weeks after injection (49). In a parallel experiment, mice of both genotypes were exposed to 8 Gy of γ-rays (whole-body exposure). PARP-deficient mice were also found to be exquisitely sensitive to γ-irradiation. Most of the treated animals died by 4–6 days, while half of the control mice were still alive 3 weeks after irradiation (Fig. 3.1). This last result has been confirmed with the PARP-deficient mice developed by Wagner's group (54).

At 72 hours' postirradiation, autopsies were performed on both the irradiated and control mice of both genotypes. In the irradiated mutant mice, the small intestine was found to be distended by an accumulation of intraluminal fluid. The wild-type mice did not show this reaction. Transverse sections

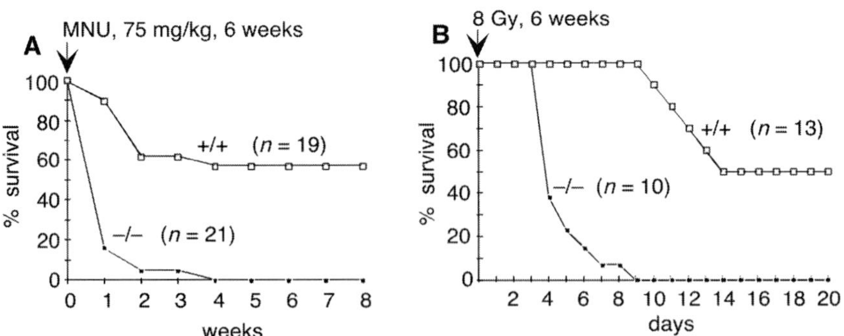

Fig. 3.1   Survival of *PARP*$^{+/+}$ and *PARP*$^{-/-}$ mice after (A) an i.p. injection of MNU at 75 mg/kg body weight at 6 weeks of age; (B) whole body γ-radiation with 8 Gy at 6–8 weeks of age. The percentage of mice alive is plotted against weeks (A) or days (B) post-treatment. (Taken from ref. 49: Ménissier-de Murcia *et al.* (1997). Requirement of poly(ADP-ribose)polymerase in recovery from DNA damage in mice and in cells. *Proc. Natl Acad. Sci. USA*, **94**, 7303, with permission. Copyright (1997) National Academy of Sciences, U.S.A.)

Table 3.1 Phenotypes of the various PARP knockout mice and derived cell lines

Constructs	Phenotype	Refs.
KO 1 (disruption of the 2nd exon)	Mice healthy and fertile. Impaired fibroblast and thymocyte proliferation after γ-irradiation. Decreased ability to repair DNA damage induced by MNNG.	48
	Hypersensitivity to whole-body γ-radiation. Increased genomic instability (SCE, micronuclei).	54
	Telomere length deregulation.	156
	p53: low basal level, defective post damage induction, normal transactivation.	157
	Accumulation of DNA breaks following streptozotocin treatment.	65
	Pancreatic islet cells are resistant to the toxicity of NO and ROI.	50
	Protection against peroxynitrite-induced arthritis.	159
	Protection against MPTP-induced parkinsonism.	158
	Protection against streptozotocin-induced diabetes.	65, 160
	Defective induction of NF-κB.	106
	Resistance to cerebral ischamia.	162, 163
	Resistance to LPS-induced septic shock.	161
KO 2 (disruption of the 4th exon)	Extreme sensitivity and high genomic instability of mice (SCE, chromatid, and chromosome breaks) following MNU exposure and γ-irradiation. Acute radiation toxicity to the small intestine. High sensitivity of splenocytes and bone-marrow cells to MNU exposure, G_2/M accumulation and apoptosis. p53 accumulation following DNA damage (MNU, MMS).	49
	Sensitization to camptothecin.	this chapter
	Growth retardation, G_2/M accumulation, chromosome instability (micronuclei), severe DNA repair defect in the BER pathway in PARP-deficient 3T3 cells exposed to MMS. Normal NER.	61
	Resistance to VP16 due to decresease expression of Topo IIβ	58
	Protection against streptozotocin-induced diabetes.	a
	Defective induction of NF-kB.	164
	Resistance to LPS-induced septic shock.	164
KO 3 (disruption of the 1st exon)	Reduced survival of PARP-deficient ES cells after MMS treatment and γ-irradiation. p53 accumulation following γ-irradiation.	46
	Protection against Streptozotocin-induced diabetes.	47

[a]Janiak *et al*. unpublished results.

through the duodenum of the non-irradiated mutant mice were histologically indistinguishable from their wild-type counterparts (Plate 9A). Irradiated wild-type mice demonstrated a shortening of the villi (Plate 9B), but intact epithelial lining. The intestinal villi of irradiated PARP-deficient mice were considerably shortened (Plate 9C), with almost all their epithelial cells undergoing necrosis (Plate 9F). These data suggest that death in mutant mice is caused by dehydration and/or endotoxicosis secondary to acute radiation toxicity.

The DNA topoisomerase I inhibitor, camptothecin, is a cytotoxic drug used in the treatment of tumours, which forms a covalent complex with DNA under torsional stress (55). This compound was administered in a single i.p. dose of 140 mg/kg body weight to mice of both genotypes. Almost 64% (7/11) of the PARP-deficient mice had died by 2 weeks' post-treatment, while all (0/11) the wild-type mice were still alive at 8 weeks (Fig. 3.2). This result shows that the absence of PARP also renders mice hypersensitive to a clinically relevant cytostatic agent that damages DNA through different mechanisms to those involving monofunctional alkylating agents or γ-irradiation, but which nevertheless also induce DNA strand breaks, notably at the vicinity of replication forks (56, 57). Similar to the PARP-deficient cell lines established by random mutagenesis (8, 12), 3T3 cells derived from PARP-deficient mice were proven to be resistant to the topoisomerase II inhibitor VP16 (58).

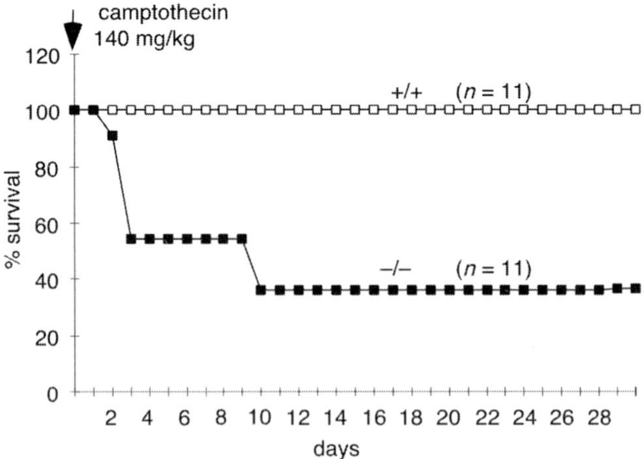

Fig. 3.2 Survival of *PARP*[+/+] and *PARP*[–/–] mice after an i.p. injection of camptothecin (CPT-11) at 140 mg/kg body weight at 6 weeks of age. The percentage of mice alive is plotted against days post-treatment.

3.5.2   Genomic instability and rearrangement in the absence of PARP

A conspicuous property of PARP is that the inhibition of its enzymatic activity by chemical inhibitors, *trans* dominant-negative mutants, or antisense RNA expression invariably results in an increased genomic instability of cultured cells injured by genotoxic stress (35, 59, 60). Indeed, a dramatic increase in sister-chromatid exchange (SCE) frequencies as well as chromatid and chromosome breaks were observed in bone marrow cells of KO mice exposed to γ-irradiation or MNU (Fig. 3.3) (49). Moreover, micronuclei formation, which is considered to be a critical indicator of chromosomal damage, was monitored in mouse embryonic fibroblasts (MEFs) derived from wild-type or PARP-deficient mice exposed to 50 μM of MMS. An increased frequency of micronuclei formation was observed in PARP-deficient MEFs in comparison with the wild-type cells (61), again indicating an important role for PARP in the maintenance of genomic integrity. Wagner and co-workers obtained similar results using PARP-deficient fibroblasts and splenocytes (54).

PARP-deficient mice exhibit unaltered developmentally programmed chromosomal rearrangements, such as immunoglobulin class-switching and V(D)J recombination (62), and do not display any difference in the hypermutation of immunoglobulin genes in memory B cells (63). To a limited extent, the introduction of the null PARP mutation into SCID mice—which accumulate V(D)J recombination intermediates—rescues V(D)J recombination in T-lymphocyte but not in B-lymphoid development (64). Interestingly, double-mutant mice develop early T-cell lymphomas, suggesting that anomalous V(D)J recombination leads directly to chromosomal abnormalities. These results suggest that PARP and DNA-PK may interact genetically in this specific recombination process to ensure chromosomal integrity during T-lymphocyte development (see Chapter 7).

3.5.3   PARP-deficient cells are severely affected in the base excision–repair pathway

Initial experiments using cells derived from PARP-deficient mice did not support the notion that PARP contributed to DNA repair (48). The experimental system used by Wang *et al.* (48) was based on the repair of an SV40-CAT plasmid damaged by MNNG that was subsequently transfected into 3T3 cells derived from PARP-deficient mice. They used the CAT enzyme activity in the recovered plasmid as an estimate of the repair capacity of wild-type and PARP-deficient cells. An untreated plasmid was used as a control and a luciferase-expressing vector was used to normalize the transfection frequency. At 12 hours' post-treatment they observed about 80% CAT activity in the wild-type but only 35% in the PARP-deficient cells, indicating a significant delay or defect in the ability of those cells to repair alkylated bases in

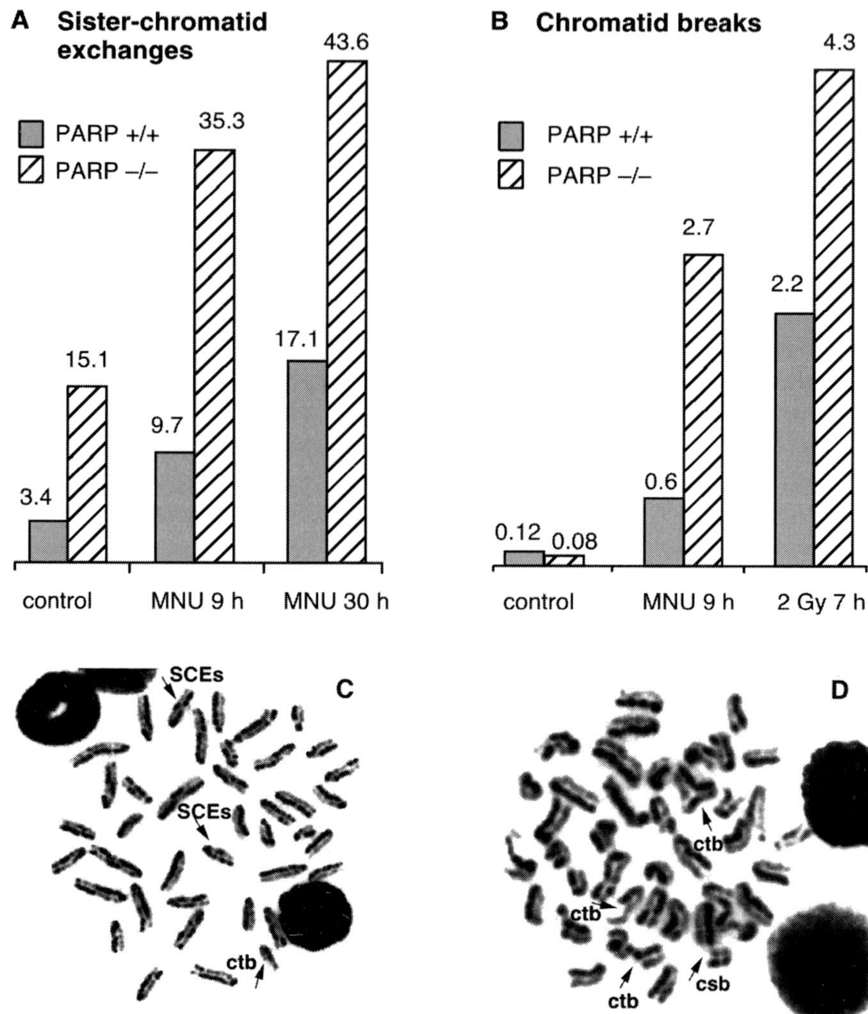

Fig. 3.3 Chromosome instability in bone marrow cells of mice mock-treated or injected with MNU at 80 mg/kg body weight, or irradiated with 2 Gy. (A) Mean number of sister-chromatid exchanges per cell in *PARP⁻/⁻* and *PARP⁺/⁺* mice, before and after exposure to MNU during 9 or 30 h. (B) Mean number of chromatid breaks before and after exposure to MNU during 9 h or 7 h after exposure to γ-rays. Microphotographs of *PARP⁻/⁻* bone marrow cells isolated 9 h after MNU injection (C) showing a high frequency of sister-chromatid exchanges (SCEs), and after irradiation with 2 Gy (D) showing chromosome breakage (ctb, chromatid breaks; csb, chromosome breaks). (Taken from ref. 49: Ménissier-de Murcia *et al.* (1997). Requirement of poly(ADP-ribose)polymerase in recovery from DNA damage in mice and in cells. *Proc. Natl Acad. Sci. USA*, **94**, 7303, with permission. Copyright (1997) National Academy of Sciences, U.S.A.)

DNA. As expected, no difference was found when the plasmid was UV-irradiated. Conclusions from this type of experiment could be biased by the fact that the transcription-coupled repair assay used in this work was not particularly relevant to PARP, since its involvement in the repair of transcribed genes is unclear (13, 14).

In another report from the de Murcia laboratory, MEFs derived from PARP-deficient mice were found to be sensitive to alkylating agents. This hypersensitivity was rescued by ectopic expression of wild-type PARP cDNA (61). Single-cell gel electrophoresis (Comet assay) clearly demonstrated that the steady-state level of DNA breaks decreased rapidly in the wild-type cell lines, while a dramatic delay in the DNA strand-break rejoining was evident in PARP-deficient cell lines (Fig. 3.4). The half-life of DNA breaks in MMS-exposed wild-type MEFs is about 1 hour, whereas mutant cells resealed half their DNA breaks in 5 hours (61), demonstrating unequivocally that PARP-deficient cell lines performed very limited DNA repair during the first 6 hours. Elevated levels of accumulated spontaneous DNA strand breaks were observed in pancreatic islet cells (65) of another PARP-deficient mouse model, thus supporting the critical role of PARP in the resolution of breaks. Alkylating agents delivered at sublethal doses for wild-type cells profoundly

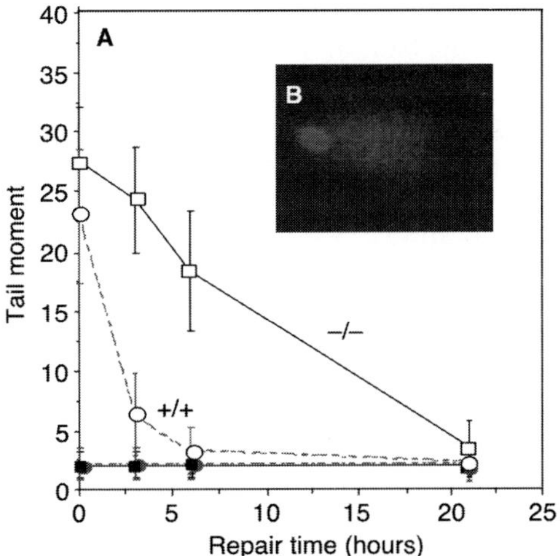

Fig. 3.4 DNA repair capacity of primary embryonic fibroblasts (MEFs) of both *PARP*[+/+] and *PARP*[–/–] genotypes as assessed by the single-cell, gel electrophoresis assay (Comet assay). (A) Distribution of the tail moment of *PARP*[–/–] cells (squares) and *PARP*[+/+] cells (circles) as a function of time following 150 μM methyl methanesulfonate (MMS) treatment (open symbols) or mock treatment (filled symbols). The data are the mean of the tail moments of 100 cells measured at each time point. (B) Example of comet tail observed in a *PARP*[–/–] damaged cell. (Taken from ref. 61; Trucco *et al.* (1998). DNA repair defect in poly(ADP-ribose)polymerase-deficient cell lines. *Nucleic Acids Res.*, **26**, 2644, with permission from OUP.)

disturbed the cell cycle of PARP-deficient cells (49, 61). The persistence of DNA strand breaks in PARP-deficient mice may well be at the origin of the genomic instability mentioned above. The accumulation of cells at the G_2/M boundary as well as the sustained induction of p53, ultimately lead to increased cell death. A similar prolongation of the p53 response was also obtained in cells depleted of PARP by antisense RNA expression following exposure to 6.3 Gy of γ-radiation (66).

The sensitivity of different wild-type and deficient cell populations with respect to several inducers of apoptosis were compared: CD95 and TNF-α in hepatocytes and MEFs; dexamethasone, ceramide, and etoposide in thymocytes; potassium withdrawal, colchicine, staurosporine, 1-methyl-4-phenylpyridinium, and peroxynitrite in neurons isolated from murine cerebellum (54, 67), and IL-3 removal (68). Under all these conditions (which are not necessarily known to trigger PARP activity) removal of the PARP pathway had no influence on the onset of apoptosis, suggesting that PARP is dispensable and that energy depletion is not an essential prerequisite for apoptosis (for an overview on this aspect see Chapter 4).

PARP-deficient ES cells recently developed by Sugimura's group (46, 47) displayed enhanced sensitivity to γ-irradiation and MMS compared to the parental heterozygote or wild-type ES cells (46), confirming the essential role of PARP in cellular recovery from DNA damage.

These results are in full agreement with those obtained with cell lines over-expressing the antisense RNA (20), as well as with the *trans*-dominant or the inhibitor approaches (69) mentioned above. They shed light on the molecular basis for the impaired survival under genotoxic stress of PARP-deficient mice, irrespective of which exon of the *PARP* gene had been disrupted. In addition, these features suggest a function for PARP in the base excision–repair (BER) pathway. Section 3.6 provides one clue as to the function of PARP in a dynamic multifactorial complex involving XRCC1, DNA polymerase-β, and DNA ligase III—three major components of the BER pathway.

3.6   Searching for partners: the multiple facets of PARP function

The identification of proteins physically associated with PARP has provided insights into a better understanding of its biological function. These PARP partners were identified by various approaches including co-purification and co-immunoprecipitation experiments and by Western blot analysis. More recently, a genetic approach using the two-hybrid system allowed the detection of proteins interacting with PARP in *S. cerevisiae* (70–72). In the de Murcia laboratory, the full-length human PARP fused to the DNA binding domain of LexA was used as bait to screen a pool of chimeric proteins, which consisted of fusions of the Gal 4 activation domain with polypeptides expressed from a HeLa cDNA library. Among the 15 independent clones obtained, two were identified: the Ubc9 ubiquitin-conjugating enzyme and the DNA repair factor XRCC1 (see below). Interactions between PARP and these protein partners were fully confirmed in mammalian cells by immuno-precipitation or by pull-down experiments using glutathione-*S*-transferase (GST) tagged to the interactor (70, 71). Table 3.2 summarizes the proteins reported to interact physically with PARP, *in vitro* and *in vivo*.

3.6.1   PARP: a DNA damage-dependent, chromatin-remodelling machine?

PARP has been described as a non-histone chromatin component that is preferentially associated with the core of the nucleosome (73, 74). A physical interaction between PARP and histone H1 has been described, and the binding sites localized to the 186–290 (C-terminal of the zinc-finger domain) and 446–525 (automodification domain) regions of PARP, and in the basic C-terminal portion of histone H1 (75). Apart from PARP, histones are the main acceptors of ADP-ribose polymers; mostly histones H1 and H2B (76–78) and, to a lesser extent, H2A, H3, and H4 are poly (ADP-ribosylated) both *in vitro* and *in vivo* (see ref. 79 for a review). In response to the treatment of cells with monofunctional alkylating agents, histones are poly (ADP-ribosylated) *in vivo* (80–82). This reaction is much more efficient when histones are folded into a nucleosomal structure, suggesting that their structural environment is critical for modification (75).

Relaxation of the chromatin superstructure (30-nm fibre) was observed both by sedimentation velocity and electron microscopy (see Chapter 2, Fig. 2.8) following poly (ADP-ribosylation) of purified calf thymus polynucleo-somes. Under these experimental conditions, H1 was found to be the major histone acceptor of poly (ADP-ribose); however, it was not dissociated from chromatin (83, 84). The subsequent polymer degradation by PARG restored the condensed superstructure of chromatin (85), demonstrating a potential remodelling function of PARP acting on a damaged chromatin template *in vitro*.

Histone tails, which are key regulators of chromatin function (86, 87), were also shown to bind non-covalently to PARP-bound poly (ADP-ribose) (88). Althaus and co-workers proposed a 'histone shuttling' model, in which histones shuttle from DNA to poly (ADP-ribose) emanating from automodified PARP that is thought to act as the catalyst for nucleosomal unfolding (88–92). It seems likely that, in response to DNA damage, either the covalent poly (ADP-ribosylation) of histones or their electrostatic binding to polymers of ADP-ribose would facilitate their transient removal or displacement from the site of the lesion and the opening of the chromatin structure. This would allow repair enzymes access to DNA breaks and

Table 3.2   The various partners of PARP

Biological function of the partner	Isolation procedure	PARP domain involved	Effect of PARP on the partner activity (*)	Ref.
Chromatin structure				
Histone H1	Cross-linking	A–D		75
Histone H2B	Cross-linking	A–D		75
PARP	Cross-linking	A–D	dimerization	75
PARP	Far Western/GST	A–D	dimerization	130 *a*
DNA replication				
DNA polymerase α-primase	IP	A	(see text)	118, 119
Multiprotein replication complex (MRC)	co-purification			23
Topoisomerase I	co-purification		stimulation	111, 112, 165
DNA repair, recombination, and cell-cycle control				
XRCC-1	Two-hybrid/IP/GST	A, D		71, 139
DNA ligase III	IP/GST			*b*
DNA polymerase-β	IP/GST	A, D		140
Ku 70, Ku 80	co-purification/IP			166–168
DNA-PK	co-purification			166–168
p53	IP			169–171
h Ubc9	Two-hybrid	D		70
B23, C23				149

Transcription

E47	Expression screen	A–D (aa 20–429)		102
DF1,4	co-purification		stimulation	105
TEF-1	Expression screen	A (aa 40–223)	stimulation	104
AP-2	GST		stimulation	101
RXRa	GST	A (aa 82–220)	repression	103
Oct-1	Two-hybrid/GST	D	stimulation	72
YY1	IP	D	repression	99
PC1	co-purification		stimulation	100
NF-kB (p65)	IP		stimulation	106

[a]V. Schreiber, unpublished results ; [b]C. Niedergang, unpublished results. Abbreviations: IP, immunoprecipitation; GST, glutathione-*S*-transferase pull-down experiment. * In the absence of NAD.

hence increase the overall rate of repair. Once the break was sealed, PARP would no longer be activated; although the degradation of histone-bound poly (ADP-ribose) by PARG would still occur, leading to the reconstitution of nucleosomes and recondensation of the chromatin fibre. Some aspects of PARP biology (interaction with dynamic regions of the genome such as promoters and nuclear matrix attachment regions, formation of DNA loops) are reminiscent of a remodelling function for PARP, capable of both transcriptional regulation and modulation of the basic chromatin fibre structure mediated by poly (ADP-ribosylation) of histone tails in an NAD^+-dependent manner. Alternatively, poly (ADP-ribosylation) of histone H1 has recently been correlated with the internucleosomal degradation of DNA during apoptosis, suggesting a potential means of increasing the access of apoptotic DNases to chromatin (93). In line with this, the association of PARP with Ubc9 (70) provides another link between DNA damage and proteolytic degradation.

Non-histone chromosomal proteins of the high mobility group (HMG), which are known to be structural components as well as co-activators for the transcription of important genes (notably those involved in inflammatory processes), can also be ADP-ribosylated in intact cells (94–96). This again suggests a potential capacity of poly (ADP-ribosylation) reactions for regulating chromatin function.

3.6.2   A potential role for PARP in gene expression

Several lines of evidence support a role for PARP in transcription. In the past, the enzyme, identified as the transcription factor TFIIC, has been mainly considered to be a necessary component for the elimination of the non-specific transcription that takes place at breaks *in vitro* (97, 98). The question of the contribution of PARP to several aspects of the transcriptional process has recently been re-investigated, given the frequency with which PARP was identified in several genetic and molecular screens.

Using the yeast two-hybrid system, immunoprecipitation or pull-down assays, PARP has been identified as a potential partner of the following transcription factors: YY1 (99), Oct-1 (72), PC1 (100), AP-2 (101), E47 (102), RXRα (103), TEF-1 (104), DF1–4 (105) and NF-κB (106).

Remarkably, in most of these cases (see Table 3.2) the effect of PARP, as an interacting protein component, was shown to stimulate the activity of its partner by increasing its loading to the cognate target sequence or to an enhancer region. In many instances, this interaction involves the automodification domain (domain D) containing a BRCT motif (see Chapter 2, Fig 2.2).

This characteristic binding property strongly suggests that the contact between PARP and its partner may be regulated by automodification, implying *de facto* a link with the cellular response to DNA damage. In a few other cases (TEF-1 and RXRα), a region corresponding to the second zinc finger is necessary and sufficient for the interaction with the transcription factor. In several cases, the target sequence on DNA contains potentially bent regions (TEF-1, NF-κB).

Regarding the effect of the PARP protein on its partners, there is no direct correlation between the structure (interacting domain) and the functional consequences (stimulation/repression) on transcriptional activity. However, when we consider PARP enzymatic activity, modification of the factors described above, reverses, as usual, the positive effect of PARP on the function of its partner *in vitro*.

An NAD$^+$-dependent silencing of RNA polymerase II-dependent transcription has been described in a cell-free context (107, 108). In this case, the TATA-binding protein (TBP), as well as the transcription factor YY1, could be negatively regulated by poly (ADP-ribosyl)ation but only prior to their binding to DNA. Moreover, several components of the basal transcription apparatus, RAP 30 and RAP 74, which are TFIIF subunits, have been identified as polymer acceptors (109), supporting a possible involvement of poly (ADP-ribosylation) reactions in the regulation of transcription, at least in *in vitro* assays.

On the basis of the existing data, it is difficult at the moment to evaluate the biological relevance of PARP in such a fundamental process. As far as we know, and given the obligatory dependence of its activity upon the presence of DNA strand breaks, it is almost impossible to separate this positive or negative regulation of transcription by PARP from its key role in the maintenance of genomic integrity. Therefore it should be interesting to use the recently developed PARP-deficient models to try to unravel the complexity of these novel and exciting aspects of PARP biology in living cells.

3.6.3   PARP links the DNA damage-surveillance network to the replication apparatus

Numerous observations indicate that PARP is associated with the DNA replication machinery (see Table 3.2). PARP was shown to co-purify with the DNA enriched in replication forks (110) (Plate 10) and topoisomerase I (111, 112). More recently, PARP was identified as a component of a multiprotein DNA replication complex (MRC) containing: replicative enzymes for leading- and lagging-strand synthesis: DNA polymerase α-primase, DNA polymerase-δ, as well as accessory proteins such as RPA, RFC, RNase H, PCNA, topoisomerase I and II, and DNA ligase I (23, 113). Among these proteins, DNA polymerase-α, RPA, topoisomerase I, PCNA, and DNA ligase I are potential targets for poly (ADP-ribosyl)ation *in vivo* (23, 80, 112, 114–117). Confocal analysis revealed that PARP and DNA polymerase-α are co-localized, close to the nuclear membrane (118) (Plate 10).

A direct interaction between PARP and DNA polymerase-α has been demonstrated through the catalytic subunit (p180) of DNA polymerase-α and the PARP DNA binding domain. DNA was not necessary to this physical association (22, 23, 118, 119). Although PARP and DNA polymerase-α are coexpressed during the entire cell cycle, their association is maximal during S and G₂/M phases (118).

However, the consequences of the interaction between PARP and DNA polymerase-α have been the subject of controversy. In the absence of NAD⁺, purified PARP was first shown to suppress, *in vitro,* the replication of DNA driven by DNA polymerase-α (120) and by DNA polymerases-α, -δ, and -ε (121, 122). In contrast, PARP was reported to stimulate, *in vitro*, the activity of DNA polymerase-α (118, 119). Inhibition of PARP by 3-aminobenzamide impaired the ability of MRC purified from HeLa cells to replicate SV40 DNA (23). In addition, the total DNA polymerase activity is reduced in mouse fibroblasts lacking PARP, and the rate of ongoing S phase is retarded following DNA damage (118, 123). The apparent discrepancy between the two results may well be due to the dual property of PARP: (1) its ability to bind to breaks, blocks or slows down a number of DNA transactions; (2) in contrast, binding cooperatively to DNA crossovers (see Chapter 2, Fig. 2.6) locally increases the density of the replication template, which in turn increases the activity of DNA polymerase-α and hence the kinetics of DNA synthesis *in vitro.*

Altogether, the data support a model (Fig. 3.5) in which PARP, as an element of the DNA damage-surveillance network and the replication apparatus, is involved in the coordination of cellular responses to DNA damage with the ongoing DNA synthesis. Under genotoxic stress, the presence of single-strand breaks in the template constitute a strong block for replication (124); their detection by PARP molecules associated with DNA polymerase α-primase would immobilize the replication complex at the lesion thus preventing priming downstream of the damage. At the same time, the presence of PARP at the break would ensure the rapid recruitment of DNA repair enzymes (see below) to restore DNA continuity, in a time-frame compatible with normal completion of the cell cycle. As a consequence of the persistence of unreplicated sequences containing lesions, the cell cycle may be delayed in G₂/M, leading eventually to apoptosis. This replication-linked repair pathway prevents cells from entering mitosis unless their genome has been repaired.

3.6.4 PARP is a key element of the base excision–repair pathway

The involvement of PARP in base excision repair (BER) was disputed until a direct link between PARP and other proteins involved in this pathway was discovered (see Table 3.2). Using immunoprecipitation and co-purification

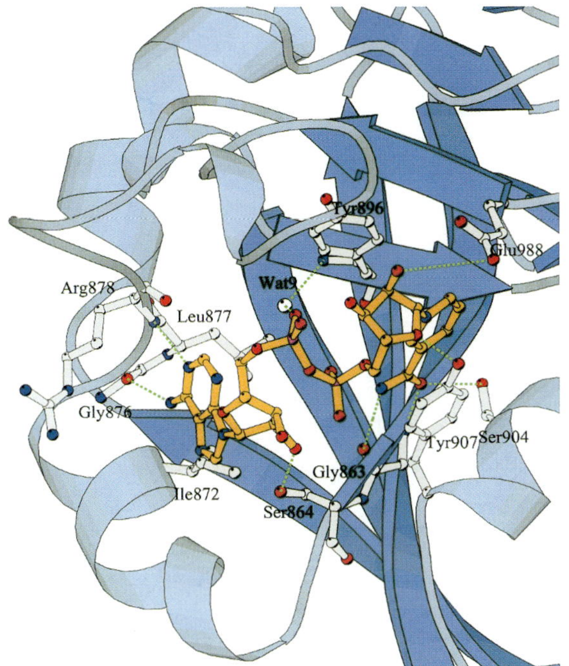

Plate 5 Model of NAD⁺ bound to PARP-CF. Putative hydrogen bonds are given by dotted lines.

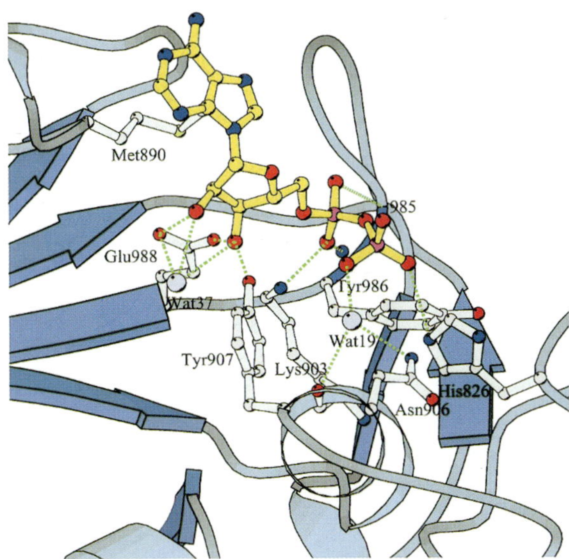

Plate 6 (ADP-ribose) acceptor binding site with bound ADP-moiety of carba-NAD. Hydrogen bonds are shown as dotted lines. The pyrophosphates are clearly the most tightly bound part of the ligand. The 2'-hydroxyl, which is the attacking nucleophile in the PARP elongation reaction, is hydrogen-bonded to the catalytic base Glu988. There are no specific interactions between the adenine and PARP-CF.

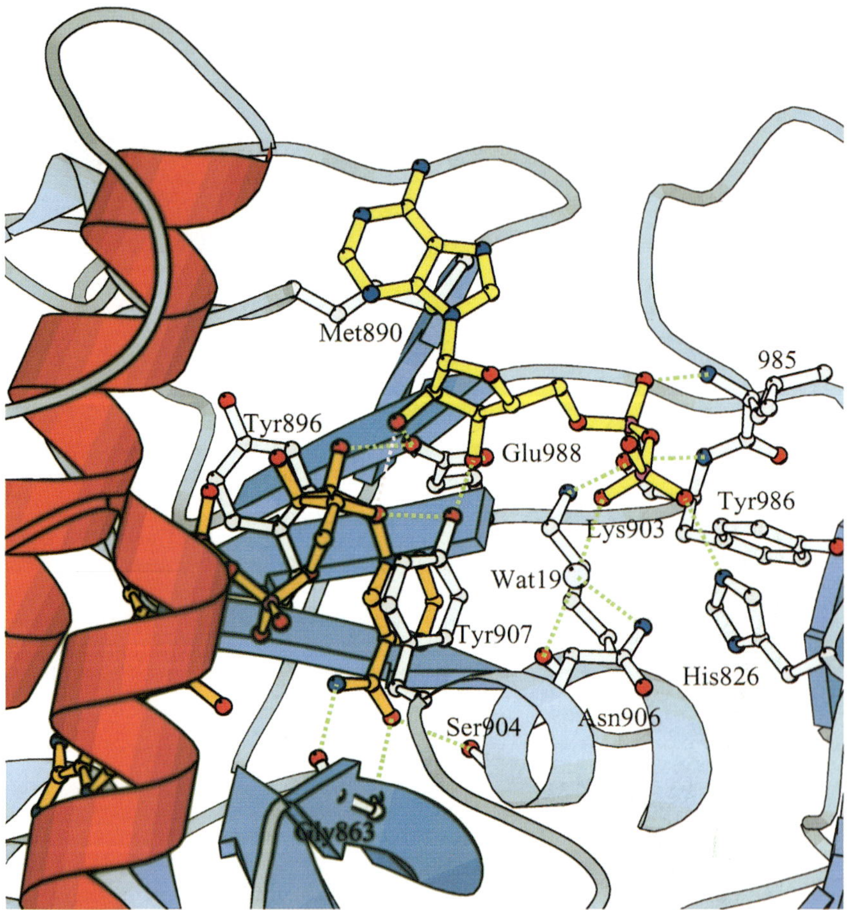

Plate 7 The reaction geometry in the PARP ADP-ribosyl transfer reaction (elongation). Shown is the acceptor ADP-moiety of a poly (ADP-ribose) chain as bound to PARP-CF together with the donor NAD⁺. The acceptor position is known from the X-ray structure of a PARP-CF/carba-NAD complex. The nicotinamide and ribose positions of NAD⁺ are derived from analogue-inhibitor complex structures. Hydrogen bonds are depicted as green dotted lines. The nucleophilic attack is marked by a yellow dotted line.

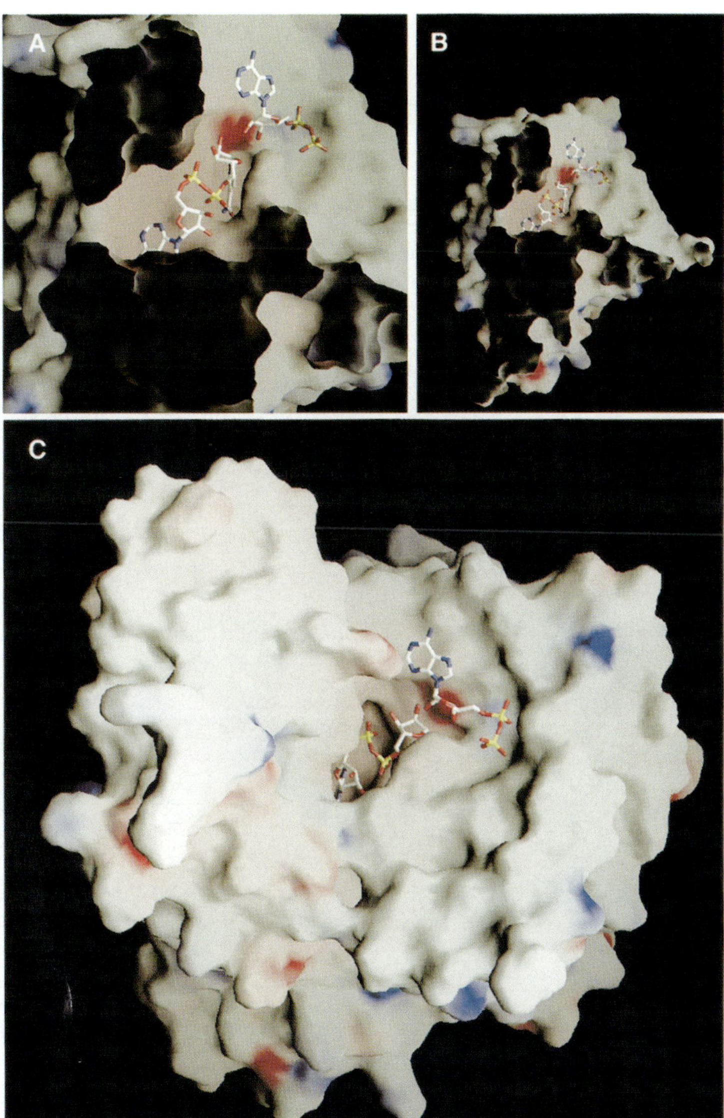

Plate 8 Molecular surface of PARP-CF coloured according to its electrostatic potential (red, negative; blue, positive). NAD^+ is bound to the deep pocket formed at the interface of the N- and C-terminal domains of PARP-CF, whereas the ADP-ribose acceptor is bound to a shallow crevice next to the NAD^+ pocket. (A) PARP-CF in approximately the same orientation as in Plate 3. The adenine and nicotinamide groups of NAD^+ are hidden by the N-terminal domain and by helix L. Further residues of poly (ADP-ribose) could possibly bind to the shallow crevices south of the binding site of the observed ADP-ribose acceptor. (B) The PARP-CF molecular surface is cut open to show the NAD^+ pocket. The negative surface potential located between the two ligands corresponds to the catalytic base Glu988. This figure was prepared using the program GRASP (183, Chapter 2).

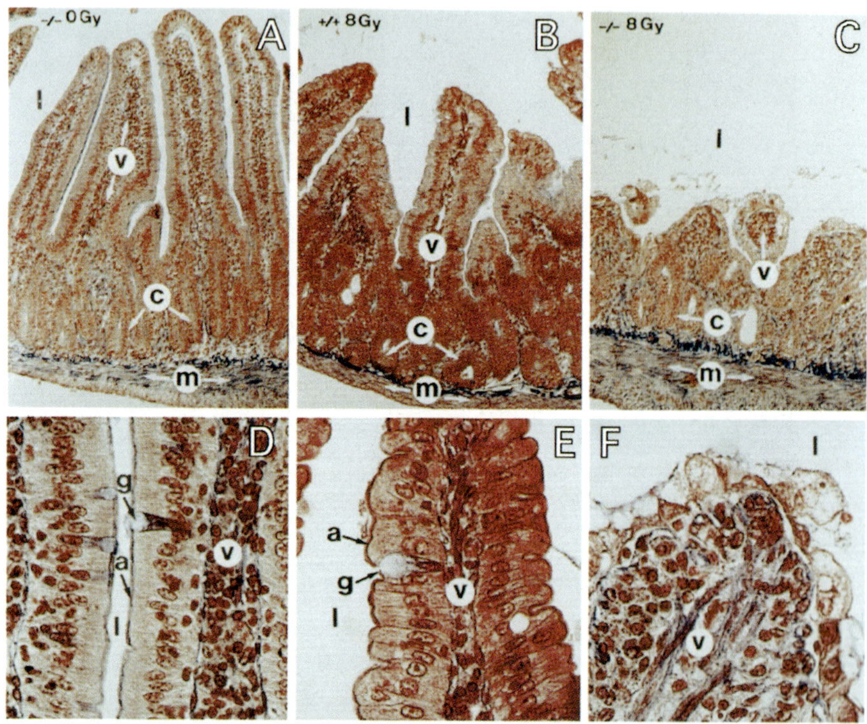

Plate 9 Transverse histological sections through the duodenum of an untreated *PARP*^{-/-} mouse (A and D) and an 8 Gy irradiated *PARP*^{+/+} mouse (B and E) or an irradiated *PARP*^{-/-} mouse (C and I). (A–C) show the full thickness of the duodenal wall and (D–F) the details of the epithelium near the tips of the villi. The untreated *PARP*^{-/-} duodenum (A and D) is histologically indistinguishable from its wild-type counterpart (not shown). Abbreviations: a, absorptive cell; c, crypt; g, goblet cell; l, lumen of the small intestine; m, muscularis; v, villi. Original magnifications: 170 × (A–C) and 750 × (D–F). (Taken from Ménissier-de Murcia *et al.* (49, Chapter 3) with permission.)

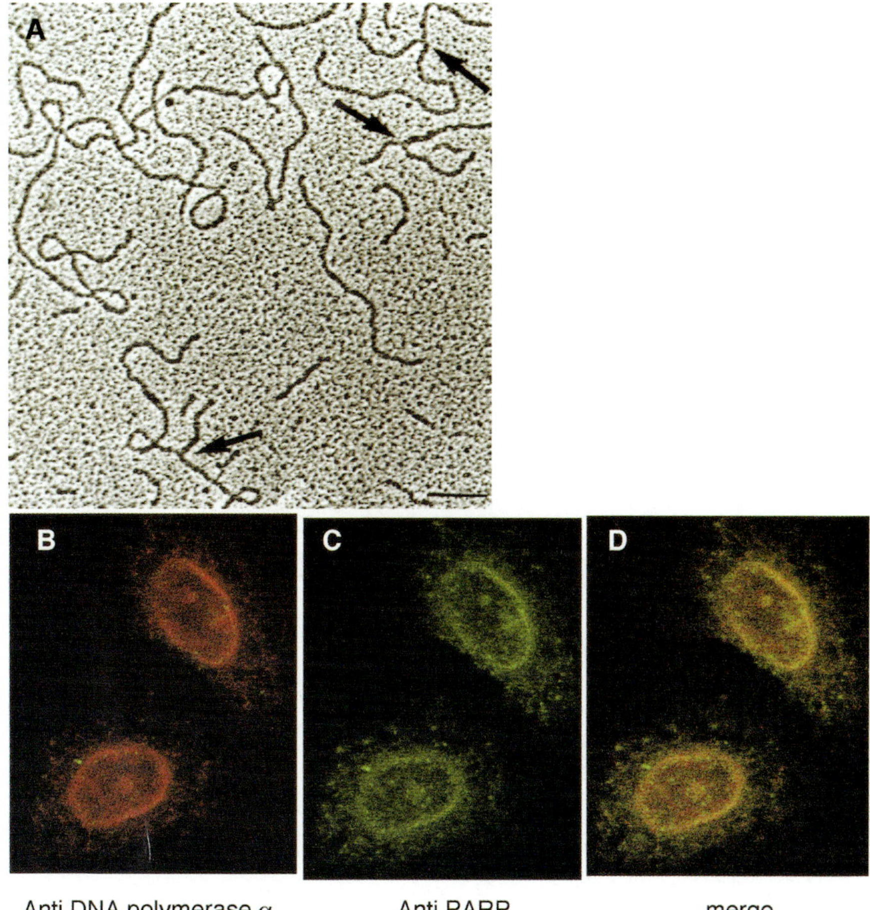

Plate 10   Confocal analysis of PARP and DNA polymerase-a double staining in HeLa cells. Panel A, fluorescein-labelled PARP (green); panel B, rhodamine-labelled DNA polymerase-a (red); panel C, merged image (regions of overlap are in yellow). (Taken from Dantzer *et al.* (118, Chapter 3) with permission.)

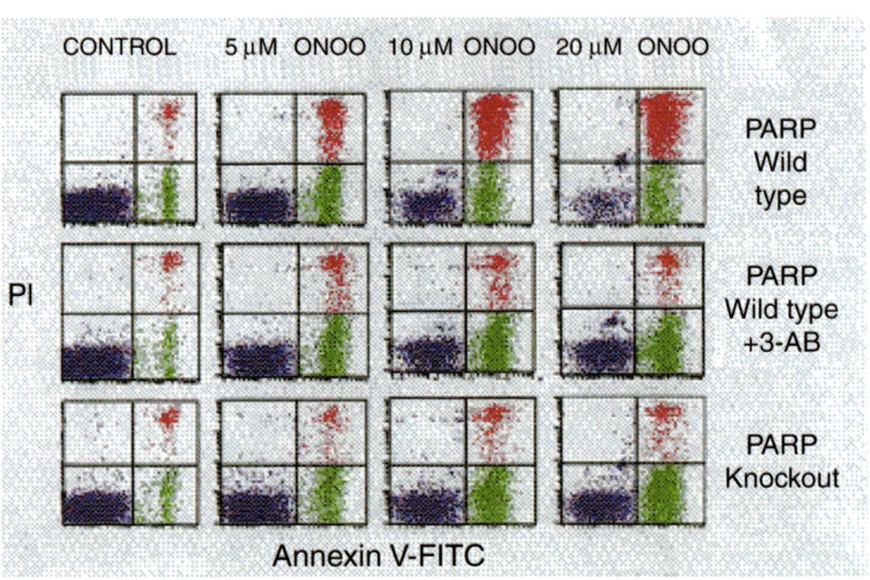

Plate 11   PARP activation in the development of cell necrosis and apoptosis in thymocytes challenged with peroxynitrite. Thymocytes were treated for 4 h with 10 and 50 μM peroxynitrite. Cells were then stained with Annexin V-FITC and propidium iodide (PI) and two-colour analysis was performed by flow cytometry. An increase in the number of apoptotic (Annexin V-FITC positive) cells was observed in response to 5 μM peroxynitrite treatment, whereas necrotic (stained by both Annexin V-FITC and PI) cells dominated in response to 20 μM peroxynitrite. PARP-deficient cells were protected against the necrotic loss of membrane integrity, as indicated by decreased PI uptake. Inhibition or genetic inactivation of PARP increased the number of cells in the normal population, as well as in the apoptotic population. (Reprinted, in modified form, with permission, from Virág, L., Scott, G.S., Salzman, A.L., and Szabó, C. (1998). Peroxynitrite-induced thymocyte apoptosis: the role of caspases and poly-(ADP-ribose) synthetase (PARS) activation. *Immunology*, **94**, 345–355, with permission of Blackwell Science Ltd.)

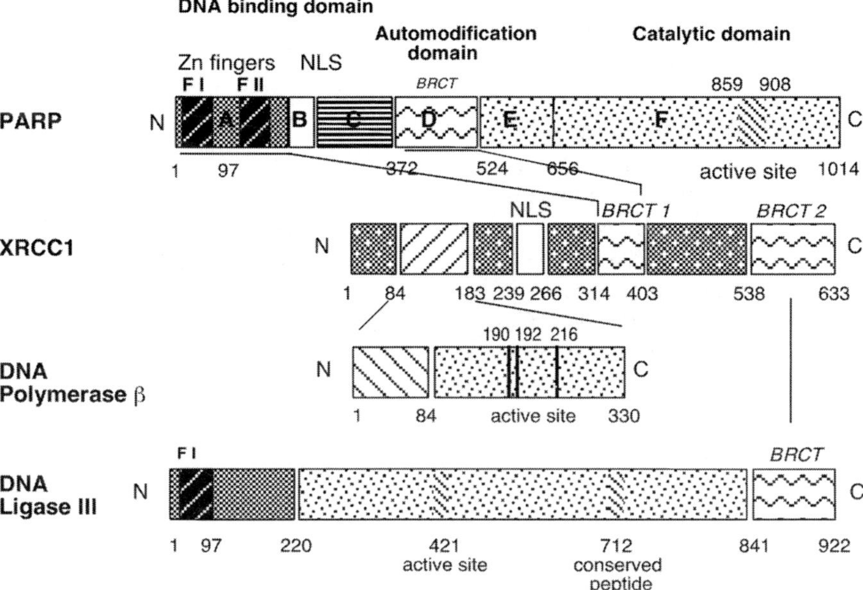

Fig. 3.6   Network of interactions between enzymes and factors participating in the BER pathway. BRCT modules involved in PARP–XRCC1 (71) and in XRCC1–DNA ligase III (127, 129, 134) interactions are indicated. XRCC1 interacts with DNA polymerase-β via its N-terminal region (128). PARP and DNA ligase III share the same nick-detection motif (zinc finger FI, aa 1–97) (132, 135). (Taken from ref. 71: Masson, M., *et al.* (1998). XRCC1 is specifically associated with poly(ADP-ribose) polymerase and negatively regulates its activity following DNA damage. *Mol. Cell Biol.*, **18**, 3563, with permission of the American Society for Microbiology.)

PARP is involved in base excision repair

Accumulating observations demonstrate that the base excision repair of DNA is a multistep process, consisting of at least two distinct, but partially overlapping, pathways (see ref. 141 for a review). The major 'short-patch repair' pathway involves a single-nucleotide replacement, mediated by DNA polymerase-β and DNA ligase III in the presence of XRCC1. The minor 'long-patch repair' pathway involves the replacement of several nucleotides, is mediated mostly by DNA polymerases-δ and -ε, and is dependent on PCNA. DNA polymerase-β is also able to carry out long-patch repair synthesis (137, 138). Reconstitution of a DNA base excision–repair assay *in vitro* with purified proteins showed that XRCC1 suppresses strand displacement by DNA polymerase-β during repair, allowing for more efficient ligation (128). In addition cell-free extracts from XRCC1-defective cells (EM-C11) showed an impaired ligation step in the short-patch repair pathway (142). However, the authors noticed that long-patch DNA repair synthesis was decreased by half in cell extracts from EM-9 cells (XRCC1-deficient). This observation was confirmed in the de Murcia laboratory (F. Dantzer, unpublished results), thus suggesting a possible involvement of XRCC1 in the long-patch base excision–repair pathway as well.

The contribution of PARP in these two repair pathways was investigated by evaluating the ability of PARP-deficient cell extracts to repair single abasic sites (AP) present on a covalently closed, circular duplex plasmid in a standard *in vitro* repair assay (143). Two types of AP sites were generated: by removing either a uracil residue with the monofunctional uracil DNA glycosylase (UDG) or an 8-oxoguanine with the bifunctional DNA glycosylase AP lyase, OGG1. Results indicated that PARP is directly involved in the base excision repair of these two lesions, with a comparable implication (140–144). PARP-deficient cell extracts are half as efficient as wild-type cells in performing short-patch repair, whereas the long-patch repair is totally impaired. The defect is associated with the polymerization step for both pathways, and both the protein and the catalytic activity take part in the repair process.

Why is the long-patch repair so much affected in PARP-deficient cells? PARP could take part in the decision of strand displacement leading to long-patch repair synthesis. One possibility is that the enzyme recognizes the 5'-dRP that is generated, and therefore could participate in the separation of the DNA strands thus allowing the progression of the DNA polymerase (-β, -δ, or -ε). The size of the released dRP-oligonucleotide was shown to be controlled by DNA polymerase-β (138).

PARP is able to interact with DNA polymerase-β, but not with DNA polymerases-δ or -ε (140). In order to gain insight into the functional relationship between these two proteins, mouse fibroblasts cell lines were generated that lacked both the PARP and the DNA polymerase-β genes. These cells were derived from mouse embryos resulting from intercrosses between $PARP^{-/-}$ (49) with $pol\beta^{+/-}$ mice (136) and between $PARP^{-/-}/pol\beta^{+/-}$ mice (140). The cell-free extract repair assay described above was used to monitor the base excision–repair efficiency of these cells. Cells lacking DNA polymerase-β were affected in both the short- and the long-patch repair pathways, whereas cells lacking both PARP and DNA polymerase-β were almost totally inefficient in both repair pathways. First, these results confirm that DNA polymerase-β is involved in both the short- and the long-patch repair pathways (137, 138). Second, in the absence of both PARP and DNA polymerase-β, repair is so inefficient that we may speculate on a possible synergy between the two proteins in base excision–repair processes.

It is tempting to propose a model for the base excision–repair pathway that now includes PARP (Fig. 3.7). In this model, XRCC1 would be considered as

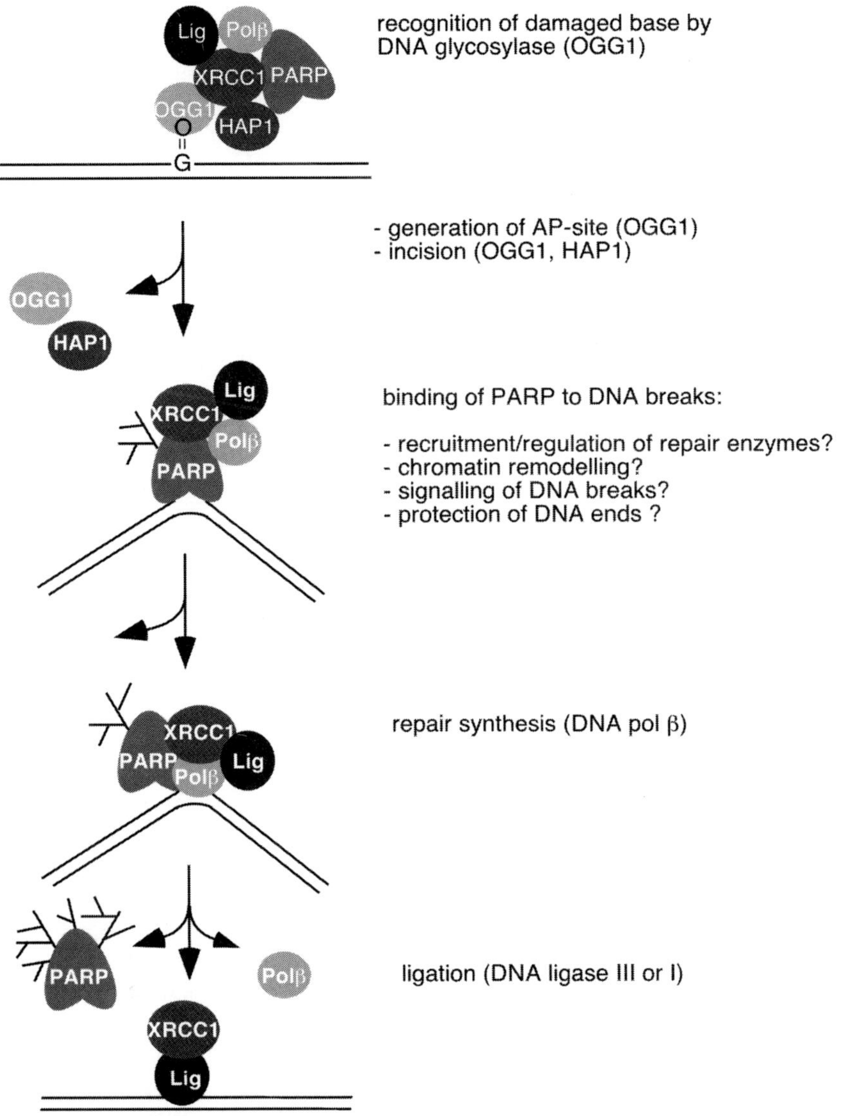

Fig. 3.7 Scheme depicting the involvement of PARP in the base excision–repair process. XRCC1 can be considered as the central adaptor to form a complex containing OGG1, HAP1, PARP, Polβ, and ligase III. PARP is likely to bind to the DNA interruption only after incision by OGG1 or HAP1. At this step, its activity may consist in either the recruitment or regulation of the subsequent repair enzymes. This scenario does not exclude a possible contribution of PARP in chromatin remodelling, signalling of the DNA breaks or in protection of the DNA ends from recombination. After DNA repair synthesis, PARP is displaced from the nick to allow access for DNA ligase III.

the central player of the BER process, sequentially recruiting each BER protein to the site of the lesion to fulfil its task. The first player would be the DNA lesion-recognizing protein, DNA glycosylase, which removes the damaged base. This protein may recruit XRCC1 at the site of the lesion. We know that this is true at least for OGG1 (A. Vidal, S. Boiteux, and P. Radicella, personal communication). However, it is not yet known whether this interaction has functional consequences for OGG1, or if this step constitutes the first link between the DNA lesion and the repair complex. The AP site generated would be incised either 5' by HAP1, or 3' by the lyase activity of a bifunctional DNA glycosylase. PARP could then be recruited by XRCC1 at this step on the incised AP site and forced to reside on the DNA break or gap, since XRCC1 limits PARP automodification (71). The next step would consist in DNA polymerization, either via long-patch or short-patch repair synthesis, mainly conducted by DNA polymerase-β (128, 137, 138) which is loaded by XRCC1.

In the case of long-patch repair synthesis, the displaced strand would be cleaved by FEN-1 in association with PCNA (137, 145). In both BER pathways, the remaining nick would then be sealed by either DNA ligase I or III (128). XRCC1 is therefore present throughout all the repair process, since it interacts with the first (e.g. OGG1) and the last (DNA ligase III) actor in this process.

From a mechanistic point of view, what would be the *precise* role of PARP in base excision repair? PARP in concert with XRCC1, settled in the nick/gap, could take part in several biological processes:

1. *Facilitating the repair process.* It might detect the DNA strand termini at the site of the lesion and recruit the other actors of the BER reaction whose activity could be regulated by poly (ADP-ribosylation). As mentioned above, four different proteins (XRCC1 and the three proteins it binds) have now been reported to bind to nicked DNA independently. Two of them, XRCC1 and DNA polβ, can simultaneously bind to this site (131). The presence of both DNA damage-detection modules (zinc fingers) and interaction interfaces (BRCT motifs) in this complex may contribute to increase the efficiency of the recruitment and the coordination of the functions of BER actors at DNA interruptions, to improve the kinetics of the overall repair pathway. The redundancy in their DNA binding property suggests a possible competition for the same target, DNA breaks, and possibly a displacement of PARP from the nick by DNA ligase III at the end of the repair process. Moreover, it also ensures that no nicked DNA could be left unbound.

2. *Protecting DNA from accidental recombination events.* In this model, initially proposed by Satoh and Lindahl (146), the binding of PARP to DNA ends protects them from nuclease attack and prevents non-homologous recombination. It is difficult at present to evaluate the relevance of this model, since no clear picture has yet emerged from the data obtained with

chemical inhibitors that show either a stimulation (147) or an inhibition of homologous recombination (148). However, the fact that the absence of PARP rescues V(D)J recombination in SCID mice argues for an antirecombinogenic role in this recombination process in T-lymphocytes (62). Moreover, the identification in B-cells of a recombination complex containing three polymer acceptors: numatrin (B23), nucleolin (C23), and PARP, further supports a possible involvement of PARP, at least in B-cell specific recombination (149).

3. *Remodelling of chromatin.* The contrasting requirement between storage of the genome in a compact ordered structure and the efficiency of its function, is met by the use of molecular machines specialized in the reversible disruption of chromatin. As mentioned above, PARP is an appropriate candidate for such a task, since it modifies H1 in chromatin. This reversible modification immediately triggers a local disruption of chromatin structure at a DNA break, which in turn allows access to multifactorial DNA repair complexes. Both the covalent modification of histone tails (74, 150) or their non-covalent binding to PARP-bound poly (ADP-ribose) (92) could be responsible for their release from the neighbouring nucleosome, leading to the transient decondensation of the 30-nm chromatin fibre.

4. *Signalling the presence of DNA strand interruptions to the cell.* Being a DNA damage-dependent polymerase, PARP is, by definition, well suited to functioning in DNA damage signalling. Depending on the intensity of this informative signal reflecting the amount of DNA damage and/or repair areas, the cell would probably balance between two options: repair with or without affecting cell-cycle progression, or no repair but opening an apoptotic programme. However, in PARP-deficient cells, treatment with genotoxic agents leads to a more pronounced G_2/M arrest and apoptosis, without efficient DNA repair, reflecting an accumulation of non-repaired DNA lesions (61). Therefore, other DNA damage-sensing proteins must be involved in this strand-break signalling; possibly DNA-PK or ATM, even though these factors are mostly involved in double-strand break repair. PARP, p53, DNA-PK, and ATM could have complementary functions in repair, recombination, cell-cycle control, and apoptosis. Recent publications describing a clear connection between PARP and several effectors of the DNA damage-signalling cascade are reviewed in Chapter 7, where emphasis has been put on NF-κB and p53, which now appear as sensors of oxidative stress (151–154).

3.7 Future directions

A variety of molecular and genetic approaches have been employed during this last decade to provide new insights into the complex functional role of

PARP in living cells. From the knockout model, we have learned that even if life and development without PARP are possible, survival under stress conditions is rather limited due to the dramatically slow repair phenotype that, in turn, generates genomic instability. The four hypotheses presented above, to explain why PARP needs to be activated in response to DNA breaks, are not mutually exclusive. Clearly, they reflect the necessity to design new experimental approaches aimed at a better understanding of the role of PARP in the BER pathway. It is now crucial to clarify the timing and the order with which the different partners interact with the DNA lesion and with each other. The importance of chromatin structure, as the natural environment of PARP, needs to be assessed in future DNA repair assays in which a nucleosomal structure will contain the lesion to be repaired. It will be also important to evaluate the function of PARP in the double-strand break pathway, since a link with the DNA-PK/Ku has been repeatedly observed (see Table 3.2 and Chapter 7). Moreover, the recently discovered PARP homologues, particularly those like PARP-2 that are activated by DNA damage (155), will have to enter into the game too, perhaps to bring another level of regulation in the positive, but potentially dangerous, function of PARP (the classical one) as a caretaker of the genome.

Finally, one of the challenges facing us in the future is to figure out a unified view of the role of PARP and poly (ADP-ribosylation) reactions in general, that now appear to cover various important but apparently unrelated biological processes. For example, what is there in common between the repair of an abasic site, the activation of NF-κB, and the maintenance of telomere length (156) (see Chapter 7)—the protection of the genome against oxidative stress only?

Acknowledgements

We wish to thank J.-L. Vonesch and D. Plotton for their fine contributions to confocal microscopy imaging. The work resulting in the original data shown in this chapter was supported by CNRS, Association pour la Recherche contre le Cancer, La Ligue contre le Cancer, Electricité de France, le Commissariat à l'Energie Atomique, and la Fondation pour la Recherche Médicale.

References

1. Milam, K. M. and Cleaver, J. E. (1984). Inhibitors of poly(adenosine diphosphate ribose)synthesis: effect on other metabolic processes. *Science*, **223**, 589.
2. Milam, K. M., Thomas, G. H., and Cleaver, J. E. (1986). Disturbances in DNA precursor metabolism associated with exposure to an inhibitor of poly(ADP-ribose) synthetase. *Exp. Cell Res.*, **165**, 260.

3. Moses, K., Harris, A. L., and Durkacz, B. W. (1988). Synergistic enhancement of 6-thioguanine cytotoxicity by ADP-ribosyltransferase inhibitors. *Cancer Res.*, **48**, 5650.

4. Rankin, P. W., Jacobson, E. L., Benjamin, R. C., Moss, J., and Jacobson, M. K. (1989). Quantitative studies of inhibitors of ADP-ribosylation *in vitro* and *in vivo*. *J. Biol. Chem.*, **264**, 4312.

5. Chatterjee, S., Petzold, S. J., Berger, S. J., and Berger, N. A. (1987). Strategy for selection of cell variants deficient in poly (ADP-ribose)polymerase. *Exp. Cell Res.*, **172**, 245.

6. Chatterjee, S., Hirschler, N. V., Petzold, S. J., Berger, S. J., and Berger, N. A. (1989). Mutant cells defective in poly(ADP-ribose) synthesis due to stable alterations in enzyme activity or substrate availability. *Exp. Cell Res.*, **184**, 1.

7. Chatterjee, S., Trivedi, D., Petzold, S. J., and Berger, N. A. (1990). Mechanism of epipodophyllotoxin-induced cell death in poly(adenosine diphosphate-ribose) synthesis-deficient V79 Chinese hamster cell lines. *Cancer Res.*, **50**, 2713.

8. Chatterjee, S., Cheng, M. F., and Berger, N. A. (1990). Hypersensitivity to clinically useful alkylating agents and radiation in poly(ADP-ribose) polymerase-deficient cell lines. *Cancer Commun.*, **2**, 401.

9. Chatterjee, S., Cheng, M. F., Berger, S. J., and Berger, N. A. (1991). Alkylating agent hypersensitivity in poly(adenosine diphosphate-ribose) polymerase deficient cell lines. *Cancer Commun.*, **3**, 71.

10. Chatterjee, S., Cheng, M. F., Berger, S. J., and Berger, N. A. (1994). Induction of M(r) 78,000 glucose-regulated stress protein in poly(adenosine diphosphate-ribose) polymerase- and nicotinamide adenine dinucleotide-deficient V79 cell lines and its relation to resistance to the topoisomerase II inhibitor etoposide. *Cancer Res.*, **54**, 4405.

11. Chatterjee, S., Cheng, M. F., Berger, R. B., Berger, S. J., and Berger, N. A. (1995). Effect of inhibitors of poly(ADP-ribose) polymerase on the induction of GRP78 and subsequent development of resistance to etoposide. *Cancer Res.*, **55**, 868.

12. Chatterjee, S., Cheng, M. F., Trivedi, D., Petzold, S. J., and Berger, N. A. (1989). Camptothecin hypersensitivity in poly(adenosine diphosphate-ribose) polymerase-deficient cell lines. *Cancer Commun.*, **1**, 389.

13. Ray, L. S., Chatterjee, S., Berger, N. A., Grishko, V. I., LeDoux, S. P., and Wilson, G. L. (1996). Catalytic activity of poly(ADP-ribose)polymerase is necessary for repair of *N*-methylpurines in nontranscribed, but in transcribed, nuclear DNA sequences. *Mutat. Res.*, **363**, 105.

14. Stevnsner, T., Ding, R., Smulson, M., and Bohr, V. A. (1994). Inhibition of gene-specific repair of alkylation damage in cells depleted of poly(ADP-ribose) polymerase. *Nucleic Acids Res.*, **22**, 4620.

15. Whitacre, C. M., Hashimoto, H., Tsai, M. L., Chatterjee, S., Berger, S. J., and Berger, N. A. (1995). Involvement of NAD-poly(ADP-ribose) metabolism in p53 regulation and its consequences. *Cancer Res.*, **55**, 3697.

16. Wright, S. C., Wei, Q. S., Kinder, D. H., and Larrick, J. W. (1996). Biochemical pathways of apoptosis: nicotinamide adenine dinucleotide-deficient cells are resistant to tumor necrosis factor or ultraviolet light activation of the 24-kD apoptotic protease and DNA fragmentation. *J. Exp. Med.*, **183**, 463.

17. MacLaren, R. A., Witmer, M. V., Richardson, E., and Stamato, T. D. (1990).

Isolation of Chinese hamster ovary cells with reduced poly(ADP-ribose) polymerase activity. *Mutat. Res.*, **231**, 265.

18. Witmer, M. V., Aboul, E. N., Jacobson, M. K., and Stamato, T. D. (1994). Increased sensitivity to DNA-alkylating agents in CHO mutants with decreased poly(ADP-ribose) polymerase activity. *Mutat. Res.*, **314**, 249.

19. Yoshihara, K., Itaya, A., Hironaka, T., Sakuramoto, S., Tanaka, Y., Tsuyuki, M., Inada, Y., Kamiya, T., Ohnishi, K., Honma, M., *et al.* (1992). Poly(ADP ribose) polymerase-defective mutant cell clone of mouse L1210 cells. *Exp. Cell Res.*, **200**, 126.

20. Ding, R., Pommier, Y., Kang, V. H., and Smulson, M. (1992). Depletion of poly(ADP-ribose) polymerase by antisense RNA expression results in a delay in DNA strand break rejoining. *J. Biol. Chem.*, **267**, 12804.

21. Ding, R. and Smulson, M. (1994). Depletion of nuclear poly(ADP-ribose) polymerase by antisense RNA expression: influences on genomic stability, chromatin organization, and carcinogen cytotoxicity. *Cancer Res.*, **54**, 4627.

22. Smulson, M. E., Kang, V. H., Ntambi, J. M., Rosenthal, D. S., Ding, R., and Simbulan, C. M. (1995). Requirement for the expression of poly(ADP-ribose) polymerase during the early stages of differentiation of 3T3-L1 preadipocytes, as studied by antisense RNA induction. *J. Biol. Chem.*, **270**, 119.

23. Simbulan-Rosenthal, C. M., Rosenthal, D. S., Hilz, H., Hickey, R., Malkas, L., Applegren, N., Wu, Y., Bers, G., and Smulson, M. E. (1996). The expression of poly(ADP-ribose)polymerase during differentiation linked DNA replication reveals that it is a component of the Multiprotein DNA Replication Complex. *Biochemistry*, **35**, 11622.

24. Rosenthal, D. S., Shima, T. B., Celli, G., De, L. L., and Smulson, M. E. (1995). Engineered human skin model using poly(ADP-ribose) polymerase antisense expression shows a reduced response to DNA damage. *J. Invest. Dermatol.*, **105**, 38.

25. Taniguchi, T., Yamauchi, K., Yamamoto, T., Toyoshima, K., Harada, N., Tanaka, H., Takahashi, S., Yamamoto, H., and Fujimoto, S. (1988). Depression in gene expression for poly(ADP-ribose) synthetase during the interferon-gamma-induced activation process of murine macrophage tumor cells. *Eur. J. Biochem.*, **171**, 571.

26. Qu, Z., Fujimoto, S., and Taniguchi, T. (1994). Enhancement of interferon-γ-induced major histocompatibility complex class II gene expression by expressing an antisense RNA of poly(ADP-ribose)synthetase. *J. Biol. Chem.*, **269**, 5543.

27. Taniguchi, T., Ota, K., Qu, Z., and Morisawa, K. (1995). Effect of poly(ADP-ribose) synthetase on the expression of major histocompatibility complex (MHC) class II genes. *Biochimie*, **77**, 472.

28. Tanaguchi, T., Takahashi, S., Yamamoto, H., and Fujimoto, S. (1991). Requirements of down-regulation of NAD$^+$ ADP-ribosyltransferase for the interferon-gamma-induced activation process of murine macrophage tumor cells. *Eur. J. Biochem.*, **195**, 557.

29. Nomura, I., Kurashige, T., and Taniguchi, T. (1991). Inhibitory effect of interferon-gamma-dependent induction of major histocompatibility complex class II antigen by expressing exogenous poly(ADP-ribose) synthetase gene. *Biochem. Biophys. Res. Commun.*, **175**, 685.

30. Tomoda, T., Kurashige, T., and Taniguchi, T. (1992). Inhibition of interfereon-

gamma- and phorbol ester-induced HLA-DR and interleukin-1 production by the expression of a transfected poly(ADP-ribose) synthetase gene in human leukemia THP-1 cells. *Biochem. Biophys. Acta*, **1135**, 79.

31. Gäken, J. A., Tavassoli, M., Gan, S. U., Vallian, S., Giddings, I., Darling, D. C., Galea, L. J., Thomas, M. G., Abedi, H., Schreiber, V., Ménissier-de Murcia, J., Collins, M. K., Shall, S., and Farzaneh, F. (1996). Efficient retroviral infection of mammalian cells is blocked by inhibition of poly(ADP-ribose) polymerase activity. *J. Virol.*, **70**, 3992.

32. Petersen, J., Dandri, M., Bürkle, A., Zhang, L., and Rogler, C. (1997). Increase in the frequency of hepadnavirus DNA integrations by oxidative DNA damage and inhibition of DNA repair. *J. Virol.*, **71**, 5455.

33. Küpper, J. H., de Murcia, G., and Bürkle, A. (1990). Inhibition of poly(ADP-ribosyl)ation by overexpressing the poly(ADP-ribose) polymerase DNA-binding domain in mammalian cells. *J. Biol. Chem.*, **265**, 18721.

34. Molinete, M., Vermeulen, W., Bürkle, A., Ménissier de Murcia, J., Küpper, J. H., Hoeijmakers, J. H., and de Murcia, G. (1993). Overproduction of the poly(ADP-ribose) polymerase DNA-binding domain blocks alkylation-induced DNA repair synthesis in mammalian cells. *EMBO J.*, **12**, 2109.

35. Schreiber, V., Hunting, D., Trucco, C., Gowans, B., Grunwald, D., de Murcia, G., and Ménissier de Murcia, J. (1995). A dominant-negative mutant of human poly(ADP-ribose) polymerase affects cell recovery, apoptosis, and sister chromatid exchange following DNA damage. *Proc. Natl Acad. Sci. USA*, **92**, 4753.

36. Küpper, J. H., Müller, M., Jacobson, M. K., Tatsumi, M. J., Coyle, D. L., Jacobson, E. L., and Bürkle, A. (1995). *Trans*-dominant inhibition of poly(ADP-ribosyl)ation sensitizes cells against gamma-irradiation and *N*-methyl-*N'*-nitro-*N*-nitrosoguanidine but does not limit DNA replication of a polyomavirus replicon. *Mol. Cell Biol.*, **15**, 3154.

37. Küpper, J. H., Müller, M., and Bürkle, A. (1996). *Trans*-dominant inhibition of poly(ADP-ribosyl)ation potentiates carcinogen-induced gene amplification in SV40-transformed Chinese Hamster cells. *Cancer Res.*, **56**, 2715.

38. Bürkle, A., Meyer, T., Hilz, T., and Zur Hausen, H. (1987). Enhancement of *N*-methyl-nitro-*N*-nitrosoguanidine-induced DNA amplification in a Simian virus transformed Chinese cell line by 3-aminobenzamide. *Cancer Res.*, **47**, 3632.

39. Bürkle, A., Heilbronn, R., and Zur Hausen, H. (1990). Potentiation of carcinogen-induced methotrexate resistance and dihydrofolate reductase gene amplification by inhibitors of poly(adenosine diphosphate-ribose) polymerase. *Cancer Res.*, **50**, 5756.

40. Tatsumi-Miyajima, J., Küpper, J.-H., Takebe, H., and Bürkle A. (1997). *trans*-Dominant inhibition of poly(ADP-ribosyl)ation potentiates alkylation-induced shuttle vector mutagenesis in Chinese hamster cells. *Mol. Cell. Biochem.*, **193**, 31.

41. Kaiser, P., Auer, B., and Schweiger, M. (1992). Inhibition of cell proliferation in *Saccharomyces cerevisiae* by expression of human NAD+ ADP-ribosyltransferase requires the DNA binding domain ('zinc fingers'). *Mol. Gen. Genet.*, **232**, 231.

42. Avila, M. A., Velasco, J. A., Smulson, M. E., Dritschilo, A., Castro, R., and Notario, V. (1994). Functional expression of human poly(ADP-ribose) polymerase in *Schizosaccharomyces pombe* results in mitotic delay at G1, increased mutation rate, and sensitization to radiation. *Yeast*, **10**, 1003.

43. Fritz, G., Auer, B., and Kaina, B. (1994). Effect of transfection of human poly(ADP-ribose)polymerase in Chinese hamster cells on mutagen resistance. *Mutat. Res.*, **308**, 127.

44. van Gool, L., Meyer, R., Tobiasch, E., Cziepluch, C., Jauniaux, J.-C., Mincheva, A., Lichter, P., Poirier, G. G., Bürkle, A., and Küpper, J.-H. (1997). Overexpression of human poly(ADP-ribose) polymerase in transfected hamster cells leads to increased poly(ADP-ribosyl)ation and cellular sensitization to γ-irradiation. *Eur. J. Biochem.*, **244**, 15.

45. Bernges, F., Bürkle, A., Küpper, J. H., and Zeller, W. J. (1997). Functional over-expression of human poly(ADP-ribose) polymerase in transfected rat tumor cells. *Carcinogenesis*, **18**, 663.

46. Masutani, M., Nozaki, T., Nishiyama, E., Shimokawa, T., Tachi, Y., Suzuki, H., Nakagama, H., Wakabayashi, K., and Sugimura, M. (1999). Function of poly(ADP-ribose) polymerase in response to DNA damage: gene-disruption study in mice. *Mol. Cell. Biochem.*, **193**, 149.

47. Masutani, M., Suzuki, H., Kamada, N., Watanabe, M., Ueda, O., Nozaki, T., Jishage, K., Watanabe, T., Sugimoto, T., Nakagama, H., Ochiya, T., and Sugimura, T. (1999). Poly(ADP-ribose) polymerase gene disruption conferred mice resistant to streptozotocin-induced diabetes. *Proc. Natl Acad. Sci. USA*, **96**, 2301.

48. Wang, Z. Q., Auer, B., Stingl, L., Berghammer, H., Haidacher, D., Schweiger, M., and Wagner, E. W. (1995). Mice lacking ADPRT and poly(ADP-ribosyl)ation develop normally but are susceptible to skin disease. *Genes Dev.*, **9**, 509.

49. Ménissier de Murcia, J., Niedergang, C., Trucco, C., Ricoul, M., Dutrillaux, B., Mark, M., Olivier, F. J., Masson, M., Dierich, A., LeMeur, M., Walztinger, C., Chambon, P., and de Murcia, G. (1997). Requirement of poly(ADP-ribose)polymerase in recovery from DNA damage in mice and in cells. *Proc. Natl Acad. Sci. USA*, **94**, 7303.

50. Heller, B., Wang, Z. Q., Wagner, E. F., Radons, J., Bürkle, A., Fehsel, K., Burkart, V., and Kolb, H. (1995). Inactivation of the poly(ADP-ribose) polymerase gene affects oxygen radical and nitric oxide toxicity in islet cells. *J. Biol. Chem.*, **270**, 11176.

51. Le Rhun, Y., Kirkland, J. B., and Shah, G. M. (1998). Cellular responses to DNA damage in the absence of Poly(ADP-ribose) polymerase. *Biochem. Biophys. Res. Commun.*, **245**, 1.

52. D'Amours, D., Desnoyers, S., D'Silva, I., and Poirier, G. G. (1999). Poly(ADP-ribosyl)ation reactions in the regulation of nuclear functions. *Biochem J.*, **342**, 249.

53. Shall, S. and de Murcia, G. (2000). Poly(ADP-ribose) polymerase-1: what have we learned from the deficient mouse model? *Mutat. Res.* 460, 1.

54. Wang, Z. Q., Stingl, L., Morrison, C., Jantsch, M., Los, M., Schulze-Osthoff, K., and Wagner, E. F. (1997). PARP is important for genomic stability but dispensable in apoptosis. *Genes Dev.*, **11**, 2347.

55. Horowitz, R. W., Wadler, S., and Wiernik, P. H. (1997). A review of the clinical experience with irinotecan (Cpt-11). *Am. J. Therapeut.*, **4**, 203.

56. Avemann, K., Knippers, R., Koller, T., and Sogo, J. M. (1988). Camptothecin, a specific inhibitor of type I DNA topoisomerase, induces DNA breakage at replication forks. *Mol. Cell Biol.*, **8**, 3026.

57. Johnson, R. T., Gotoh, E., Mullinger, A. M., Ryan, A. J., Shiloh, Y., Ziv, Y., and Squires, S. (1999). Targeting double-strand breaks to replicating DNA identifies a subpathway of DSB repair that is defective in ataxia-telangiectasia cells. *Biochem. Biophys. Res. Commun.*, **261**, 317.

58. Canitrot, Y., de Murcia, G., and Salles, B. (1998). Decreased expression of topoisomerase IIbeta in poly(ADP-ribose) polymerase-deficient cells. *Nucleic Acids Res.*, **26**, 5134.

59. Oikawa, A., Tohda, H., Kanai, T., Miwa, M., and Sugimura, T. (1980). Inhibitors of poly(adenosine diphosphate ribose) polymerase induce sister chromatid exchanges. *Biochem. Biophys. Res. Commun.*, **97**, 1311.

60. Natarajan, A. T., Csuka, I., and van Zeeland, A. A. (1981). Contribution of incorporated 5-bromodeoxyuridine in DNA to the frequencies of sister-chromatid exchanges induced by inhibitors of poly(ADP-ribose)polymerase. *Mutat. Res.*, **84**, 125.

61. Trucco, C., Oliver, F. J., de Murcia, G., and Ménissier-de Murcia, J. (1998). DNA repair defect in poly(ADP-ribose)polymerase-deficient cell lines. *Nucleic Acids Res.*, **26**, 2644.

62. Morrison, C., Smith, G. C., Stingl, L., Jackson, S. P., Wagner, E. F., and Wang, Z. Q. (1997). Genetic interaction between PARP and DNA-PK in V(D)J recombination and tumorigenesis. *Nature Genet.*, **17**, 479.

63. Jacobs, H., Fukita, Y., van der Horst, G. T., de Boer, J., Weeda, G., Essers, J., de Wind, N., Engelward, B. P., Samson, L., Verbeek, S., de Murcia, J. M., de Murcia, G., te Riele, H., and Rajewsky, K. (1998). Hypermutation of immunoglobulin genes in memory B cells of DNA repair-deficient mice. *J. Exp. Med.*, **187**, 1735.

64. Morrison, C. and Wagner, E. (1996). Extrachromosomal recombination occurs efficiently in cells defective in various DNA repair systems. *Nucleic Acids Res.*, **24**, 2053.

65. Pieper, A. A., Brat, D. J., Krug, D. K., Watkins, C. C., Gupta, A., Blackshaw, S., Verma, A., Wang, Z. Q., and Snyder, S. H. (1999). Poly(ADP-ribose) polymerase-deficient mice are protected from streptozotocin-induced diabetes. *Proc. Natl Acad. Sci. USA*, **96**, 3059.

66. Simbulan-Rosenthal, C. M., Rosenthal, D. S., Ding, R., Bhatia, K., and Smulson, M. E. (1998). Prolongation of the p53 response to DNA strand breaks in cells depleted of PARP by antisense RNA expression. *Biochem. Biophys. Res. Commun.*, **253**, 864.

67. Leist, M., Single, B., Künstle, G., Volbracht, C., Hentze, H., and Nicotera, P. (1997). Apoptosis in the absence of poly(ADP-ribose) polymerase. *Biochem. Biophys. Res. Commun.*, **233**, 518.

68. Oliver, F. J., de la Rubia, G., Rolli, V., Ruiz-Ruiz, M. C., de Murcia, G., and Ménissier-de Murcia, J. (1998). Importance of poly(ADP-ribose) polymerase and its cleavage in apoptosis. Lesson from an uncleavable mutant. *J. Biol. Chem.*, **273**, 33533.

69. Shall, S. (1984). ADP-ribose in DNA repair: a new component of DNA excision repair. *Adv. Radiat. Biol.*, **11**, 1.

70. Masson, M., Ménissier-de Murcia, J., Mattei, M. J., de Murcia, G., and Niedergang, C. P. (1997). Poly(ADP-ribose)polymerase interacts with a novel human ubiquitin conjugating enzyme: hUbc9. *Gene*, **190**, 287.

71. Masson, M., Niedergang, C., Schreiber, V., Ménissier-de Murcia, J., and de

Murcia, G. (1998). XRCC1 is specifically associated with poly(ADP-ribose) polymerase and negatively regulates its activity following DNA damage. *Mol. Cell Biol.*, **18**, 3563.

72. Nie, J., Sakamoto, S., Song, D., Qu, Z., Ota, K., and Taniguchi, T. (1998). Interaction of Oct-1 and automodification domain of poly(ADP-ribose) synthetase. *FEBS Lett.*, **424**, 27.

73. Leduc, Y., Lawrence, J. J., de Murcia, G., and Poirier, G. G. (1986). Visualization of poly(ADP-ribose) synthetase associated with polynucleosomes by immunoelectron microscopy. *Biochim. Biophys. Acta*, **875**, 248.

74. de Murcia, G., Huletsky, A., and Poirier, G. G. (1988). Modulation of chromatin structure by poly(ADP-ribosyl)ation. *Biochem. Cell Biol.*, **66**, 625.

75. Buki, K. G., Bauer, P. I., Hakam, A., and Kun, E. (1995). Identification of domains of poly(ADP-ribose) polymerase for protein binding and self-association. *J. Biol. Chem.*, **270**, 3370.

76. Aubin, R. J., Dam, V. T., Miclette, J., Brousseau, Y., Huletsky, A., and Poirier, G. G. (1982). Hyper(ADP-ribosyl)ation of histone H1. *Can. J. Biochem.*, **60**, 1085.

77. Aubin, R. J., Frechette, A., de Murcia, G., Mandel, P., Lord, A., Grondin, G., and Poirier, G. G. (1983). Correlation between endogenous nucleosomal hyper(ADP-ribosyl)ation of histone H1 and the induction of chromatin relaxation. *EMBO J.*, **2**, 1685.

78. Huletsky, A., Niedergang, C., Frechette, A., Aubin, R., Gaudreau, A., and Poirier, G. G. (1985). Sequential ADP-ribosylation pattern of nucleosomal histones. ADP-ribosylation of nucleosomal histones. *Eur. J. Biochem.*, **146**, 277.

79. Althaus, F. R., Bachmann, S., Hofferer, L., Kleczkowska, H. E., Malanga, M., Panzeter, P. L., Realini, C., and Zweifel, B. (1995). Interactions of poly(ADP-ribose) with nuclear proteins. *Biochimie*, **77**, 423.

80. Krupitza, G. and Cerutti, P. (1988). Poly(ADP-ribosyl)ation of histones in intact human keratinocytes. *Biochemistry*, **28**, 4054.

81. Amietz, P. and Rudolph, A. (1984). ADP-ribosylation of nuclear proteins *in vivo*. Identification of histone H2B as a major acceptor for mono and poly(ADP-ribose) in dimethyl sulfate treated AH 7974 cells. *J. Biol. Chem.*, **259**, 6841.

82. Adamietz, P. (1985). Isolation and identification of mono and poly(ADP-ribosyl)proteins formed in intact cells in association with DNA repair. In *ADP-ribosylation of proteins* (ed. F. R. Althaus, H. Hilz, and S. Shall), p. 264. Springer-Verlag, Heidelberg.

83. Poirier, G. G., de Murcia, G., Jongstra-Bilen, J., Niedergang, C., and Mandel, P. (1982). Poly(ADP-ribosyl)ation of polynucleosomes cause relaxation of chromatin structure. *Proc. Natl Acad. Sci. USA*, **79**, 3423.

84. Niedergang, C., de Murcia, G., Ittel, M. E., Pouyet, J., and Mandel, P. (1985). Time course of polynucleosome relaxation and ADP ribosylation: correlation between relaxation and histone H1 hyper-ADP-ribosylation. *Eur. J. Biochem.*, **146**, 185.

85. de Murcia, G., Huletski, A., Lamarre, D., Gaudreau, A., Pouyet, J., Daune, M., and Poirier, G. G. (1986). Modulation of chromatin superstructure induced poly(ADP-ribose) synthesis and degradation. *J. Biol. Chem.*, **261**, 7011.

86. Wolffe, A. P. and Hayes, J. J. (1999). Chromatin disruption and modification. *Nucleic Acids Res.*, **27**, 711.

87. Kornberg, R. D. and Lorch, Y. (1999). Twenty-five years of the nucleosome, fundamental particle of the eukaryote chromosome. *Cell*, **98**, 285.

88. Panzeter, P. L., Realini, C. A., and Althaus, F. R. (1992). Noncovalent interactions of poly(adenosine diphosphate ribose) with histones. *Biochemistry*, **31**, 1379.

89. Althaus, F. R., Naegeli, H., Realini, C., Mathis, G., Loetscher, P., and Mattenberger, M. (1990). The poly-ADP-ribosylation system of higher eukaryotes: a protein shuttle mechanism in chromatin? *Acta Biol. Hung.*, **41**, 9.

90. Althaus, F. R. (1992). Poly ADP-ribosylation: a histone shuttle mechanism in DNA excision repair. *J. Cell Sci.*, **102**, 663.

91. Althaus, F. R., Hofferer, L., Kleczkowska, H. E., Malanga, M., Naegeli, H., Panzeter, P., and Realini, C. (1993). Histone shuttle driven by the automodification cycle of poly(ADP-ribose)polymerase. *Environ. Mol. Mutagen.*, **22**, 278.

92. Panzeter, P. L., Zweifel, B., Malanga, M., Waser, S. H., Richard, M., and Althaus, F. R. (1993). Targeting of histone tails by poly(ADP-ribose). *J. Biol. Chem.*, **268**, 17662.

93. Yoon, Y. S., Kim, J. W., Kang, K. W., Kim, Y. S., Choi, K. H., and Joe, C. O. (1996). Poly(ADP-ribosyl)ation of histone H1 correlates with internucleosomal DNA fragmentation during apoptosis. *J. Biol. Chem.*, **271**, 9129.

94. Johnson, G. S. and Ralhan, R. (1986). Glucocorticoid agonists as well as antagonists are effective inducers of mouse mammary tumor virus RNA in mouse mammary tumor cells treated with inhibitors of ADP-ribosylation. *J. Cell Physiol.*, **129**, 36.

95. Tanuma, S., Johnson, L. D., and Johnson, G. S. (1983). ADP-ribosylation of chromosomal proteins and mouse mammary tumor virus gene expression. Glucocorticoids rapidly decrease endogenous ADP- ribosylation of nonhistone high mobility group 14 and 17 proteins. *J. Biol. Chem.*, **258**, 15371.

96. Wong, N. C., Poirier, G. G., and Dixon, G. H. (1977). Adenosine diphosphoribosylation of certain basic chromosomal proteins in isolated trout testis nuclei. *Eur. J. Biochem.*, **77**, 11.

97. Slattery, E., Dignam, J. D., Matsui, T., and Roeder, R. G. (1983). Purification and analysis of a factor which suppresses nick-induced transcription by RNA polymerase II and its identity with poly(ADP- ribose) polymerase. *J. Biol. Chem.*, **258**, 5955.

98. Kurl, R. N. and Jacob, S. T. (1985). Characterization of a factor that can prevent random transcription of cloned rDNA and its probable relationship to poly(ADP-ribose) polymerase. *Nucleic Acids Res.*, **13**, 89.

99. Oei, S. L., Griesenbeck, J., Schweiger, M., Babich, V., Kropotov, A., and Tomilin, N. (1997). Interaction of the transcription factor YY1 with human poly(ADP-ribosyl) transferase. *Biochem. Biophys. Res. Commun.*, **240**, 108.

100. Meisterernst, M., Stelzer, G., and Roeder, R. G. (1997). Poly(ADP-ribose) polymerase enhances activator-dependent transcription *in vitro*. *Proc. Natl Acad. Sci. USA*, **94**, 2261.

101. Kannan, P., Yu, Y., Wankhade, S., and Tainsky, M. A. (1999). PolyADP-ribose polymerase is a coactivator for AP-2-mediated transcriptional activation. *Nucleic Acids Res.*, **27**, 866.

102. Dear, T. N., Hainzl, T., Follo, M., Nehls, M., Wilmore, H., Matena, K., and Boehm, T. (1997). Identification of interaction partners for the basic-helix–loop–helix protein E47. *Oncogene*, **14**, 891.

103. Miyamoto, T., Kakizawa, T., and Hashizume, K. (1999). Inhibition of nuclear receptor signalling by poly(ADP-ribose) polymerase. *Mol. Cell. Biol.*, **19**, 2644.

104. Butler, A. J. and Ordahl, C. P. (1999). Poly(ADP-ribose) polymerase binds with transcription enhancer factor 1 to MCAT1 elements to regulate muscle-specific transcription. *Mol. Cell. Biol.*, **19**, 296.

105. Plaza, S., Aumercier, M., Bailly, M., Dozier, C., and Saule, S. (1999). Involvement of poly (ADP-ribose)-polymerase in the Pax-6 gene regulation in neuroretina. *Oncogene*, **18**, 1041.

106. Hassa, P. O. and Hottiger, M. O. (1999). A role of poly (ADP-ribose) polymerase in NF-kappaB transcriptional activation. *Biol. Chem.*, **380**, 953.

107. Oei, S. L., Griesenbeck, J., Ziegler, M., and Schweiger, M. (1998). A novel function of poly(ADP-ribosyl)ation: silencing of RNA polymerase II-dependent transcription. *Biochemistry*, **37**, 1465.

108. Oei, S. L., Griesenbeck, J., Schweiger, M., and Ziegler, M. (1998). Regulation of RNA polymerase II-dependent transcription by poly(ADP-ribosyl)ation of transcription factors. *J. Biol. Chem.*, **273**, 31644.

109. Rawling, J. M. and Alvarez-Gonzalez, R. (1997). TFIIF, a basal eukaryotic transcription factor, is a substrate for poly(ADP-ribosyl)ation. *Biochem. J.*, **324**, 249.

110. de Murcia, G., Jongstra-Bilen, J., Ittel, M., Mandel, P., and Delain, E. (1983). Poly(ADP-ribose)polymerase automodification and interaction with DNA: electronic microscopic visualization. *EMBO J.*, **2**, 543.

111. Ferro, A. M., Higgins, N. P., and Olivera, B. M. (1983). Poly(ADP-ribosyl)ation of DNA topoisomerase I. *J. Biol. Chem.*, **258**, 6000.

112. Jongstra-Bilen, J., Ittel, M. E., Niedergang, C., Vosberg, H. P., and Mandel, P. (1983). DNA topoisomerase I from calf thymus is inhibited *in vitro* by poly(ADP-ribosyl)ation. *Eur. J. Biochem.*, **136**, 391.

113. Coll, J. M., Hickey, R. J., Cronkey, E. A., Jiang, H. Y., Schnaper, L., Lee, M. Y., Uitto, L., Syvaoja, J. E., and Malkas, L. H. (1997). Mapping specific protein–protein interactions within the core component of the breast cell DNA synthesome. *Oncol. Res.*, **9**, 629.

114. Darby, M. K., Schmitt, B., Jongstra-Bilen, J., and Vosberg, H. P. (1985). Inhinition of calf thymus type II DNA topoisomerase by poly(ADP-ribosyl)ation. *EMBO J.*, **4**, 2129.

115. Yoshihara, K., Itaya, A., Tanaka, Y., Ohashi, Y., Ito, K., Teraoka, H., Tsukada, H., Matzukage, A., and Kamya, T. (1985). Poly(ADP-ribosyl)ation of nuclear enzymes. In *ADP-ribosylation of proteins* (ed. A. H. Althaus, H. Hilz, and S. Shall), p. 82. Springer Verlag, Heidelberg.

116. Eki, T. and Hurwitz, J. (1991). Influence of poly(ADP-ribose) polymerase on the enzymatic synthesis of SV40 DNA. *J. Biol. Chem.*, **266**, 3087.

117. Scovassi, A. I., Mariani, C., Negroni, M., Negri, C., and Bertazzoni, U. (1993). ADP-ribosylation of non histone proteins in HeLa cells: modification of DNA topoisomerase II. *Exp. Cell Res.* **206**, 177.

118. Dantzer, F., Nasheuer, H. P., Vonesch, J.-L., de Murcia, G., and Ménissier-de Murcia, J. (1998). Functional association of poly(ADP-ribose)polymerase with DNA polymerse alpha-primase complex: a link between DNA strand break detection and DNA replication. *Nucleic Acids Res.*, **26**, 1891.

119. Simbulan, C. M. G., Suzuki, M., Izuta, S., Sakurai, T., Savoysky, E., Kojima,

K., Miyahara, K., Shizuta, Y., and Yoshida, S. (1993). Poly(ADP-ribose) polymerase stimulates DNA polymerase alpha by physical association. *J. Biol. Chem.*, **268**, 93.

120. Nobori, T., Yamanaka, H., and Carson, D. A. (1989). Poly(ADP-ribose) polymerase inhibits DNA synthesis initiation in the absence of NAD. *Biochem. Biophys. Res. Commun.*, **163**, 1113.

121. Eki, T. (1994). Poly (ADP-ribose) polymerase inhibits DNA replication by human replicative DNA polymerase alpha, delta and epsilon *in vitro*. *FEBS Lett.*, **356**, 261.

122. Eki, T., Matsumoto, T., Murakami, Y., and Hurwitz, J. (1992). The replication of DNA containing the simian virus 40 origin by the monopolymerase and dipolymerase systems. *J. Biol. Chem.*, **267**, 7284.

123. Simbulan-Rosenthal, C. M., Rosenthal, D. S., Luo, R., and Smulson, M. E. (1999). Poly(ADP-ribose) polymerase upregulates E2F-1 promoter activity and DNA pol alpha expression during early S phase. *Oncogene*, **18**, 5015.

124. Stillman, B. (1994). Smart machines at the DNA replication fork. *Cell*, **78**, 725.

125. Thompson, L. H., Brookman, K. W., Jones, N. J., Allen, S. A., and Carrano, A. V. (1990). Molecular cloning of the human XRCC1 gene, which corrects defective DNA strand break repair and sister chromatid exchange. *Mol. Cell Biol.*, **10**, 6160.

126. Zdzienicka, M. Z., van der Schans, G. P., Natarajan, A. T., Thompson, L. H., Neuteboom, I., and Simons, J. W. I. M. (1992). A Chinese hamster ovary cell mutant (EM-C11) with sensitivity to simple alkylating agents and a very high level of sister chromatid exchanges. *Mutagenesis*, **7**, 265.

127. Caldecott, K. W., McKeown, C. K., Tucker, J. D., Ljungquist, S., and Thompson, L. H. (1994). An interaction between the mammalian DNA repair protein XRCC1 and DNA ligase III. *Mol. Cell Biol.*, **14**, 68.

128. Kubota, Y., Nash, R. A., Klungland, A., Barnes, D., and Lindahl, T. (1996). Reconstitution of DNA base excision repair with purified human proteins: interaction between DNA polymerase beta and the XRCC1 protein. *EMBO J.*, **23**, 6662.

129. Nash, R. A., Caldecott, K. W., Barnes, D. E., and Lindahl, T. (1997). XRCC1 protein interacts with one of two distinct forms of DNA ligase III. *Biochemistry*, **36**, 5207.

130. Griesenbeck, J., Oei, S. L., Mayer-Kuckuk, P., Ziegler, M., Buchlow, G., and Schweiger, M. (1997). Protein–protein interaction of the human poly(ADP-ribose)polymerase depends on the functional state of the enzyme. *Biochemistry*, **36**, 7297.

131. Marintchev, A., Mullen, M. A., Maciejewski, M. W., Pan, B., Gryk, M. R., and Mullen, G. P. (1999). Solution structure of the single-strand break repair protein XRCC1 N-terminal domain. *Nature Struct. Biol.*, **6**, 884.

132. Wei, Y. F., Robins, P., Carter, K., Caldecott, K., Pappin, D. J. C., Yu, G. L., Wang, R. P., Shell, B. K., Nash, R. A., Schär, P., Barnes, D. E., Haseltine, W. A., and Lindahl, T. (1995). Molecular cloning and expression of human cDNAs encoding a novel DNA ligase IV and DNA ligase III, an enzyme active in DNA repair and genetic recombination. *Mol. Cell. Biol.*, **15**, 3206.

133. Chen, J., Tomkinson, A. E., Ramos, W., Mackey, Z. B., Danehower, S., Walter, C. A., Schultz, R. A., Besterman, J. M., and Husain, I. (1995). Mammalian

DNA ligase III: Molecular cloning, chromosomal localization, and expression in spermatocytes undergoing meiotic recombination. *Mol. Cell. Biol.*, **15**, 5412.

134. Mackey, Z. B., Ramos, W., Levin, D. S., Walter, C. A., Mc Carrey, J. R., and Tomkinson, A. E. (1997). An alternative splicing event which occurs in mouse pachytene spermatocytes generates a form of DNA ligase III with distinct biochemical properties that may function in meiotic recombination. *Mol. Cell. Biol.*, **17**, 989.

135. Mackey, Z., Niedergang, C., Ménissier-de Murcia, J., Leppard, J., Au, K., Chen, J., de Murcia, G., and Tomkinson, A. (1999). DNA ligase III is recruited to DNA strand breaks by a zinc finger motif homologous to that of poly (ADP-ribose) polymerase. *J. Biol. Chem.*, **274**, 21679.

136. Sobol, R. W., Horton, J. K., Kuhn, R., Gu, H., Singhal, R. K., Prasad, R., Rajewsky, K., and Wilson, S. H. (1996). Requirement of mammalian DNA polymerase beta in base excision repair. *Nature*, **379**, 183.

137. Klungland, A. and Lindahl, T. (1997). Second pathway for completion of human DNA base excision-repair: reconstitution with purified proteins and requirement for DNase IV (FEN1). *EMBO J.*, **16**, 3341.

138. Dianov, G. L., Prasad, R., Wilson, S. H., and Bohr, V. A. (1999). Role of DNA polymerase beta in the excision step of long patch mammalian base excision repair. *J. Biol. Chem.*, **274**, 13741.

139. Caldecott, K., Aoufouchi, S., Johnson, P., and Shall, S. (1996). XRCC1 polypeptide interacts with DNA polymerase β and possibly poly(ADP-ribose)polymerase, and DNA ligase III is a novel molecular 'nick sensor' *in vitro. Nucleic Acids. Res.*, **24**, 4387.

140. Dautzer, F., de la Rubia, G., Ménissier-de Murcia, J., Hostomsky, Z., de Murcia, G. and Schreiber, V. (2000). Base excision repair is impaired in mammalian cells lacking poly (ADP-ribose) polymerase-1. *Biochemistry* **39**, 7559.

141. Wilson III, D. M. and Thompson, L. H. (1997). Life without DNA repair. *Proc. Natl Acad. Sci. USA*, **94**, 12754.

142. Cappelli, E., Taylor, R., Cevasco, M., Abbondandolo, A., Caldecott, K., and Frosina, G. (1997). Involvement of XRCC1 and DNA ligase III gene products in DNA base excision repair. *J. Biol. Chem.*, **272**, 23970.

143. Frosina, G., Fortini, P., Rossi, O., Carrozzino, F., Raspaglio, G., Cox, L. S., Lane, D. P., Abbondandolo, A., and Dogliotti, E. (1996). Two pathways for base excision repair in mammalian cells. *J. Biol. Chem.*, **271**, 9573.

144. Dantzer, F., Schreiber, V., Niedergang, C., Trucco, C., Flatter, E., De La Rubia, G., Oliver, J., Rolli, V., Menissier-de Murcia, J., and de Murcia, G. (1999). Involvement of poly(ADP-ribose) polymerase in base excision repair. *Biochimie*, **81**, 69.

145. Matsumoto, Y., Kim, K., and Bogenhagen, D. F. (1994). Proliferating cell nuclear antigen-dependent abasic site repair in *Xenopus laevis* oocytes: an alternative pathway of base excision DNA repair. *Mol. Cell. Biol.*, **14**, 6187.

146. Satoh, M. S. and Lindahl, T. (1992). Role of poly(ADP-ribose) formation in DNA repair. *Nature*, **356**, 356.

147. Waldman, A. S. and Waldman, B. C. (1991). Stimulation of intrachromosomal homologous recombination in mammalian cells by an inhibitor of poly (ADP-ribosyl)ation). *Nucleic Acids Res.*, **19**, 5943.

148. Semionov, A., Cournoyer, D., and Chow, T. Y. (1999). Inhibition of poly(ADP-

ribose)polymerase stimulates extrachromosomal homologous recombination in mouse Ltk-fibroblasts. *Nucleic Acids Res.*, **27**, 4526.

149. Borggrefe, T., Wabl, M., Akhmedov, A. T., and Jessberger, R. (1998). A B-cell-specific DNA recombination complex. *J. Biol. Chem.*, **273**, 17025.

150. de Murcia, G., Huletsky, A., Lamarre, D., Gaudreau, A., Pouyet, J., Daune, M., and Poirier, G. G. (1986). Modulation of chromatin superstructure induced by poly(ADP-ribose) synthesis and degradation. *J. Biol. Chem.*, **261**, 7011.

151. Ambs, S., Ogunfusika, M. O., Merriam, W. G., Bennett, W. P., Billiar, T. R., and Harris, C. C. (1998). Up-regulation of inducible nitric oxide synthase expression in cancer-prone p53 knockout mice. *Proc. Natl Acad. Sci. USA*, **95**, 8823.

152. Ambs, S., Hussain, S. P., and Harris, C. C. (1997). Interactive effects of nitric oxide and the p53 tumor suppressor gene in carcinogenesis and tumor progression. *FASEB J.*, **11**, 443.

153. Li, N. and Karin, M. (1999). Is NF-kappaB the sensor of oxidative stress? *FASEB J.*, **13**, 1137.

154. Barnes, P. J. and Karin, M. (1997). Nuclear factor-kappaB: a pivotal transcription factor in chronic inflammatory diseases. *N. Engl. J. Med.*, **336**, 1066.

155. Amé, J.-C., Rolli, V., Schreiber, V., Niedergang, C., Apiou, F., Decker, P., Muller, S., Höger, T., Ménissier-de Murcia, J., and de Murcia, G. (1999). PARP-2, a novel mammalian DNA-damage dependent poly(ADP-ribose) polymerase. *J. Biol. Chem.*, **274**, 17860.

156. d'Adda di Fagagna, F., Hande, M. P., Tong, W. M., Lansdorp, P. M., Wang, Z. Q., and Jackson, S. P. (1999). Functions of poly(ADP-ribose) polymerase in controlling telomere length and chromosomal stability. *Nature Genet.*, **23**, 76

157. Agarwal, M. L., Agarwal, A., Taylor, W. R., Wang, Z. Q., Wagner, E. F., and Stark, G. R. (1997). Defective induction but normal activation and function of p53 in mouse cells lacking poly-ADP-ribose-polymerase. *Oncogene*, **15**, 1035.

158. Mandir, A. S., Przedborski, S., Jackson-Lewis, V., Wang, Z. Q., Simbulan-Rosenthal, C. M., Smulson, M. E., Hoffman, B. E., Guastella, D. B., Dawson, V. L., and Dawson, T. M. (1999). Poly(ADP-ribose) polymerase activation mediates 1-methyl-4-phenyl-1, 2, 3, 6-tetrahydropyridine (MPTP)-induced parkinsonism. *Proc. Natl Acad. Sci. USA*, **96**, 5774.

159. Szabo, C., Virag, L., Cuzzocrea, S., Scott, G. S., Hake, P., O'Connor, M. P., Zingarelli, B., Salzman, A., and Kun, E. (1998). Protection against peroxynitrite-induced fibroblast injury and arthritis development by inhibition of poly(ADP-ribose) synthase. *Proc. Natl Acad. Sci. USA*, **95**, 3867

160. Burkart, V., Wang, Z. Q., Radons, J., Heller, B., Herceg, Z., Stingl, L., Wagner, E. F., and Kolb, H. (1999). Mice lacking the poly(ADP-ribose) polymerase gene are resistant to pancreatic beta-cell destruction and diabetes development induced by streptozocin. *Nature Med.*, **5**, 314

161. Kuhnle, S., Nicotera, P., Wendel, A., and Leist, M. (1999). Prevention of endotoxin-induced lethality, but not of liver apoptosis in poly(ADP-ribose) polymerase-deficient mice. *Biochem. Biophys. Res. Commun.*, **263**, 433.

162. Endres, M., Wang, Z. Q., Namura, S., Waeber, C., and Moskowitz, M. A. (1997). Ischemic brain injury is mediated by the activation of poly(ADP-ribose)polymerase. *J. Cereb Blood Flow Metab.*, **17**, 1143

163. Eliasson, M. J., Sampei, K., Mandir, A. S., Hurn, P. D., Traystman, R. J., Bao, J., Pieper, A., Wang, Z. Q., Dawson, T. M., Snyder, S. H., and Dawson, V. L.

(1997). Poly(ADP-ribose) polymerase gene disruption renders mice resistant to cerebral ischemia. *Nature Med.*, **3**, 1089

164. Oliver, F. J., Menissier-de Murcia, J., Nacci, C., Decker, P., Andriantsitohaina, R., Muller, S., de la Rubia, G., Stoclet, J. C., and de Murcia, G. (1999). Resistance to endotoxic shock as a consequence of defective NF-kappaB activation in poly (ADP-ribose) polymerase-1 deficient mice. *EMBO J.*, **18**, 4446

165. Ferro, A. M., McElwain, M. C., and Olivera, B. M. (1984). Poly (ADP-ribosyl)ation) of DNA topoisomerase I: a nuclear response to DNA-strand interruptions. *Cold Spring Harb. Symp. Quant. Biol.*, **49**, 683

166. Ruscetti, T., Lehnert, B. E., Halbrook, J., Le Trong, H., Hoekstra, M. F., Chen, D. J., and Peterson, S. R. (1998). Stimulation of the DNA-dependent protein kinase by poly(ADP-ribose) polymerase. *J. Biol. Chem.*, **273**, 14461

167. Galande, S. and Kohwi-Shigematsu, T. (1999). Poly(ADP-ribose) polymerase and Ku autoantigen form a complex and synergistically bind to matrix attachment sequences. *J. Biol. Chem.*, **274**, 20521

168. Ariumi, Y., Masutani, M., Copeland, T. D., Mimori, T., Sugimura, T., Shimotohno, K., Ueda, K., Hatanaka, M., and Noda, M. (1999). Suppression of the poly(ADP-ribose) polymerase activity by DNA-dependent protein kinase *in vitro*. *Oncogene*, **18**, 4616

169. Vaziri, H., West, M. D., Allsopp, R. C., Davison, T. S., Wu, Y. S., Arrowsmith, C. H., Poirier, G. G., and Benchimol, S. (1997). ATM-dependent telomere loss in aging human diploid fibroblasts and DNA damage lead to the post-translational activation of p53 protein involving poly(ADP-ribose) polymerase. *EMBO J.*, **16**, 6018

170. Wesierska-Gadek, J., Wang, Z. Q., and Schmid, G. (1999). Reduced stability of regularly spliced but not alternatively spliced p53 protein in PARP-deficient mouse fibroblasts. *Cancer Res.*, **59**, 28

171. Wesierska-Gadek, J., Schmid, G., and Cerni, C. (1996). ADP-ribosylation of wild-type p53 *in vitro*: binding to specific p53 consensus sequence prevents its modification. *Biochem. Biophys. Res. Commun.*, **224**, 96.

4

Role of poly (ADP-ribose) polymerase in cell death

El Bachir Affar, Marc Germain, and Guy G. Poirier

4.1 Introduction

Cells are continuously exposed to a variety of potentially genotoxic endogenous and exogenous DNA damaging agents. To deal with such an insult, cells are provided with a complex and hierarchical network of proteins responsible for the detection, signalling, and repair of DNA strand breaks. Among these proteins, a chromatin associated protein, poly (ADP-ribose) polymerase (PARP) appears to be involved in the maintenance of genomic integrity. This enzyme is stimulated by DNA strand breaks to catalyse the synthesis of poly (ADP-ribose) (pADPr), which interacts with many nuclear proteins to possibly regulate their functions (see Chapter 1). The pADPr was recently found to be synthesized by other enzymes with homologies to the catalytic domain of PARP (see Chapter 2). The catabolism of the pADPr is achieved by the poly (ADP-ribose) glycohydrolase (PARG), a 111-kDa cytoplasmic enzyme (1), see Chapter 1).

PARP has been involved in many cellular functions related to DNA metabolism such as DNA repair, transcription, and replication (2). The rapid activation of PARP following DNA damage has led to the belief that this enzyme could be involved in the balance between survival and cell death. Thus, many authors have explored the potential involvement of PARP in these processes following moderate or excessive exposure to DNA damaging agents. Some of these studies have suggested that PARP acts as a survival-promoting molecule, permitting the recovery of the cell from DNA damage. This enzyme has also been involved as a mediator of cell death, allowing the elimination of cells harbouring large amounts of DNA damage. In addition, several studies have implicated PARP in the initiation and/or the execution of apoptosis, whereas others have found no effect. The cleavage of PARP by the apoptotic proteases, the caspases, has also been suggested to be an important event during apoptosis.

The aim of this review is to summarize the current knowledge on the role of PARP in cell survival and death. Based on the results obtained using different approaches, we propose a model in which PARP would be, depending on the amount of DNA damage, either a survival-promoting factor or a death-inducing protein.

4.2 PARP as a survival factor for the cellular recovery from DNA damage

4.2.1 Effects of PARP inhibition

The dependence of PARP activity on DNA breaks has led to a considerable amount of work towards understanding the role of PARP in DNA repair. Many investigations have correlated the effects of competitive inhibitors of PARP with both DNA repair and cell recovery following DNA damage (Table 4.1). In these studies, PARP inhibitors alone did not significantly alter cell survival, but greatly potentiated cell death after treatment with various DNA damaging agents (e.g. alkylation, oxidation, irradiation). Similar results were obtained using the 46-kDa DNA binding domain of PARP (DBD) as a *trans*-dominant inhibitor of the full-length enzyme (3, 4) (Table 4.1). The inhibition or delay of DNA repair, the occurrence of genomic alterations, and the potentiation of cell death by these inhibitors have led several authors to postulate that PARP plays an important role in the recovery from DNA damage. However, these effects may not directly reflect the involvement of PARP in these processes. In fact, both chemical inhibitors and DBD may provoke a continuous fixation of PARP to the DNA breaks, which could interfere with DNA repair and other genomic processes. Also, PARP inhibitors may provoke some metabolic side-effects (5–7). In addition, the other ADP-ribose polymerizing activities (8–10) which are inhibitable by PARP inhibitors, could be involved in processes that were claimed to be mediated by PARP itself.

These limitations have been overcome by the use of antisense PARP RNA and knockout mice (Table 4.1). Cells expressing the antisense RNA were shown to exhibit decreased DNA repair activity and a reduction in cell survival following exposure to the DNA damaging agents methyl methanesulfonate (MMS) (11, 12) and nitrogen mustard (HN_2) (13). In concordance with these results, PARP-deficient cells exhibit an increased mortality following DNA damage (14–16) even though the effects on DNA repair are still unclear. At the level of the whole animal, $PARP^{-/-}$ phenotypical mice are highly sensitive to γ-irradiation (14, 17) and to *N*-methyl-*N*-nitrosourea (MNU) (14). These findings support the notion that PARP is an important molecule allowing the recovery and survival of cells following DNA damage.

Table 4.1 Effect of PARP inhibition on cell survival

Cell type	DNA damaging agent[a]	PARP inhibition approach[b]	Effect of the inhibition	Ref.
Chinese hamster ovary	Bleomycin	3-AB (5 mM) 3-MB (2.5 mM)	Decreased survival and DNA repair	111
Chinese hamster ovary	MNNG EMS	3-AB (0–10 mM)	Decreased survival, increased mutations and chromosomal aberrations	112
V79-B310H	X-ray	3-AB (0–20 mM) NA (0–20 mM)	Decreased survival	113
C3H10T$^1/_2$	MNNG	3-MB (1 mM)	Decreased survival, decreased DNA synthesis	114
C3H10T$^1/_2$	X/XO	DHQ (1 mM)	Decreased survival	115
Human fibroblasts	DMS γ-radiation	3-AB (5 mM)	Decreased survival, delayed rejoining of strand breaks No effect on survival, slight effect on strand breaks	116
Human fibroblasts	MMS	3-AB (4 mM)	Decreased survival, decreased DNA repair	117
BALB/3T3	MNNG	3-AB (3 mM)	Decreased survival, increased transformation frequency	118
L1210	DMS	3-AB (2 mM)	Decreased survival, decreased DNA repair	102
HeLa S3	MNNG	Overproduction of DBD	Decreased survival, increased SCE[c]	4
CO60	MNNG γ-radiation	Overproduction of DBD	Decreased survival	3
HeLa S3	HN$_2$	Antisense RNA	Decreased survival, decreased gene-specific repair	13
HeLa S3	MMS	Antisense RNA	Decreased survival, delayed DNA strand break rejoining	11, 12
ES embryonic cells	MMS	Knockout (Sugimura)	Decreased survival	16
Bone-marrow cells	MNU	Knockout (de Murcia)	Decreased survival, increased SCE	15

Table 4.1 Effect of PARP inhibition on cell survival—*continued*

Cell type	DNA damaging agent[a]	PARP inhibition approach[b]	Effect of the inhibition	Ref.
Fibroblast	MMS			
Mice	MNU γ-radiation	Knockout (de Murcia)	Decreased survival	14
Splenocytes	MNU			
Mice	γ-radiation	Knockout (Wang)	Decreased survival	17
Embryonic fibroblasts	Adriamicin γ-radiation	Knockout (Wang)	No difference	21

[a]DMS, dimethyl sulfate; EMS, ethyl methanesulfonate; HN$_2$, nitrogen mustard; MMS, methyl methanesulfonate; MNNG, 1-methyl-3-nitro-1-nitrosoguanidine; MNU, *N*-methyl-*N*-nitrosourea; X/XO, xanthine/xanthine oxidase.
[b]3-AB, 3-aminobenzamide; 3-MB, 3-methoxybenzamide; DBD, DNA binding domain; DHQ, 1,5-(dihydroxy)isoquinoline; NA, nicotinamide.
[c]SCE, sister-chromatid exchange.

4.2.2 Possible molecular basis

The survival function of PARP may be mediated by its direct interaction with other proteins involved in DNA repair. Alternatively, the pADPr could covalently and/or non-covalently bind to these proteins to regulate their functions. PARP, in conjunction with other molecules, could promote the maintenance of genomic integrity by signalling DNA damage. Agarwal *et al.* and Szumiel have proposed a model in which PARP, as a DNA break-sensing molecule, plays a role along with DNA-PK, the ataxia–telangiectasia mutated protein (ATM), and p53 in a DNA damage-signalling network (18, 19). Some evidence suggests a link between PARP and the p53 pathway of cell-cycle arrest and apoptosis. Whitacre *et al.* have shown that V79-derived cell lines defective in poly (ADP-ribosylation) have lower basal p53 levels than the parental V79 cell line and that they also fail to induce p53 in response to etoposide (20). In agreement with this study, Agarwal *et al.* have reported a defective induction, but a normal activation of p53 in PARP$^{-/-}$ fibroblasts treated with DNA damaging agents (21). Furthermore, normally spliced p53 protein, but not its alternatively spliced form, was shown to be unstable in PARP-deficient cells (22). However, a prolonged p53 accumulation following DNA damage was observed in cells where PARP was depleted using antisense RNA expression (23). Using splenocytes from another PARP knockout mouse, Ménissier-de Murcia *et al.* have found that following DNA damage, the induction of p53 and apoptosis was faster in the PARP$^{-/-}$ cells compared to wild-type cells. The authors suggest that a defect in DNA repair would prompt these cells to an apoptotic death (14). At the molecular level, co-immunoprecipitation studies have indicated a physical association between PARP and p53 (24–26). p53 was also shown to be a pADPr acceptor protein (24, 25, 27) and to possess three conserved pADPr binding motifs that could bind free pADPr (27). Still, the exact nature and consequences of these interactions on the cellular response to DNA damage remain to be elucidated. However, since PARP-deficient mice do not develop spontaneous tumours, it seems that PARP is not absolutely required for p53 functions.

On the other hand, recent studies have reported that PARP co-purifies with DNA-PK and Ku proteins, indicating an interaction of these proteins (28, 29). *In vitro* experiments have shown that PARP poly (ADP-ribosylates) the catalytic subunit of DNA-PK to increase its activity (28). In addition, the double-mutant SCID-PARP$^{-/-}$ mice develop a high frequency of T-cell lymphomas, unlike mice defective in only one of the two enzymes (30). This suggests a cooperation between PARP and DNA-PK in the maintenance of genomic integrity. PARP could also promote the survival of the cells in a p53-independent manner following DNA damage. For example, it has been shown that PARP binds specifically to the repair protein XRCC1 (31). Finally, the synthesis of pADPr might be an emergency signal that modulates the function of proteins involved in the maintenance of DNA integrity.

4.3 Role of PARP in apoptosis

4.3.1 General considerations on apoptosis

Apoptosis is a conserved mechanism of cell death controlling the development and homeostasis of multicellular organisms (32). This programmed cell death is involved in organogenesis, maturation of the immune system, and elimination of potentially dangerous cells (33). Defects in cell death by apoptosis are associated with cancer, while its inappropriate occurrence is associated with neurodegenerative diseases (32).

Apoptosis can be induced by DNA damage, changes in growth conditions, chemical agents such as inhibitors of protein or nucleic acid synthesis, or by activation of death receptors (32, 34, 35). During the apoptotic process, all components of the cell are carefully dismantled: a set of proteins implicated in cellular homeostasis are specifically cleaved by a family of cysteine proteases named caspases, DNA becomes condensed and is degraded into internucleosomal-size fragments (DNA ladder), cells are then disassembled into small vesicles and marked for engulfment by other nearby cells (32, 36). In this way, the death of one cell does not damage surrounding tissues. In contrast, necrosis is characterized by an early disruption of membrane integrity and leakage of the cellular contents into the extracellular medium (36, 37). Thus, certain pathologies linked to necrosis, like chronic inflammation, could be avoided by apoptosis (32, 36, 37). Also, some apoptotic inducers, such as H_2O_2 (38), peroxynitrite (39), heat shock (38), N-methyl-D-aspartate (NMDA) (40), and glutamate (41), can induce necrosis at high doses. On the other hand, necrosis can be prevented by antiapoptotic members of the BCL-2 family of proteins (42). This indicates that these two modes of cell death can be linked and depend on the intensity of the stimulus.

Apoptosis is an active mode of cell death requiring energy to proceed, as cells that are triggered to undergo apoptosis but lack sufficient energy die by necrosis (36, 43, 44). In particular, the activation of caspase-9 requires the presence of its cofactors APAF-1, cytochrome c, and ATP or dATP (45). Other energy-requiring steps of the apoptotic process are those of chromatin condensation and the formation of the apoptotic bodies, which are blocked in the absence of ATP (46).

At the molecular level, the caspase family of cysteine proteases, highly related to the interleukin-1β-converting enzyme (ICE, caspase-1) and the proapoptotic CED-3 gene of *Caenorhabditis elegans*, are important mediators of the apoptotic process (47–49). Caspases are implicated both in the induction and in the execution of cell death. All caspases cleave after an aspartic acid, for which they have an absolute specificity. Caspases exist in the cytoplasm as inactive proenzymes which are processed to a large and a small subunit to form the active tetrameric enzyme. The processing sites within procaspases match the recognition sequences of caspases, an observation that has led to the suggestion that caspases are activated in a proteolytic cascade, the activation of initiator caspases resulting in the processing and activation of execution caspases.

The caspase cascade, shown in Fig. 4.1, illustrates the different pathways

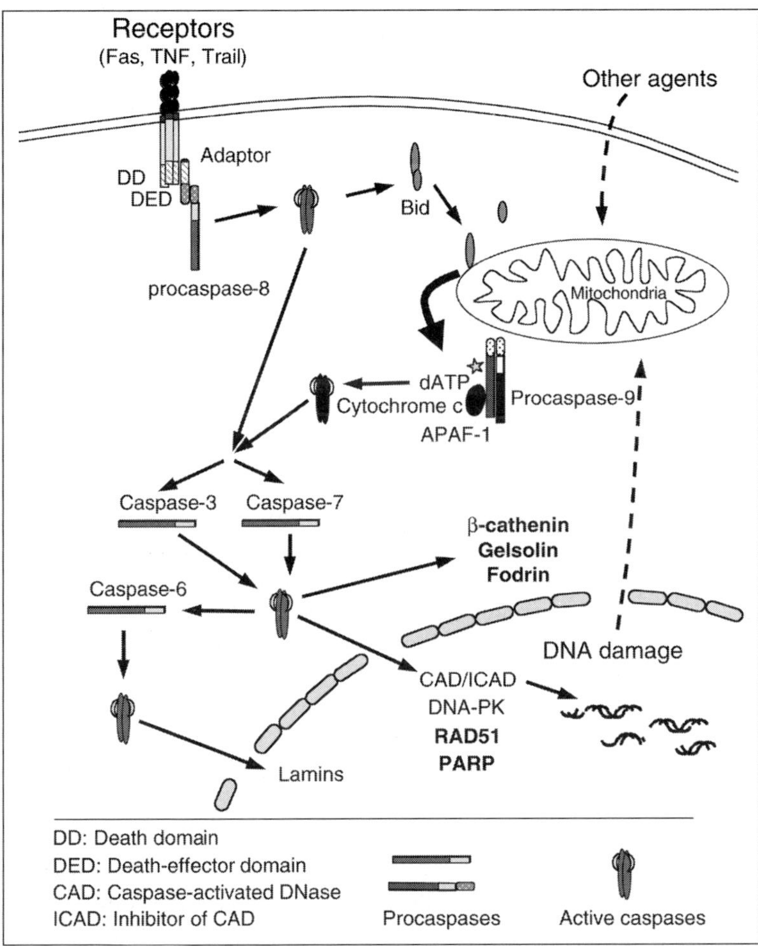

Fig. 4.1 Molecular pathways of apoptosis. Apoptosis is characterized by the activation of cysteine proteases (caspases) which trigger the cell-death process. Caspases' proteolytic cascade can be induced upon the fixation of specific ligands to their death receptors (Fas, TNF, Trail) which recruit an adaptor protein and procaspase-8. This results in the autocatalytic activation of this protease, which in turn processes downstream proteases, procaspase-3, -6, and -7 resulting in their activation. On the other hand, DNA damage and other apoptotic inducers act via still unknown pathways to stimulate the release of cytochrome c from the mitochondria. In the presence of cytochrome c, dATP (or ATP), and APAF-1 the procaspase-9 is activated by autocatalysis resulting in the processing and activation of caspase-3, -6, and -7. Death-receptor pathways could also activate effector caspases via the caspase-8 cleavage of Bid. Once cleaved, this factor induces the release of cytochrome c and caspase-9 activation.

which trigger apoptosis. It can be induced directly by activation of cell-death receptors, like Fas, which leads to procaspase-8 processing, while DNA damaging agents and other chemotherapeutic drugs act via still unknown pathways to induce cytochrome c release from mitochondria and the activation of caspase-9. Both pathways result in the activation of the effector proteases, caspase-3, -6, and –7, responsible for the degradation of most of the proteins cleaved during apoptosis. To date, about 60 proteins, which can be divided into eight categories, including structural proteins and enzymes implicated in the maintenance of cellular homeostasis (Table 4.2), are known to be cleaved by caspases (see ref. 50 for a detailed list).

4.3.2   PARP activation and apoptosis

A cellular drop in NAD content, which is not observed in the presence of PARP inhibitors, has been reported to occur during apoptosis (38, 51–53), thus suggesting the activation of PARP. More direct evidence for PARP activation during apoptosis has come from studies measuring pADPr levels in permeabilized cells (54–56). Yoon *et al.* found that poly (ADP-ribosyl)ation of histone H1 correlated with the internucleosomal degradation of DNA, which suggests a role for PARP in permitting the access of the apoptotic DNase to DNA (56). By measuring NAD levels and NAD incorporation into permeabilized cells, some authors have found a correlation between PARP activation and the occurrence of the DNA ladder (38, 51, 54, 56, 57). Using an antibody against pADPr, Smulson's group also showed that PARP activation is an early event during apoptosis of osteosarcoma cells grown to confluence (58), fibroblasts treated with anti-Fas, and HL-60 cells treated with camptothecin (59). Very recently, our laboratory has shown, using a new immunological technique to measure endogenous levels of pADPr, that PARP is activated concomitantly with, but not before, the appearance of the DNA ladder in HL60 cells treated with VP-16 (60). Thus, most evidence suggests that PARP activation during apoptosis is the consequence of the internucleosomal DNA degradation, although in some cases (e.g. the use of DNA damaging agents) apoptosis could be preceded by an earlier activation of the enzyme, linked to the primary response to DNA damage (38).

Chemical PARP inhibitors have been widely used in order to understand the role of PARP activation during apoptosis (Table 4.3). Some of these studies have shown that the presence of inhibitors can prevent cell death, as well as some (61) or all of the morphological changes associated with apoptosis (38, 55). However, some studies have shown the opposite effect (62, 63), while others found no influence of PARP inhibitors on the apoptotic process (51, 64). These results are difficult to reconcile but, as discussed before, some care should be taken in drawing conclusions from the use of chemical inhibitors, since they can cause side-effects and also inhibit other pADPr polymerizing activities.

Table 4.2 Caspase substrates

Substrate category	Examples[a]	Caspases involved	Effect of the cleavage
Cytoskeletal and structural proteins	Fodrin, gelsolin Lamins	Caspase-3-like Caspase-6	Morphological features of apoptosis and disassembly of the cytoskeleton
Transcription and translation	Sp1, SREBPs, U1–70 kDa sRNP	Caspase-3-like	Halt of processes implicated in the normal homeostasis of the cell and inhibition of repair mechanisms
Cell cycle and replication	Rb, p21, RFC140		Inhibition of antiapoptotic signals
DNA repair	PARP, DNA-PK, RAD51		Activation of proapoptotic signals
Signal transduction	PKCδ, MEKK-1, Ras GAP		
Apoptotic signaling	Procaspases, Bid	Upstream caspases	Activation of the apoptotic cascade
Cytokines	Pro-interleukin-1β	Caspase-1 subfamily	Inflammatory response
Others	Huntington, hsp90, presenilins	Effector caspases	Roles of cleavage not understood

[a] See reference 50 for an exhaustive list of caspases substrates.

Table 4.3 Effect of loss of PARP activity on apoptosis

Cell type	Apoptosis-inducing agent[a]	PARP inhibition approach[b]	Effect of the inhibition	Ref.
U937 HL60	TNF, UV	NAD depletion 3-AB (0–1 mM)	Resistance to apoptosis as determined by DNA fragmentation and percent of cell death	119
U937	H_2O_2, heat shock	3-AB (2.5 mM)	Prevention of the NAD depletion in apoptotic cells, inhibition of apoptosis as determined by nuclear fragmentation	38
HL60	actinomycin D	3-AB, BA (1 mM)	Suppression of apoptotic bodies formation, no difference for DNA ladder and PARP cleavage	61
HL60 COLO320	Camptothecin, VM-26	3-AB (0–10 mM)	Inhibition of DNA fragmentation	120
HL60	VP-16	3-AB (200 µM)	Partial prevention of NAD loss, but not of DNA laddering and PARP cleavage	51
HT-29	H_2O_2	3-AB, NA (1–40 mM)	Inhibition of necrosis, but not apoptosis as determined by cell count and flow cytometry	64
HL60	UV, adriamycin, mitomycin, cisplatin	3-AB (5 mM)	Prevents internucleosomal DNA degradation and morphological changes	56
HL60	VP-16	3-AB (5 mM)	Prevents NAD and ATP depletion, and morphological changes	55
Jurkat rat colon	NaDOC	3-AB (1 mM)	Enhances morphological features of apoptosis	62
3T3-L1 Fibroblasts	Anti-Fas + CHX Anti-Fas + CHX	Antisense RNA Knockout (Wang)	Inhibits DNA ladder, activation of caspase-3-like, and morphological features of apoptosis	59
Fibroblasts Thymocytes	TNF-α, anti-FAS γ-Ray, Dex	Knockout (Wang)	No difference as determined by MTT reduction and propidium iodide staining	17

Thymocytes	VP-16, Dex, ionomycin, ceramides	Knockout (Wang)	No difference at the end of the treatment	65
Hepatocytes	TNF, anti-Fas	Knockout (Wang)	No difference at the end of the treatment	
Neurons	Staurosporine, colchicine, MPP, peroxinitrite	Knockout (Wang)	No difference at the end of the treatment	
Thymocytes	Peroxynitrite, Dex, anti-Fas	Knockout (Wang), 3-AB (1 mM)	No difference for anti-Fas, Dex, and low doses of DNA damaging agents, switch from apoptosis to necrosis at high doses of DNA damaging agents in presence of PARP	39
Mouse splenocytes	MNU	Knockout (de Murcia)	Acceleration of apoptosis as seen by DNA laddering and p53 accumulation	14
Bone marrow, fibroblasts	MNU, MMS, anti-Fas	Knockout (de Murcia)	Acceleration of apoptosis for DNA damaging agents, no difference for anti-Fas	15

[a]CHX, cycloheximide; Dex, dexamethasone; MMS, methyl methanesulfonate; N-methyl-N-nitrosourea; MPP, N-methyl-4-phenylpiridine; NaDOC, sodium deoxycholate; NO, nitric oxide; SNAP, S-nitroso-N-acetyl-DL,-penicillamine; SNP, sodium nitroprusside; VP-16, etoposide[b] 3-AB, 3-aminobenzamide; BA, benzamide; NA, nicotinamide.

Two PARP knockout mice have been used to study the implication of PARP in apoptosis. de Murcia's group have found that cells from *PARP*[-/-] mice are more sensitive to alkylating agent-induced apoptosis (14, 15). The knockout mice developed by Wang *et al.* were shown to be as sensitive to apoptosis as their wild-type counterpart (17, 39, 65), even in cells treated with DNA damaging agents known to activate PARP (γ-radiation in thymocytes (65); peroxynitrite in neurons (65) and thymocytes (39). These results suggest that, although PARP has been implicated in the p53 network as a DNA nick sensor, it does not participate in the induction of apoptosis following DNA damage. Nevertheless, Smulson and colleagues recently showed that PARP[-/-] fibroblasts, or fibroblasts with depleted PARP due to antisense RNA expression, were completely resistant to apoptosis induced by anti-Fas and that these cells returned to normal sensitivity when PARP was reintroduced (59). They concluded that PARP activation, as an apoptotic signal, is required to activate effector caspases. However, the discrepancy between their results and those of other groups, who found no difference in Fas-mediated apoptosis between wild-type and mutant cells (15, 17, 39, 65), cannot be explained. The notion of PARP activation as an apoptotic signal is also difficult to reconcile with the current knowledge of the Fas-induction pathway (see Fig. 4.1). Upon activation, the Fas receptor is known to recruit procaspase-8 via an adapter protein, named FADD, which promotes the autocatalytic activation of caspase-8. This protease then directly activates effector caspases, resulting in the execution of the apoptotic programme (35, 66). It is thus difficult to place PARP activation, which requires DNA strand breaks, within this pathway. Indeed, PARP activation is likely to occur as a consequence of the internucleosomal degradation of the DNA, mediated by the caspase-activated DNase (CAD) (67, 68). In normal cells, CAD is complexed with its inhibitor ICAD. Upon induction of apoptosis, ICAD is cleaved by caspase-3 and CAD is activated. DNA degradation is therefore the consequence of the apoptotic process, which explains the absence of a role for PARP activation in inducing apoptosis. The existence of healthy and fertile PARP[-/-] mice has also discarded a critical role for PARP in programmed cell death during development (14, 69, 70). However, the acute sensitivity of PARP[-/-] mice to γ-irradiation (14, 17) and of their cells to alkylating agents (14–16) suggests that, as discussed before, PARP could be a survival factor rather than a death-promoting molecule. With this in mind, it is interesting to note that PARP is specifically cleaved and inactivated by caspases during apoptosis, abrogating its survival-promoting function.

4.3.3   PARP cleavage during apoptosis

PARP is one of the first substrates found to be cleaved by caspases during apoptosis (51, 71, 72). PARP cleavage occurs at the conserved sequence

DEVD-/G to generate a 89-kDa fragment containing the catalytic site and the automodification domain, and a 24-kDa fragment containing the zinc fingers responsible for DNA binding activity (51, 71). The 24-kDa fragment binds irreversibly to DNA breaks (73, 77) and can no longer be poly (ADP-ribosylated) (unpublished results). The 89-kDa fragment still retains a low basal activity but can no longer be stimulated by DNA strand breaks (51).

PARP cleavage has been shown in almost all forms of apoptosis, including apoptosis induced by irradiation, chemotherapeutic agents, and the activation of death receptors. This cleavage occurs early in the execution phase of apoptosis, concomitant with the internucleosomal degradation of DNA, but before the appearance of the apoptotic bodies and the externalization of phosphatidylserines (51). Caspases-3 and -7 are probably responsible for PARP cleavage *in vivo*, since they have the highest affinity for the DEVD cleavage site (74, 75). Indeed, we recently found that caspase-7 is even more efficient in PARP cleavage than caspase-3. Furthermore, only caspase-7 cleaves poly (ADP-ribosylated) PARP more efficiently than the non-activated substrate (60). Since PARP cleavage coincides with its activation by the internucleosomal DNA degradation, these results suggest that caspase-7 is the endogenous PARP-cleaving caspase.

4.3.4   Role of PARP cleavage during apoptosis

A recent study by Virag *et al.* showed that, while low doses of peroxynitrite caused apoptosis independent of PARP status, higher doses caused necrosis in the presence of PARP but caused apoptosis in knockout cells and in the presence of PARP inhibitors (39). Thus, although PARP does not influence the apoptotic death, its presence could result in a switch from apoptosis to necrosis in the presence of large amounts of DNA damage, possibly following the depletion of energy stocks (Section 4.4, see also Chapter 5). Since DNA is extensively degraded during apoptosis, the overactivation of PARP could lead to a switch towards necrosis. Inactivation of the full-length enzyme by caspases would prevent this overactivation and necrotic death. The apoptotic fragments of PARP could also inhibit PARP molecules that have not been cleaved. Studies using the complete 46-kDa DBD have shown that this domain can bind to DNA and prevent the activation of PARP (3, 4, 76, 77). In apoptotic cells, the 24-kDa fragment could thus bind tightly to strand breaks and prevent the activation of the uncleaved PARP molecules. This apoptotic fragment could also act as a *trans*-dominant inhibitor of DNA base excision repair, as was shown for the complete 46-kDa DBD (77). Indeed, recent work from our laboratory has shown that the 24-kDa apoptotic fragment of PARP can efficiently inhibit both PARP activity and DNA repair *in vitro* (unpublished results). This would prevent futile attempts by the cell to repair its DNA during apoptosis.

Generally speaking, PARP is believed to act as a survival factor by signalling DNA damage and promoting cell recovery (see Table 4.1). PARP would therefore belong to a set of proteins implicated in cellular homeostasis by signalling stressful situations and promoting subsequent cellular responses (18, 19). Since apoptosis results in the complete dismantling of the cell, it is likely that these proteins would stimulate a response to such an insult, unless they were rapidly inactivated. Indeed, it has been observed that the activation of caspases results in the proteolysis and inhibition of proteins linked to cellular homeostasis and repair (see Table 4.2). In particular, a set of protein kinases implicated in different survival responses are cleaved during apoptosis, resulting in their inactivation (78). In addition, the presence of large numbers of DNA breaks in apoptotic cells could result in desperate attempts to repair the damaged genome, which would subsequently delay apoptosis or cause a switch to necrosis. As a rapid apoptotic death is probably preferable for the organism, inactivation of enzymes implicated in the signalling of DNA damage is primordial. It is probably most efficient to target the early events of repair in order to rapidly shut off repair mechanisms. Proteins other than PARP are implicated in the response to DNA strand breaks. For example, p21 (79, 80), DNA-PK (81–83), and RAD51 (84, 85) are also cleaved and inactivated during apoptosis. The cleavage of p21 would lead to the inhibition of the G_1 checkpoint and subsequent DNA repair. Interestingly, the cleavage of p21 disrupts its interaction with PCNA, which may have a direct effect on DNA repair (80). The cleavage of DNA-PK also leads to the inhibition of its catalytic activity (82, 83). Furthermore, it has been reported that the catalytically active form of DNA-PK is preferentially cleaved by caspase-3 (81). This could cause the clustering of DNA-PKcs–Ku complexes on DNA strand breaks resulting in a dominant-negative function and interference with DNA repair (86). Altogether, these results suggest that caspase cleavage can selectively inactivate enzymes responsible for the first steps of the stress response. This process would ensure a fast and efficient programmed cell death.

For PARP, the proposed roles for its cleavage during apoptosis are not mutually exclusive and could cooperate to ensure the completion of the process. At the present time however, there is not sufficient data to determine the exact role of PARP cleavage during apoptosis. Nevertheless, since PARP$^{-/-}$ mice do not show any apoptotic defects, some authors concluded that PARP cleavage is an accidental event during apoptosis (50, 65). However, knockout mice, while allowing the study of the role of PARP activation during apoptosis, do not permit the elucidation of the role of its cleavage. Since PARP is likely to be cleaved to ensure that it will not interfere with the apoptotic process, one would expect that apoptosis is not influenced by the absence of PARP. Experiments using PARP$^{-/-}$ cells transfected with uncleavable PARP should therefore provide answers on the role of PARP cleavage. A first report using an uncleavable PARP (bearing a DEVD to

DEVA mutation in the caspase cleavage site) has recently been published (15). The introduction of this mutant in PARP$^{-/-}$ cells delayed cell death in response to anti-Fas and, less significantly, in response to the alkylating agent MMS. Studies using other uncleavable mutants, such as Rb (87), p21 (79), or RAD51 (84), have shown similar results: the absence of cleavage delays cell death but ultimately does not prevent it. On the other hand, Herceg and Wang have reported that the transfection of PARP-null cells with an uncleavable PARP (DEVD to DEVN) results in both enhanced apoptosis and induction of necrosis in response to tumour necrosis factor (TNF) (88). Their results support the hypothesis that PARP is cleaved to prevent its over-activation and subsequent energy depletion leading to necrosis.

Thus, there is evidence supporting both hypotheses for PARP cleavage during apoptosis (inactivation of survival-factor functions or a switch to necrosis). Despite the fact that the exact function of PARP cleavage remains to be elucidated, the results from the uncleavable mutants clearly indicate that PARP cleavage is an important event for an efficient apoptotic cell death.

4.3.5   PARP cleavage during necrosis

As discussed above, PARP cleavage occurs early in almost all forms of apoptosis, which makes it a sensitive marker for apoptosis detection. However, PARP cleavage has also been found during necrosis (89, 90). This proteolysis results in the generation of a major fragment of 50 kDa (89, 90) and another fragment with the same apparent molecular weight as the 89-kDa apoptotic fragment, but which is inactive (89). This necrotic cleavage occurred later in the cell-death process than the apoptotic cleavage and was not related to caspase activity. The different pattern of PARP cleavage during these two types of cell death could provide a useful tool for discriminating between apoptotic and necrotic cell death.

4.4   PARP as a mediator of necrosis following excessive DNA damage

4.4.1   PARP overactivation leads to a necrotic cell death

In several cell types, cell death induced by various DNA damaging agents (directly or via the production of an oxidant in the cell)—including 1-methyl-3-nitro-1-nitrosoguanidine (MNNG) (91), H_2O_2 (92, 93), peroxynitrite (93, 94), and *N*-methyl-D-aspartate (NMDA) (95–97)—could be prevented by PARP inhibitors. This cytotoxicity in response to excessive DNA damage was prevented in the PARP-deficient cells, emphasizing the involvement of PARP in this process (39, 97–99). This cell death occurring following excessive DNA damage seems to be necrotic rather than apoptotic. In fact, the 'accidental' overactivation of PARP is not linked to the physiological processes of cell death involving specialized molecules such as death receptors, p53, and caspases. Cells rapidly lose their cytoplasmic membrane integrity during this cell-death process, as determined by the rapid release of lactate dehydrogenase and the inclusion of viability dyes (93; Affar *et al.*, in preparation). In addition, this mode of cell death is not associated with apoptotic characteristics such as the internucleosomal DNA fragmentation (39) or the typical PARP cleavage (92, Affar *et al.*, in preparation).

4.4.2 Mechanism of cell death: NAD^+- and ATP-depletion concepts

Previous work has reported that cellular NAD^+ is rapidly decreased in cells treated with DNA damaging agents. During moderate DNA damage this decrease in NAD^+, reaching more than 50% of basal level, can be restored by cell metabolism (100–102). In contrast, an irreversible NAD^+ depletion occurs during excessive DNA damage (52, 91, 103, 104). The existence of a close temporal association between the generation of DNA strand breaks, pADPr synthesis, and NAD^+ decrease has been described (105–108), suggesting that PARP may be responsible for the excessive utilization of its substrate. In addition, both polymer synthesis and NAD^+ depletion are inhibited by chemical inhibitors of PARP (107, 108). This cellular depletion of NAD^+ is therefore considered to be caused by the overactivation of PARP following excessive DNA damage. As NAD^+ is an enzymatic cofactor for glycolysis, a consequence of its depletion could be the arrest of this metabolic pathway resulting in the inhibition of ATP production (91, 108, 109). ATP depletion was indeed shown to occur following the treatment of cells with DNA damaging agents. This was followed by an arrest of DNA, RNA, and protein synthesis, and subsequently cell death (91, 108).

The causal role of PARP activation in DNA damage-induced NAD^+ depletion has been investigated in a few studies using cells derived from PARP knockout mice (9, 99, 110). Some reported the prevention of this NAD^+ decrease in knockout cells following treatment with streptozotocin (99) and the oxidant hypoxanthine/xanthine oxidase (110). However, a study by Shieh *et al.* showed similar decreases in wild-type as well as knockout cells in response to MNNG (9), and that this decrease could be inhibited by benzamide.

Although most evidence suggests the involvement of PARP in necrotic cell death in response to excessive DNA damage, the molecular basis is not well understood. In fact, considering the recent findings indicating that MNNG induces a similar NAD^+ depletion in wild-type and PARP-deficient cells (9), it is difficult to establish a direct correlation between the loss of NAD^+ and the decrease of ATP reserves and subsequent cell death. Further studies on the kinetics of both NAD^+ and ATP levels in wild-type and PARP-deficient cells in response to excessive DNA damage will give more information about their interrelation and about the possible involvement of other pADPr-synthesizing enzymes.

In the past few years, a role for PARP overactivation in the induction of necrosis has been emphasized in diverse experimental pathophysiological models like cerebral ischaemia, cardiac ischaemia, inflammation, and diabetes. In fact, resistance to cell death and protection from oxidant-induced organ injuries and dysfunction were observed using PARP inhibitors and PARP-deficient mice. This suggests that PARP inhibition may be an important avenue for therapeutic treatment (see Chapter 5).

4.5   Conclusions

PARP, as a DNA break sensor, is activated early during the cellular response to DNA damage. Our proposed model summarizing the current knowledge on the role of PARP in the cell survival and death following DNA damage is illustrated in Fig. 4.2. The involvement of PARP in these processes seems to be dependent on the number of DNA strand breaks. In the presence of small amounts of DNA damage, PARP may act as a survival-promoting molecule. This may be due to a direct interaction with proteins or via the synthesis of pADPr, which could bind covalently and/or non-covalently to proteins in order to modify their function. Other proteins of the DNA damage-signalling network could be involved in conjunction with PARP to re-establish the genomic integrity, thus permitting an accurate repair of the DNA and preventing the occurrence of genomic alterations. During excessive DNA damage, overactivation of PARP induces necrotic cell death, which could be mediated by the depletion of NAD^+ and ATP stocks. This cytotoxicity would predominate and mask the survival function of PARP. Apoptosis induced by various pathways including the stimulation of death receptors, DNA damage, and growth-factor starvation, causes the activation of caspase proteases which trigger the cleavage of many cellular substrates involved in cellular homeostasis. PARP cleavage by caspases results in its inactivation, which inhibits its survival function. This cleavage could also suppress a futile overactivation of the enzyme by the internucleosomal DNA fragments which would deplete the energy reserve of the cell and force the switch to a necrotic cell death. The 24-kDa apoptotic fragment of PARP may irreversibly bind to DNA breaks and competitively inhibit both uncleaved PARP molecules and DNA repair processes to avoid wasting energy.

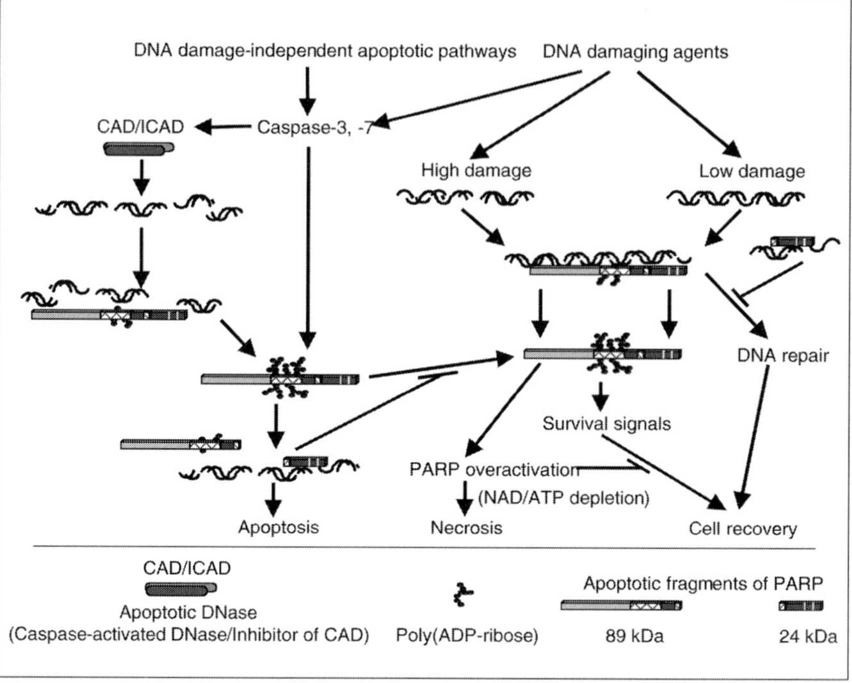

Fig. 4.2 Schematic model showing the involvement of PARP in cell survival and death. The involvement of PARP in the cellular response to DNA damage seems to be dependent on the amount of DNA breaks. In the presence of small amounts of DNA damage, PARP allows cell recovery, possibly by promoting the maintenance of genomic integrity. The overactivation of PARP following excessive DNA damage induces a necrotic cell death, most probably as the consequence of the depletion of the cellular energy. During apoptosis, PARP is cleaved by caspases to inhibit its survival function and/or to prevent a switch to necrosis.

References

1. Winstall, E., Affar, E. B., Shah, R., Bourassa, S., Scovassi, A. I., and Poirier, G. G. (1999). Preferential perinuclear localization of poly(ADP-ribose) glycohydrolase. *Exp. Cell Res.*, **251**, 372.

2. D'Amours, D., Desnoyers, S., D'Silva, I., and Poirier, G. G. (1999). Poly(ADP-ribosyl)ation reactions in the regulation of nuclear functions. *Biochem. J.*, **342**, 249.

3. Kupper, J. H., Muller, M., Jacobson, M. K., Tatsumi Miyajima, J., Coyle, D. L., Jacobson, E. L., and Burkle, A. (1995). Trans-dominant inhibition of poly(ADP-ribosyl)ation sensitizes cells against gamma-irradiation and *N*-methyl-*N*'-nitro-*N*-nitrosoguanidine but does not limit DNA replication of a polyomavirus replicon. *Mol. Cell. Biol.*, **15**, 3154.

4. Schreiber, V., Hunting, D., Trucco, C., Gowans, B., Grunwald, D., de Murcia, G., and Ménissier-de Murcia, J. (1995). A dominant-negative mutant of human

poly(ADP-ribose) polymerase affects cell recovery, apoptosis, and sister chromatid exchange following DNA damage. *Proc. Natl Acad. Sci. USA*, **92**, 4753.

5. Cleaver, J. E., Bodell, W. J., Morgan, W. F., and Zelle, B. (1983). Differences in the regulation by poly(ADP-ribose) of repair of DNA damage from alkylating agents and ultraviolet light according to cell type. *J. Biol. Chem.*, **258**, 9059.

6. Milam, K. M. and Cleaver, J. E. (1984). Inhibitors of poly(adenosine diphosphate-ribose) synthesis: effect on other metabolic processes. *Science*, **223**, 589.

7. Cleaver, J. E. (1984). Differential toxicity of 3-aminobenzamide to wild-type and 6-thioguanine-resistant Chinese hamster cells by interference with pathways of purine biosynthesis. *Mutat. Res.*, **131**, 123.

8. Smith, S., Giriat, I., Schmitt, A., and de Lange, T. (1998). Tankyrase, a poly(ADP-ribose) polymerase at human telomeres. *Science*, **282**, 1484.

9. Shieh, W. M., Ame, J. C., Wilson, M. V., Wang, Z. Q., Koh, D. W., Jacobson, M. K., and Jacobson, E. L. (1998). Poly(ADP-ribose) polymerase null mouse cells synthesize ADP-ribose polymers. *J. Biol. Chem.*, **273**, 30069.

10. Amé, J. C., Rolli, V., Schreiber, V., Niedergang, C., Apiou, F., Decker, P., Muller, S., Hoger, T., Menissier-de Murcia, J., and de Murcia, G. (1999). PARP-2, A novel mammalian DNA damage-dependent poly(ADP-ribose) polymerase. *J. Biol. Chem.*, **274**, 17860.

11. Ding, R., Pommier, Y., Kang, V. H., and Smulson, M. (1992). Depletion of poly(ADP-ribose) polymerase by antisense RNA expression results in a delay in DNA strand break rejoining. *J. Biol. Chem.*, **267**, 12804.

12. Ding, R. and Smulson, M. (1994). Depletion of nuclear poly(ADP-ribose) polymerase by antisense RNA expression: influences on genomic stability, chromatin organization, and carcinogen cytotoxicity. *Cancer Res.*, **54**, 4627.

13. Stevnsner, T., Ding, R., Smulson, M., and Bohr, V. A. (1994). Inhibition of gene-specific repair of alkylation damage in cells depleted of poly(ADP-ribose) polymerase. *Nucleic Acids Res.*, **22**, 4620.

14. Ménissier-de Murcia, J., Niedergang, C., Trucco, C., Ricoul, M., Dutrillaux, B., Mark, M., Oliver, F. J., Masson, M., Dierich, A., LeMeur, M., Walztinger, C., Chambon, P., and de Murcia, G. (1997). Requirement of poly(ADP-ribose) polymerase in recovery from DNA damage in mice and in cells. *Proc. Natl Acad. Sci. USA*, **94**, 7303.

15. Oliver, F. J., de la Rubia, G., Rolli, V., Ruiz-Ruiz, M. C., de Murcia, G., and Ménissier-de Murcia, J. (1998). Importance of poly(ADP-ribose) polymerase and its cleavage in apoptosis. Lesson from an uncleavable mutant. *J. Biol. Chem.*, **273**, 33533.

16. Masutani, M., Nozaki, T., Nishiyama, E., Shimokawa, T., Tachi, Y., Suzuki, H., Nakagama, H., Wakabayashi, K., and Sugimura, T. (1999). Function of poly(ADP-ribose) polymerase in response to DNA damage: gene disruption study in mice. *Mol. Cell. Biochem.*, **193**, 149.

17. Wang, Z. Q., Stingl, L., Morrison, C., Jantsch, M., Los, M., Schulze-Osthoff, K., and Wagner, E. F. (1997). PARP is important for genomic stability but dispensable in apoptosis. *Genes Dev.*, **11**, 2347.

18. Agarwal, M. L., Taylor, W. R., Chernov, M. V., Chernova, O. B., and Stark, G. R. (1998). The p53 network. *J. Biol. Chem.*, **273**, 1.

19. Szumiel, I. (1998). Monitoring and signaling of radiation-induced damage in mammalian cells. *Radiat. Res.*, **150**, 92.

20. Whitacre, C. M., Hashimoto, H., Tsai, M. L., Chatterjee, S., Berger, S. J., and Berger, N. A. (1995). Involvement of NAD-poly(ADP-ribose) metabolism in p53 regulation and its consequences. *Cancer Res.*, **55**, 3697.

21. Agarwal, M. L., Agarwal, A., Taylor, W. R., Wang, Z. Q., Wagner, E. F., and Stark, G. R. (1997). Defective induction but normal activation and function of p53 in mouse cells lacking poly-ADP-ribose polymerase. *Oncogene*, **15**, 1035.

22. Wesierska-Gadek, J., Wang, Z. Q., and Schmid, G. (1999). Reduced stability of regularly spliced but not alternatively spliced p53 protein in PARP-deficient mouse fibroblasts. *Cancer Res.*, **59**, 28.

23. Simbulan-Rosenthal, C. M., Rosenthal, D. S., Ding, R., Bhatia, K., and Smulson, M. E. (1998). Prolongation of the p53 response to DNA strand breaks in cells depleted of PARP by antisense RNA expression. *Biochem. Biophys. Res. Commun.*, **253**, 864.

24. Wesierska-Gadek, J., Bugajska-Schretter, A., and Cerni, C. (1996). ADP-ribosylation of p53 tumor suppressor protein: mutant but not wild- type p53 is modified. *J. Cell. Biochem.*, **62**, 90.

25. Wesierska-Gadek, J., Schmid, G., and Cerni, C. (1996). ADP-ribosylation of wild type p53 *in vitro*: binding of p53 protein to specific p53 consensus sequence prevents its modification. *Biochem. Biophys. Res. Commun.*, **224**, 96.

26. Vaziri, H., West, M. D., Allsopp, R. C., Davison, T. S., Wu, Y.-S., Arrowsmith, C. H., Poirier, G. G., and Benchimol, S. (1997). ATM-dependent telomere loss in aging human diploid fibroblasts and DNA damage lead to the post-translational activation of p53 protein involving poly(ADP-ribose) polymerase. *EMBO J.*, **16**, 6018.

27. Malanga, M., Pleschke, J. M., Kleczkowska, H. E., and Althaus, F. R. (1998). Poly(ADP-ribose) binds to specific domains of p53 and alters its DNA binding functions. *J. Biol. Chem.*, **273**, 11839.

28. Ruscetti, T., Lehnert, B. E., Halbrook, J., Le Trong, H., Hoekstra, M. F., Chen, D. J., and Peterson, S. R. (1998). Stimulation of the DNA-dependent protein kinase by poly(ADP-ribose) polymerase. *J. Biol. Chem.*, **273**, 14461.

29. Galande, S. and Kohwi-Shigematsu, T. (1999). Poly(ADP-ribose) polymerase and Ku autoantigen form a complex and synergistically bind to matrix attachment sequences. *J. Biol. Chem.*, **274**, 20521.

30. Morrison, C., Smith, G. C., Stingl, L., Jackson, S. P., Wagner, E. F., and Wang, Z. Q. (1997). Genetic interaction between PARP and DNA-PK in V(D)J recombination and tumorigenesis. *Nature Genet.*, **17**, 479.

31. Masson, M., Niedergang, C., Schreiber, V., Muller, S., Menissier-de Murcia, J., and de Murcia, G. (1998). XRCC1 is specifically associated with poly(ADP-ribose) polymerase and negatively regulates its activity following DNA damage. *Mol. Cell. Biol.*, **18**, 3563.

32. Steller, H. (1995). Mechanisms and genes of cellular suicide. *Science*, **267**, 1445.

33. Jacobson, M. D., Weil, M., and Raff, M. C. (1997). Programmed cell death in animal development. *Cell*, **88**, 347.

34. Green, D. R. and Reed, J. C. (1998). Mitochondria and apoptosis. *Science*, **281**, 1309.

35. Ashkenazi, A. and Dixit, V. M. (1998). Death receptors: signaling and modulation. *Science*, **281**, 1305.

36. Tsujimoto, Y. (1997). Apoptosis and necrosis: intracellular ATP levels as a determinant for cell death modes. *Cell Death Diff.*, **4**, 429.

37. Savill, J. (1997). Apoptosis in resolution of inflammation. *J. Leukoc. Biol.*, **61**, 375.

38. Nosseri, C., Coppola, S., and Ghibelli, L. (1994). Possible involvement of poly(ADP-ribosyl) polymerase in triggering stress-induced apoptosis. *Exp. Cell Res.*, **212**, 367.

39. Virag, L., Scott, G. S., Cuzzocrea, S., Marmer, D., Salzman, A. L., and Szabo, C. (1998). Peroxynitrite-induced thymocyte apoptosis: the role of caspases and poly (ADP-ribose) synthetase (PARS) activation. *Immunology*, **94**, 345.

40. Bonfoco, E., Krainc, D., Ankarcrona, M., Nicotera, P., and Lipton, S. A. (1995). Apoptosis and necrosis: two distinct events induced, respectively, by mild and intense insults with N-methyl-d-aspartate or nitric oxide/superoxide in cortical cell cultures. *Proc. Natl Acad. Sci. USA*, **92**, 7162.

41. Nicotera, P. and Leist, M. (1997). Energy supply and the shape of cell death in neurons and lymphoid cells. *Cell Death Differ.*, **4**, 435.

42. Shimizu, S., Eguchi, Y., Kamiike, W., Waguri, S., Uchiyama, Y., Matsuda, H., and Tsujimoto, Y. (1996). Retardation of chemical hypoxia-induced necrotic cell death by Bcl-2 and ICE inhibitors: possible involvement of common mediators in apoptotic and necrotic signal transductions. *Oncogene*, **12**, 2045.

43. Leist, M. and Nicotera, P. (1997). The shape of cell death. *Biochem. Biophys. Res. Commun.*, **236**, 1.

44. Eguchi, Y., Shimizu, S., and Tsujimoto, Y. (1997). Intracellular ATP levels determine cell death fate by apoptosis or necrosis. *Cancer Res.*, **57**, 1835.

45. Li, P., Nijhawan, D., Budihardjo, I., Srinivasula, S. M., Ahmad, M., Alnemri, E. S., and Wang, X. (1997). Cytochrome c and dATP-dependent formation of Apaf-1/caspase-9 complex initiates an apoptotic protease cascade. *Cell*, **91**, 479.

46. Kass, G. E., Eriksson, J. E., Weis, M., Orrenius, S., and Chow, S. C. (1996). Chromatin condensation during apoptosis requires ATP. *Biochem. J.*, **318**, 749.

47. Salvesen, G. S. and Dixit, V. M. (1997). Caspases: intracellular signaling by proteolysis. *Cell*, **91**, 443.

48. Nicholson, D. W., and Thornberry, N. A. (1997). Caspases: killer proteases. *Trends Biochem. Sci.*, **22**, 299.

49. Thornberry, N. A. and Lazebnik, Y. (1998). Caspases: enemies within. *Science*, **281**, 1312.

50. Stroh, C. and Schulze-Osthoff, K. (1998). Death by a thousand cuts: an ever increasing list of caspase substrates. *Cell Death Differ.*, **5**, 997.

51. Kaufmann, S. H., Desnoyers, S., Ottaviano, Y., Davidson, N. E., and Poirier, G. G. (1993). Specific proteolytic cleavage of poly(ADP-ribose) polymerase: an early marker of chemotherapy-induced apoptosis. *Cancer Res.*, **53**, 3976.

52. Coppola, S., Nosseri, C., Maresca, V., and Ghibelli, L. (1995). Different basal NAD levels determine opposite effects of poly(ADP-ribose) polymerase inhibitors on H2O2-induced apoptosis. *Exp. Cell Res.*, **221**, 462.

53. Ubol, S., Suk, P., Budihardjo, I., Desnoyers, S., Montrose, M. H., Poirier, G. G., Kaufmann, S. H., and Griffin, D. E. (1996). Temporal changes in chromatin, intracellular calcium, and poly(adp-ribose) polymerase during sindbis virus-induced apoptosis of neuroblastoma-cells. *J. Virol.*, **70**, 2215.

54. Shimizu, T., Kubota, M., Tanizawa, A., Sano, H., Kasai, Y., Hashimoto, H.,

Akiyama, Y., and Mikawa, H. (1990). Inhibition of both etoposide-induced DNA fragmentation and activation of poly(ADP-ribose) synthesis by zinc ion. *Biochem. Biophys. Res. Commun.*, **169**, 1172.

55. Tanizawa, A., Kubota, M., Hashimoto, H., Shimizu, T., Takimoto, T., Kitoh, T., Akiyama, Y., and Mikawa, H. (1989). VP-16-induced nucleotide pool changes and poly(ADP-ribose) synthesis: the role of VP-16 in interphase cell death. *Exp. Cell Res.*, **185**, 237.

56. Yoon, S. Y., Kim, J. W., Kang, K. W., Kim, Y. S., Choi, K. H., and Joe, C. O. (1996). Poly(ADP-ribosyl)ation of histone H1 correlates with internucleosomal DNA fragmentation during apoptosis. *J. Biol. Chem.*, **271**, 9129.

57. Guano, F., Bernardi, R., Negri, C., Donzelli, M., Prosperi, E., Ricotti, G. A., and Scovassi, A. I. (1994). Dose-dependent zinc inhibition of DNA ladder in apoptotic HeLa cells regulates the activity of poly(ADP-ribose) polymerase and does not protect from death induced by VP-16. *Cell Death Differ.*, **1**, 101.

58. Rosenthal, D. S., Ding, R., Simbulan-Rosenthal, C. M., Vaillancourt, J. P., Nicholson, D. W., and Smulson, M. (1997). Intact cell evidence for the early synthesis, and subsequent late apopain-mediated suppression, of poly(ADP-ribose) during apoptosis. *Exp. Cell Res.*, **232**, 313.

59. Simbulan-Rosenthal, C. M., Rosenthal, D. S., Iyer, S., Boulares, A. H., and Smulson, M. E. (1998). Transient poly(ADP-ribosyl)ation of nuclear proteins and role of poly(ADP-ribose) polymerase in the early stages of apoptosis. *J. Biol. Chem.*, **273**, 13703.

60. Germain, M., Affar, E. B., D'Amours, D., Dixit, V. M., Salvesen, G. S., and Poirier, G. G. (1999). Cleavage of automodified poly(ADP-ribose) polymerase during apoptosis: evidence for involvement of caspase-7. *J. Biol. Chem.*, **274**, 28379.

61. Shiokawa, D., Maruta, H., and Tanuma, S. (1997). Inhibitors of poly(ADP-ribose) polymerase suppress nuclear fragmentation and apoptotic-body formation during apoptosis in HL-60 cells. *FEBS Lett.*, **413**, 99.

62. Payne, C. M., Crowley, C., Washo-Stultz, D., Briehl, M., Bernstein, H., Bernstein, C., Beard, S., Holubec, H., and Warneke, J. (1998). The stress-response proteins poly(ADP-ribose) polymerase and NF-kappaB protect against bile salt-induced apoptosis. *Cell Death Differ.*, **5**, 623.

63. Saldeen, J. and Welsh, N. (1998). Nicotinamide-induced apoptosis in insulin producing cells is associated with cleavage of poly(ADP-ribose) polymerase. *Mol. Cell. Endocrinol.*, **139**, 99.

64. Watson, A. J.M., Askew, J. N., and Benson, R. S.P. (1995). Poly(adenosine diphosphate ribose) polymerase inhibition prevents necrosis induced by H2O2 but not apoptosis. Gastroenterology, **109**, 472.

65. Leist, M., Single, B., Kunstle, G., Volbracht, C., Hentze, H., and Nicotera, P. (1997). Apoptosis in the absence of poly-(ADP-ribose) polymerase. *Biochem. Biophys. Res. Commun.*, **233**, 518.

66. Varfolomeev, E. E., Schuchmann, M., Luria, V., Chiannilkulchai, N., Beckmann, J. S., Mett, I. L., Rebrikov, D., Brodianski, V. M., Kemper, O. C., Kollet, O., Lapidot, T., Soffer, D., Sobe, T., Avraham, K. B., Goncharov, T., Holtmann, H., Lonai, P., and Wallach, D. (1998). Targeted disruption of the mouse Caspase 8 gene ablates cell death induction by the TNF receptors, Fas/Apo1, and DR3 and is lethal prenatally. *Immunity*, **9**, 267.

67. Enari, M., Sakahira, H., Yokoyama, H., Okawa, K., Iwamatsu, A., and Nagata, S. (1998). A caspase-activated DNase that degrades DNA during apoptosis, and its inhibitor ICAD. *Nature*, **391**, 43.

68. Zhang, J., Liu, X., Scherer, D. C., van Kaer, L., Wang, X., and Xu, M. (1998). Resistance to DNA fragmentation and chromatin condensation in mice lacking the DNA fragmentation factor 45. *Proc. Natl Acad. Sci. USA*, **95**, 12480.

69. Wang, Z. Q., Auer, B., Stingl, L., Berghammer, H., Haidacher, D., Schweiger, M., and Wagner, E. F. (1995). Mice lacking ADPRT and poly(ADP-ribosyl)ation develop normally but are susceptible to skin disease. *Genes Dev.*, **9**, 509.

70. Masutani, M., Suzuki, H., Kamada, N., Watanabe, M., Ueda, O., Nozaki, T., Jishage, K., Watanabe, T., Sugimoto, T., Nakagama, H., Ochiya, T., and Sugimura, T. (1999). Poly(ADP-ribose) polymerase gene disruption conferred mice resistant to streptozotocin-induced diabetes. *Proc. Natl Acad. Sci. USA*, **96**, 2301.

71. Lazebnik, Y. A., Kaufmann, S. H., Desnoyers, S., Poirier, G. G., and Earnshaw, W. C. (1994). Cleavage of poly(ADP-ribose) polymerase by a proteinase with properties like ICE. *Nature*, **371**, 346.

72. Tewari, M., Quan, L. T., O'Rourke, K., Desnoyers, S., Zeng, Z., Beidler, D. R., Poirier, G. G., Salvesen, G. S., and Dixit, V. M. (1995). Yama/CPP32 beta, a mammalian homologue of CED-3, is a CrmA-inhibitable protease that cleaves the death substrate poly(ADP-ribose) polymerase. *Cell*, **81**, 801.

73. Smulson, M. E., Pang, D., Jung, M., Dimtchev, A., Chasovskikh, S., Spoonde, A., Simbulan-Rosenthal, C., Rosenthal, D., Yakovlev, A., and Dritschilo, A. (1998). Irreversible binding of poly(ADP)ribose polymerase cleavage product to DNA ends revealed by atomic force microscopy: possible role in apoptosis. *Cancer Res.*, **58**, 3495.

74. Talanian, R. V., Quinlan, C., Trautz, S., Hackett, M. C., Mankovich, J. A., Banach, D., Ghayur, T., Brady, K. D., and Wong, W. W. (1997). Substrate specificities of caspase family proteases. *J. Biol. Chem.*, **272**, 9677.

75. Thornberry, N. A., Rano, T. A., Peterson, E. P., Rasper, D. M., Timkey, T., Garcia-Calvo, M., Houtzager, V. M., Nordstrom, P. A., Roy, S., Vaillancourt, J. P., Chapman, K. T., and Nicholson, D. W. (1997). A combinatorial approach defines specificities of members of the caspase family and granzyme B. Functional relationships established for key mediators of apoptosis. *J. Biol. Chem.*, **272**, 17907.

76. Kupper, J. H., de Murcia, G., and Burkle, A. (1990). Inhibition of poly(ADP-ribosyl)ation by overexpressing the poly(ADP-ribose) polymerase DNA-binding domain in mammalian cells. *J. Biol. Chem.*, **265**, 18721.

77. Molinete, M., Vermeulen, W., Burkle, A., Menissier de Murcia, J., Kupper, J. H., Hoeijmakers, J. H., and de Murcia, G. (1993). Overproduction of the poly(ADP-ribose) polymerase DNA-binding domain blocks alkylation-induced DNA repair synthesis in mammalian cells. *EMBO J.*, **12**, 2109.

78. Widmann, C., Gibson, S., and Johnson, G. L. (1998). Caspase-dependent cleavage of signaling proteins during apoptosis. A turn-off mechanism for anti-apoptotic signals. *J. Biol. Chem.*, **273**, 7141.

79. Levkau, B., Koyama, H., Raines, E. W., Clurman, B. E., Herren, B., Orth, K., Roberts, J. M., and Ross, R. (1998). Cleavage of p21Cip1/Waf1 and p27Kip1

mediates apoptosis in endothelial cells through activation of Cdk2: role of a caspase cascade. *Mol. Cell*, **1**, 553.

80. Gervais, J. L., Seth, P., and Zhang, H. (1998). Cleavage of CDK inhibitor p21(Cip1/Waf1) by caspases is an early event during DNA damage-induced apoptosis. *J. Biol. Chem.*, **273**, 19207.

81. Han, Z., Malik, N., Carter, T., Reeves, W. H., Wyche, J. H., and Hendrickson, E. A. (1996). DNA-dependent protein kinase is a target for a CPP32-like apoptotic protease. *J. Biol. Chem.*, **271**, 25035.

82. Song, Q., Lees-Miller, S. P., Kumar, S., Zhang, Z., Chan, D. W., Smith, G. C., Jackson, S. P., Alnemri, E. S., Litwack, G., Khanna, K. K., and Lavin, M. F. (1996). DNA-dependent protein kinase catalytic subunit: a target for an ICE-like protease in apoptosis. *EMBO J.*, **15**, 3238.

83. Casciola-Rosen, L., Nicholson, D. W., Chong, T., Rowan, K. R., Thornberry, N. A., Miller, D. K., and Rosen, A. (1996). Apopain/cpp32 cleaves proteins that are essential for cellular repair—a fundamental principle of apoptotic death. *J. Exp. Med.*, **183**, 1957.

84. Huang, Y., Nakada, S., Ishiko, T., Utsugisawa, T., Datta, R., Kharbanda, S., Yoshida, K., Talanian, R. V., Weichselbaum, R., Kufe, D., and Yuan, Z. M. (1999). Role for caspase-mediated cleavage of Rad51 in induction of apoptosis by DNA damage. *Mol. Cell. Biol.*, **19**, 2986.

85. Flygare, J., Armstrong, R. C., Wennborg, A., Orsan, S., and Hellgren, D. (1998). Proteolytic cleavage of HsRad51 during apoptosis. *FEBS Lett.*, **427**, 247.

86. McConnell, K. R., Dynan, W. S., and Hardin, J. A. (1997). The DNA-dependent protein kinase catalytic subunit (p460) is cleaved during Fas-mediated apoptosis in Jurkat cells. *J. Immunol.*, **158**, 2083.

87. Tan, X., Martin, S. J., Green, D. R., and Wang, J. Y. J. (1997). Degradation of retinoblastoma protein in tumor necrosis factor- and CD95-induced cell death. *J. Biol. Chem.*, **272**, 9613.

88. Herceg, Z. and Wang, Z. Q. (1999). Failure of poly(ADP-ribose) polymerase cleavage by caspases leads to induction of necrosis and enhanced apoptosis. *Mol. Cell. Biol.*, **19**, 5124.

89. Shah, G. M., Shah, R. G., and Poirier, G. G. (1996). Different cleavage pattern for poly(ADP-ribose) polymerase during necrosis and apoptosis in HL-60 cells. *Biochem. Biophys. Res. Commun.*, **229**, 838.

90. Casiano, C. A., Ochs, R. L., and Tan, E. M. (1998). Distinct cleavage products of nuclear proteins in apoptosis and necrosis revealed by autoantibody probes. *Cell Death Differ.*, **5**, 183.

91. Berger, N. A. (1985). Poly(ADP-ribose) in the cellular response to DNA damage. *Radiat. Res.*, **101**, 4.

92. Cookson, M. R., Ince, P. G., and Shaw, P. J. (1998). Peroxynitrite and hydrogen peroxide induced cell death in the NSC34 neuroblastoma spinal cord cell line: role of poly (ADP-ribose) polymerase. *J. Neurochem.*, **70**, 501.

93. Gilad, E., Zingarelli, B., Salzman, A. L., and Szabo, C. (1997). Protection by inhibition of poly (ADP-ribose) synthetase against oxidant injury in cardiac myoblasts *in vitro*. *J. Mol. Cell. Cardiol.*, **29**, 2585.

94. Endres, M., Scott, G. S., Salzman, A. L., Kun, E., Moskowitz, M. A., and Szabo, C. (1998). Protective effects of 5-iodo-6-amino-1,2-benzopyrone, an

inhibitor of poly(ADP-ribose) synthetase against peroxynitrite-induced glial damage and stroke development. *Eur. J. Pharmacol.*, **351**, 377.

95. Zhang, J., Dawson, V. L., Dawson, T. M., and Snyder, S. H. (1994). Nitric oxide activation of poly(ADP-ribose) synthetase in neurotoxicity. *Science*, **263**, 687.

96. Said, S. I., Berisha, H. I., and Pakbaz, H. (1996). Excitoxicity in the lung: *N*-methyl-D-aspartate-induced, nitric oxide-dependent, pulmonary edema is attenuated by vasoactive intestinal peptide and by inhibitors of poly(ADP-ribose) polymerase. *Proc. Natl Acad. Sci. USA*, **93**, 4688.

97. Eliasson, M. J. L., Sampei, K., Mandir, A., Hurn, P., Traystman, R. J., Pieper, A., Wang, Z.-Q., Dawson, T. E., Snyder, S. H., and Dawson, V. L. (1997). Poly(ADP-ribose) polymerase gene disruption renders mice resistant to cerebral ischemia. *Nature Med.*, **3**, 1089.

98. Virag, L., Salzman, A. L., and Szabo, C. (1998). Poly(ADP-ribose) synthetase activation mediates mitochondrial injury during oxidant-induced cell death. *J. Immunol.*, **161**, 3753.

99. Burkart, V., Wang, Z. Q., Radons, J., Heller, B., Herceg, Z., Stingl, L., Wagner, E. F., and Kolb, H. (1999). Mice lacking the poly(ADP-ribose) polymerase gene are resistant to pancreatic beta-cell destruction and diabetes development induced by streptozocin. *Nature Med.*, **5**, 314.

100. Goodwin, P. M., Lewis, P. J., Davies, M. I., Skidmore, C. J., and Shall, S. (1978). The effect of gamma radiation and neocarzinostatin on NAD and ATP levels in mouse leukaemia cells. *Biochim. Biophys. Acta*, **543**, 576.

101. Skidmore, C. J., Davies, M. I., Goodwin, P. M., Halldorsson, H., Lewis, P. J., Shall, S., and Zia'ee, A. A. (1979). The involvement of poly(ADP-ribose) polymerase in the degradation of NAD caused by gamma-radiation and *N*-methyl-*N*-nitrosourea. *Eur. J. Biochem.*, **101**, 135.

102. Durkacz, B. W., Omidiji, O., Gray, A., and Shall, S. (1980). (ADP-ribose)n participates in DNA excision repair. *Nature*, **283**, 593.

103. Kirkland, J. B. (1991). Lipid peroxidation, protein thiol oxidation and DNA damage in hydrogen peroxide-induced injury to endothelial cells: role of activation of poly(ADP-ribose)polymerase. *Biochim. Biophys. Acta*, **1092**, 319.

104. Dypbukt, J. M., Ankarcrona, M., Burkitt, M., Sjoholm, A., Strom, K., Orrenius, S., and Nicotera, P. (1994). Different prooxidant levels stimulate growth, trigger apoptosis, or produce necrosis of insulin-secreting RINm5F cells. The role of intracellular polyamines. *J. Biol. Chem.*, **269**, 30553.

105. Jacobson, M. K., Levi, V., Juarez-Salinas, H., Barton, R. A., and Jacobson, E. L. (1980). Effect of carcinogenic *N*-alkyl-*N*-nitroso compounds on nicotinamide adenine dinucleotide metabolism. *Cancer Res.*, **40**, 1797.

106. Yamamoto, H., Uchigata, Y., and Okamoto, H. (1981). Streptozotocin and alloxan induce DNA strand breaks and poly(ADP-ribose) synthetase in pancreatic islets. *Nature*, **294**, 284.

107. Wielckens, K., Schmidt, A., George, E., Bredehorst, R., and Hilz, H. (1982). DNA fragmentation and NAD depletion: their relation to the turnover of endogenous to mono(ADP-ribosyl) and poly(ADP-ribosyl) proteins. *J. Biol. Chem.*, **257**, 12872.

108. Sims, J. L., Berger, S. J., and Berger, N. A. (1983). Poly(ADP-ribose) polymerase inhibitors preserve nicotinamide adenine dinucleotide and adenosine 5'-

triphosphate pools in DNA-damaged cells: mechanism of stimulation of unscheduled DNA synthesis. *Biochemistry*, **22**, 5188.

109. Berger, S. J., Sudar, D. C., and Berger, N. A. (1986). Metabolic consequences of DNA damage: DNA damage induces alterations in glucose metabolism by activation of poly (ADP-ribose) polymerase. *Biochem. Biophys. Res. Commun.*, **134**, 227.

110. Heller, B., Wang, Z. Q., Wagner, E. F., Radons, J., Burkle, A., Fehsel, K., Burkart, V., and Kolb, H. (1995). Inactivation of the poly(ADP-ribose) polymerase gene affects oxygen radical and nitric oxide toxicity in islet cells. *J. Biol. Chem.*, **270**, 11176.

111. Huet, J. and Laval, F. (1985). Potentiation of cell killing by inhibitors of poly(adenosine diphosphate-ribose) synthesis in bleomycin-treated Chinese hamster ovary cells. *Cancer Res.*, **45**, 987.

112. Schwartz, J. L., Morgan, W. F., Brown-Lindquist, P., Afzal, V., Weichselbaum, R. R., and Wolff, S. (1985). Comutagenic effects of 3-aminobenzamide in Chinese hamster ovary cells. *Cancer Res.*, **45**, 1556.

113. Ben-Hur, E., Chen, C. C., and Elkind, M. M. (1985). Inhibitors of poly(adenosine diphosphoribose) synthetase, examination of metabolic perturbations, and enhancement of radiation response in Chinese hamster cells. *Cancer Res.*, **45**, 2123.

114. Jacobson, E. L., Smith, J. Y., Mingmuang, M., Meadows, R., Sims, J. L., and Jacobson, M. K. (1984). Effect of nicotinamide analogues on recovery from DNA damage in C3H10T1/2 cells. *Cancer Res.*, **44**, 2485.

115. Shah, G. M., Poirier, D., Desnoyers, S., Saint-Martin, S., Hoflack, J. C., Rong, P., ApSimon, M., Kirkland, J. B., and Poirier, G. G. (1996). Complete inhibition of poly(ADP-ribose) polymerase activity prevents the recovery of C3H10T1/2 cells from oxidative stress. *Biochim. Biophys. Acta*, **1312**, 1.

116. James, M. R. and Lehmann, A. R. (1982). Role of poly(adenosine diphosphate ribose) in deoxyribonucleic acid repair in human fibroblasts. *Biochemistry*, **21**, 4007.

117. Boorstein, R. and Pardee, A. (1984). Factors modifying 3-aminobenzamide cytotoxicity in normal and repair-deficient human fibroblasts. *J. Cell Physiol.*, **120**, 335.

118. Kasid, U. N., Stefanik, D. F., Lubet, R. A., Dritschilo, A., and Smulson, M. E. (1986). Relationship between DNA strand breaks and inhibition of poly (ADP-ribosyl)ation: enhancement of carcinogen-induced transformation. *Carcinogenesis*, **7**, 327.

119. Wright, S. C., Wei, Q. S., Kinder, D. H., and Larrick, J. W. (1996). Biochemical pathways of apoptosis: nicotinamide adenine dinucleotide-deficient cells are resistant to tumor necrosis factor or ultraviolet light activation of the 24-kD apoptotic protease and DNA fragmentation. *J. Exp. Med.*, **183**, 463.

120. Bertrand, R., Solary, E., Jenkins, J., and Pommier, Y. (1993). Apoptosis and its modulation in human promyelocytic HL-60 cells treated with DNA topoisomerase I and II inhibitors. *Exp. Cell Res.*, **207**, 388.

----------------------------- **5** -----------------------------

Activation of poly (ADP-ribose) polymerase in the pathogenesis of ischaemia–reperfusion injury

Csaba Szabó

5.1   Introduction

Nitric oxide and hydroxyl radicals produce DNA damage. The PARP suicide hypothesis suggests that massive DNA damage, which occurs with severe oxidative stress, may trigger the activation of PARP to such an extent that all the cellular NAD^+ may be depleted. Thus, the cellular energy supply may become deficient and this may lead to functional and morphological alterations in the cell, which may culminate in cell death. Recent studies with PARP knockout (–/–, or null mice), PARP-deficient cell lines and novel pharmacological enzyme inhibitors confirm the role of PARP in oxidant-mediated cell death. The principal species causing DNA injury and cell death are peroxynitrite and hydroxyl radicals. When PARP activation leads to depletion of the cellular energy stores, then cell death is necrotic rather than apoptotic. Inhibition or absence of PARP protects against necrosis.

Furthermore, both peroxynitrite and hydroxyl radicals are produced during the reperfusion of ischaemic tissues. Therefore, pharmacological inhibition or genetic inactivation of PARP protects cells in animal models of stroke, myocardial and splanchnic ischaemia–reperfusion, and in haemorrhage resuscitation, which is equivalent to whole-body ischaemia–reperfusion.

PARP emerges as one of the principal, final common mediators of cellular injury in a variety of pathophysiological conditions. This chapter reviews the role of PARP activation in various forms of ischaemia–reperfusion. I also present evidence that prevention of cell necrosis is the most likely explanation of the protection elicited by PARP inhibition or absence in reperfusion injury.

5.2   The poly (ADP-ribose) polymerase suicide hypothesis

The activation of PARP results in the cleavage of NAD^+ to form poly (ADP-ribose) polymers covalently attached to chromatin proteins. It was first suggested by Nathan Berger (1–5) about two decades ago, that this process of poly (ADP-ribosyl)ation may be linked to cell death associated with DNA injury. According to this 'PARP suicide hypothesis', massive activation of PARP may rapidly deplete the intracellular NAD^+, slowing the rate of glycolysis, electron transport, and therefore of ATP formation, thereby resulting in cell dysfunction and cell death. Hence, inhibitors of PARP should protect against cell death under these conditions. The PARP suicide hypothesis was originally invoked to explain cell death caused by hydrogen peroxide and ionizing radiation. In both cases, the hydroxyl radical is the final reactive species which causes the DNA damage. The contribution of PARP varies with cell type; in endothelial cells, epithelial cells, cardiac myocytes, and fibroblasts PARP plays a greater role in the oxidant damage than, for example, in hepatocytes (4, 5). In most cell types studied, hydrogen peroxide and other agents that generate oxidants and free radicals (including nitric oxide donors, peroxynitrite generators, and streptozotocin) have been shown to induce DNA single-strand breaks and hence PARP activation, followed by cellular energy failure (NAD^+ and ATP depletion). In most of these studies, inhibition or inactivation of PARP significantly improved cellular energetics, cell membrane integrity (viability) and function (6–54) (Table 5.1).

Recent investigations have clarified the mode of oxidant-induced cell death effected by PARP. Low doses of damaging agents, with consequent low levels of DNA damage, lead either to DNA repair or apoptotic cell death. Exposure to massive amounts of DNA damaging agents would lead to correspondingly massive levels of DNA damage, with a consequent massive synthesis of poly (ADP-ribose) polymer and loss of cellular NAD^+. Thus, in contrast to the case with low levels of DNA damage, with high levels of damage one invokes cell suicide; the PARP suicide pathway, the process of 'DNA single-strand breakage $\rightarrow$ PARP activation $\rightarrow$ energy depletion' does not induce apoptosis, but rather induces necrotic cell death characterized by rapid cell injury, changes in membrane integrity, release of lactate dehydrogenase from the cells, and mitochondrial injury (25, 34, 37, 40, 48, 49, 53, 54) (Fig. 5.1). In cells challenged with high levels of peroxynitrite or hydrogen peroxide, inhibition of PARP diverts the mode of cell death from the necrotic towards the apoptotic mode, or protects against cell injury altogether (53–55) (Plate 11). It is important to point out that high levels of oxidant were used in these studies, which would induce a substantial decrease in NAD^+ and ATP levels and much necrosis; while in most earlier studies investigating the role of PARP in apoptosis, the damage was much weaker and induced a milder degree of cell injury, leading to apoptosis.

PARP activation, when associated with necrosis, has been shown to effect the mitochondrial alterations associated with cell necrosis. Mitochondrial alterations play a key role in the process of necrotic cell death. Damage to the mitochondria appears to be the primary early event in necrosis (56). Oxidative phosphorylation in the absence of the mitochondrial permeability

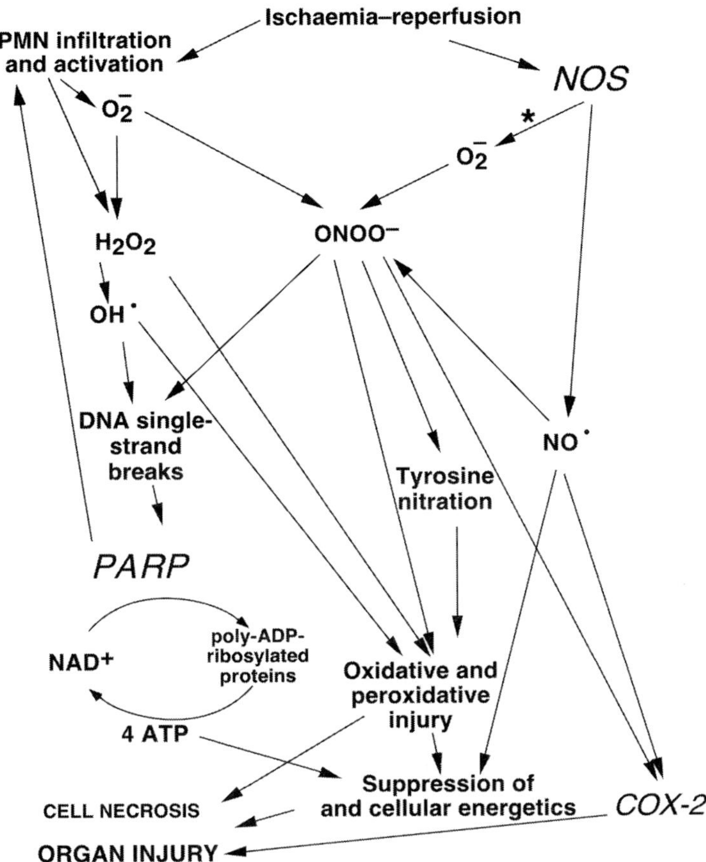

Fig. 5.1 Role of PARP activation in the development of oxidant-induced cell necrosis in during reperfusion injury. Nitric oxide (NO˙), hydroxyl radical (OH˙), superoxide (O₂), and peroxynitrite (ONOO⁻) are produced in the reperfusion stage. NMDA receptor ligands (e.g. glutamate) produced in the reperfused brain activate the constitutive neuronal NOS (bNOS). NO, in turn, combines with superoxide to yield peroxynitrite. Hydroxyl radical (produced from superoxide via the iron-catalysed Haber–Weiss reaction) and peroxynitrite or peroxynitrous acid induce the development of DNA single-strand breakage, with the consequent activation of PARP. Depletion of cellular NAD⁺ leads to inhibition of cellular ATP-generating pathways leading to cellular dysfunction. NO alone does not induce DNA single-strand breakage, but may combine with superoxide (produced from the mitochondrial chain or from other cellular sources) to yield peroxynitrite. Under conditions of low cellular L-arginine, NOS may produce both superoxide (*) and NO, which can then combine to form peroxynitrite. There are PARP-independent, parallel pathways of cellular metabolic inhibition, and these pathways can be activated by NO, hydroxyl radical, superoxide, and by peroxynitrite (alone or in combination or synergy). For instance, peroxynitrite can induce cell injury via protein tyrosine nitration. Activation of the inducible cyclo-oxygenase (COX-2) by NO or peroxynitrite may also amplify the response. PARP activation promotes neutrophil (PMN) recruitment, thereby triggering a positive feedback cycle.

Table 5.1 Role of PARP activation in oxidant-induced cell death[a]

Target cell	Inducer of injury	Mode of PARP inhibition	Effect of PARP inhibition	Reference
Murine islet cells	Streptozotocin; Alloxan	3-AB; Nicotinamide	Protection against islet cell injury and against the inhibition of proinsulin synthesis.	6–8
L1210 murine leukaemia cells; L929 fibroblasts	MNNG	3-AB; Nicotinamide	Prevention of the suppression of NAD^+ and ATP levels and the changes in cellular glucose-6-phosphate levels.	9–11
Murine macrophages, Human lymphocytes	Hydrogen peroxide	3-AB; Nicotinamide	Prevention of NAD^+ loss, ATP loss, and cell death. No effect on the hexose-monophosphate shunt.	12,13
Human umbilical-vein endothelial cells	XO/hypoxanthine; Glucose oxidase/glucose	3-AB	Inhibition of NAD^+ loss, no effect on ATP depletion.	14
Human ovarian and breast tumour lines Murine L929 fibroblasts	TNF-α	3-AB; Nicotinamide	Protection against cytotoxicity in the L9292 cells, enhancement of cytotoxicity in the oncogene-expressing lines. No effect in non-expressing human cells.	15–17
Bovine pulmonary or cardiac endothelial cells	Hydrogen peroxide; Dihydroxyfumarate; Menadione	3-AB; Nicotinamide	No effect on DNA strand breaks, but prevention of NAD^+ and ATP loss and reduction in LDH release.	18–20
Mastocytoma and lymphoma cell lines, fibroblasts	Cytotoxic T-lymphocyte attack	3-AB; Benzamide; $PARP^{-/-}$ cell line	Reduced target cell lysis.	16,17
Human T-lymphocytes	γ-Irradiation	3-AB; Nicotinamide	Protection against the loss of viability and restoration of the impaired T-cell adhesion to endothelial cells.	23

Rat cardiomyocytes, cardiac myoblast cell line	Hydrogen peroxide; Peroxynitrite	3-AB; Nicotinamide	Protection against the loss of NAD^+/ATP, maintained cardiac action potentials, inhibition of peroxidation.	24–26
Human lymphocytes	Hydrogen peroxide; γ-Irradiation; XO/hypoxanthine	3-AB; Nicotinamide	Preservation of GSH and reduced thiol content. NAD^+ depletion was not associated with ATP depletion.	27
Murine L929 fibroblasts Rat primary hepatocytes	Tert-butyl-hydroperoxide	3-AB; Benzamide	Fibroblasts, but not hepatocytes, are protected against the metabolic changes.	28
Human endothelial cells	XO/hypoxanthine; Homocysteine	3-AB	Prevention of ATP depletion, improved cellular viability.	29, 30
Mouse thymocytes	Dexamethasone	3-AB	Protection against cell death and LDH loss.	31
Rat cerebellar granule cells	Glutamate	3-AB; benzamide; 3-aminophthal-hydrazide	Protection against neuronal death, without affecting glutamate-induced calcium influx.	32
Murine pancreatic islet cells	Nitroprusside; SNAP; XO/hypoxanthine	3-AB; nicotinamide; 4-amino-1,8-naphthalimide; $PARP^{-/-}$ phenotype	Prevention of NAD^+ depletion and islet cell lysis.	33–37
Murine lung slices	Cyclophosphamide	3-AB	Protection against NAD^+ loss and LDH release.	38
Rabbit type-II pneumocytes	Hydrogen peroxide	3-AB; Nicotinamide	Maintenance of NAD^+ and ATP levels and surfactant synthesis.	39
Intestinal epithelial cell line HT-29–18-C1	Hydrogen peroxide	3-AB; Nicotinamide	Prevention of the loss of NAD^+ and ATP and protection against cell death. No effect on cell detachment.	40
Human lymphocytes	Nitrogen mustard	Nicotinamide; 3-AB	Protection against cell death, both in pretreatment and post-treatment regimens.	41

Table 5.1 Role of PARP activation in oxidant-induced cell death[a]—*continued*

Target cell	Inducer of injury	Mode of PARP inhibition	Effect of PARP inhibition	Reference
Murine macrophages; rat aortic smooth muscle cells; endothelial cells	Peroxynitrite; LPS/IFN	3-AB; Nicotinamide; INH2BP	Prevention of NAD^+- and ATP-depletion and suppression of mitochondrial respiration; maintenance of endothelial and smooth muscle functions.	42–46
YAC-1 lymphoma cells	SNAP; SNP; Hydrogen peroxide	3-AB; Nicotinamide	Protection against cell lysis.	47
Cerebral cortical culture	NMDA; Sodium nitroprusside; SNAP; SIN-1	Benzamide; 3-AB; 4-AB; 1,5-dihydro-isoquinoline; $PARP^{-/-}$ phenotype	Prevention of cell death.	48, 49
Human neuroblastoma line SH-SY5Y	SNP; SNAP	3-AB; 1,5-isoquinolinediol	Protection against cell death.	50
Pulmonary fibroblasts	Peroxynitrite; Hypoxia/reoxygenation	$PARP^{-/-}$ phenotype	Prevention of cell death.	25, 45
Human pulmonary and intestinal epithelial cells	Peroxynitrite; SIN-1	3-AB; INH2BP	Protection against NAD^+- and ATP-depletion, amelioration of epithelial hyperpermeability.	51, 52
Murine thymocytes	Peroxynitrite; Hydrogen peroxide	$PARP^{-/-}$ phenotype; 3-AB; INH2BP	Protection against cell necrosis and mitochondrial injury.	53, 54

[a]Reprinted, in modified form, with permission, from Szabó, C. and Dawson, V. L. (1998). Role of poly (ADP-ribose) synthetase activation in inflammation and reperfusion injury. *Trends Pharmacol. Sci.*, **19**, 287–298, with permission of Elsevier Science Ltd.
Abbreviations: AB, aminobenzamide; IFN, interferon-γ; INH2BP, 5-iodo-6-amino-1,2-benzopyrone; LDH, lactate dehydrogenase; LPS, bacterial lipopolysaccharide; MNNG, N-methyl-*N*'-nitro-*N*-nitrosoguanidine; NMDA, *N*-methyl-D-aspartate; SIN-1, 3-morpholino-sydnonimine; SNAP, *S*-nitroso-*N*-acetyl-DL-penicillamine; TNF, tumour necrosis factor; XO, xanthine oxidase.

transition leads to ATP depletion, ion dysregulation, and enhanced degradative hydrolase activity. If oxygen is present, toxic oxygen species may be generated and lipid peroxidation can occur. Subsequent cytoskeleton and plasma membrane damage result in plasma membrane bleb formation. If injury continues, bleb rupture and cell lysis occur. In addition, mitochondrial damage may result in the mitochondrial permeability transition, which involves the formation of proteaceous, regulated pores, probably by apposition of inner and outer mitochondrial membrane proteins, which cooperate to form the mitochondrial mega-channel, also known as the mitochondrial permeability transition pore. This latter step is thought to be irreversible and to lead to cell death (57). It is believed that the mitochondrial permeability transition occurs secondary to oxidative stress, increased intracellular Ca^{2+}, and ATP depletion. The permeability transition has important metabolic consequences, namely the collapse of the mitochondrial *trans*-membrane potential, uncoupling of the respiratory chain, hyperproduction of superoxide anions, disruption of mitochondrial biogenesis, outflow of matrix Ca^{2+} and glutathione, and release of soluble intermembrane proteins (57). These mitochondrial alterations have been described in various cell types challenged with peroxynitrite and hydrogen peroxide (54, 58). Importantly, the changes in mitochondrial membrane potential, the mitochondrial permeability transition, the increase in reactive oxygen intermediate production, the increased Ca^{2+} mobilization, and the destruction of mitochondrial structure are attenuated by the inhibition or inactivation of PARP (54). Since cellular NAD^+ and ATP are important regulators of mitochondrial functions, maintenance of energy pools in cells where PARP is inhibited may explain the improvement of the mitochondrial functions (54).

Necrotic cell death—which, according to the conventional wisdom, was not generally believed to be amenable to pharmacological or therapeutic interactions—is an important pathway of cell death, which has direct relevance for various forms of reperfusion injury. Under these conditions, overwhelming oxidant production can occur and cells may die rapidly via necrosis. For example, in the reperfused myocardium, necrotic cells lose the integrity of the plasma membrane and release their intracellular contents. Indeed, the release of intracellular enzymes from specific tissues has long been used as a clinical diagnostic tool; for example, the measurement of the plasma levels of creatine kinase isoforms specific for the heart in the diagnosis of myocardial infarction, or the detection of various hepatic transaminase enzymes in the plasma in the diagnosis of liver injury.

5.3   Tissue damage caused by nitric oxide-derived oxidant species

It was only about 10 years ago that it was discovered that ischaemia–reperfusion injury was, at least in part, mediated by free radicals. In the last decade, the free radical, nitric oxide (NO) has emerged as a major mediator in the control of the cardiovascular and immune systems, both in physiological and pathophysiological conditions. NO is produced by a family of isoenzymes called NO synthases (NOS). The NOS enzymes in the vascular endothelial cells (ecNOS) and in the neurons (bNOS) are present constitutively, and produce only small or moderate amounts of NO; whereas the inducible NOS (iNOS) isoform is expressed in response to immunological stimuli and produces high, cytotoxic concentrations of NO. NOS produces NO by the oxidation of one of the guanidino nitrogens of L-arginine. This process involves the oxidation of NADPH and the reduction of molecular oxygen, utilizing a variety of cofactors. NO diffuses in the cytosol and easily crosses lipid membranes; thus it is either a diffusible mediator of signal transduction or a cytotoxic molecule at higher concentrations (59–61).

For this review the cytotoxic aspects of NO are relevant. Much of the toxicity caused by NO is evident when NO and oxygen-derived free radicals are produced simultaneously. NO and an oxygen radical (like superoxide) react together extremely rapidly to form the peroxynitrite anion ($ONOO^-$) (62). The rate constant for the reaction of superoxide with NO, approximates the diffusion-controlled limit (62, 63). The oxidant reactivity of peroxynitrite is mediated by an intermediate with the biological activity of the hydroxyl radical. However, this product does not appear to be the hydroxyl radical itself, but peroxynitrous acid (ONOOH) or its activated isomer (ONOOH*) (63). Peroxynitrite is a highly reactive species, which causes the rapid oxidation of sulfhydryl groups and thioethers, as well as the nitration and hydroxylation of aromatic compounds, such as tyrosine and tryptophan (62, 63). Peroxynitrite also induces DNA damage; causing both DNA base modification and DNA strand breakage (42, 64).

Although compounds that generate NO have been demonstrated to induce PARP activation (Table 5.2), it is likely that the compound which most often causes DNA single-strand breaks in cells exposed to NO donors is peroxynitrite rather than NO itself. For example, exposure of thymocytes to 3-morpholino-sydnonimine, a compound which releases equimolar quantities of NO and superoxide, induces the development of DNA single-strand breaks and activates PARP. On the other hand, the 'pure' NO donor, *S*-nitroso-*N*-acetyl-DL-penicillamine, neither induces DNA strand breaks nor does it produce energy derangement by PARP activation in macrophages, smooth muscle cells, epithelial cells, fibroblasts, endothelial cells, and cardiac myocytes (25, 42, 65). In the earlier studies in brain slices, PARP was not activated by the 'pure' NO donor sodium nitroprusside, but rather by the combination of sodium nitroprusside and 3-morpholino-sydnonimine (48). In these experiments, 3-morpholino-sydnonimine presumably provided superoxide for the generation of peroxynitrite. Moreover, in pancreatic islet cells, where NO donors induce a PARP-dependent suppression of mitochondrial respiration (34, 37), scavenging of superoxide (or peroxynitrite) with vitamin E prevents the cell damage

Table 5.2 Biologically relevant reactive oxygen and nitrogen species and their relevance to PARP activation

Reactive oxygen or nitrogen species	Selected sources	Diffusible?	Relationship to PARP activation
Superoxide anion	Mitochondrial respiratory chain; Activated neutrophils, macrophages; Xanthine oxidase; Lipid peroxidation; Catechol auto-oxidation; Redox cycling	Yes	Does not induce DNA single-strand breakage of PARP activation. Although not a particularly potent oxidant, it can be cytotoxic via a number of PARP-independent mechanisms.
Hydrogen peroxide	Xanthine oxidase; Mitochondrial respiratory chain; Catechol auto-oxidation; Redox cycling	Yes	Potent trigger of DNA single-strand breakage and PARP activation via the generation of hydroxyl radicals in the vicinity of DNA. Potent oxidizing agent. Can also be cytotoxic via a number of PARP-independent mechanisms.
Hydroxyl radical	From hydrogen peroxide, via the classic Fenton reaction. Monoamine oxidase; In response to ionizing radiation; Catechol auto-oxidation; Redox cycling	No	Potent trigger of DNA single-strand breakage and PARP activation. Can also be cytotoxic via a number of PARP-independent mechanisms. The decomposition of peroxynitrite yields a hydroxyl-radical-like intermediate, which, however, does not appear to be a 'free' hydroxyl radical.
Hypochlorous acid	Activated macrophages	Yes	Does not induce DNA single-strand breakage of PARP activation. Can be cytotoxic via a number of PARP-independent mechanisms.
Nitric oxide	Nitric oxide synthases; Acidification of nitrite; Non-enzymatic pathways	Yes	Does not induce DNA single-strand breakage of PARP activation, unless it combines with superoxide to produce peroxynitrite. Although not a particularly potent oxidant, it can be cytotoxic via a number of PARP-independent mechanisms. It can also exert marked antioxidant and cytoprotective actions.

Table 5.2 Biologically relevant reactive oxygen and nitrogen species and their relevance to PARP activation—*continued*

Reactive oxygen or nitrogen species	Selected sources	Diffusible?	Relationship to PARP activation
Peroxynitrite	Reaction of nitric oxide and superoxide. Nitric oxide synthases can produce peroxynitrite under conditions of low cellular L-arginine levels	Yes	Potent trigger of DNA single-strand breakage and PARP. activation. Can also be cytotoxic via a number of PARP-independent mechanisms.
Nitroxonium anion	Nitric oxide	Yes	Trigger of DNA single- and double-strand breakage. May trigger PARP activation. Can also be cytotoxic via a number of PARP-independent mechanisms.

elicited by NO donors (66). Taken together, it seems that it is peroxynitrite and not NO itself which causes the DNA strand breaks, which in turn initiates PARP activation (42, 64). Normal cellular processes may provide small amounts of superoxide, so that low levels of peroxynitrite are formed. Furthermore, inhibition of mitochondrial respiratory enzymes by NO or peroxynitrite may induce the 'leak' of superoxide from the mitochondria, thereby leading to the generation of peroxynitrite and initiation of a positive feedback cycle. In addition, under conditions of low cellular L-arginine concentrations, NOS itself may generate cytotoxic quantities of peroxynitrite (67).

Nitric oxide can be produced by many cell types. In 1994, two groups, working on the pathogenesis of diabetes and neuronal injury, independently proposed that NO is a potential activator of PARP due to its ability to induce DNA single-strand breaks (34, 48). When pancreatic islet cells or rat cerebellar slices were incubated with high concentrations of NO gas or high concentrations of drugs that can form NO, there was an increase in PARP activity and a consequent decrease in cellular NAD^+ levels leading to cytotoxic effects resembling cellular necrosis (34, 48). In pancreatic islets of the $PARP^{-/-}$ knockout mice, the absence of functional PARP enzyme provides protection against oxidant stress or NO donor compounds (37). Similarly, the genetic inactivation of PARP protects cells in cerebellar slices and pulmonary embryonal fibroblasts exposed to a variety of oxidants (42, 49, 65) (see Table 5.1). Interestingly, however, the protection waned when extremely high concentrations of the oxidants were used (37, 65). This suggests that when cells are exposed to extreme oxidant stress, PARP-independent, parallel pathways of cytotoxicity become activated.

Studies in brain slices have demonstrated that activation of the NMDA receptor increases NO production from endogenous sources, which in turn decreases membrane integrity (48, 49). Inhibitors of NOS, as well as inhibition or inactivation of PARP, decrease the effects caused by N-methyl-D-aspartate (NMDA) receptor activation (48, 49). Another common source of reactive oxygen and nitrogen species is the immunostimulated macrophage. In cultured macrophages and fibroblasts, similar to the results observed with authentic peroxynitrite, we found that stimulating cultured macrophages with bacterial lipopolysaccharide and interferon-gamma results in DNA strand breaks and PARP activation, and that these changes parallel the onset of NO and peroxynitrite production (43, 65). Pharmacological inhibition of PARP activity or inactivation of PARP prevents the depression of mitochondrial respiration in these cells (43, 65).

5.4 Stroke and brain injury

The involvement of superoxide and the protective effect of superoxide neutralizing strategies (68, 69), as well as the involvement of NO and the protective effect of NOS inhibition (70), have been established in various forms of central nervous system injury, including the reperfusion of the ischaemic brain (stroke). Infarct volume following vascular stroke is also markedly diminished in animals treated with NOS inhibitors and in mice with bNOS gene disruption (70). Peroxynitrite, but neither NO nor superoxide separately, is the major cytotoxic mediator in neuronal injury during stroke (71). Reperfusion injury in the central nervous system is associated with the activation of NMDA receptors; these are now believed to play a key role in the pathogenesis of stroke, because they trigger the simultaneous production of superoxide and NO. There is now indirect evidence suggesting that activation of the NMDA receptor is associated with a marked increase in a hydroxyl radical-like reactivity in the brain which may be blocked by inhibition of NOS (72), and which is presumably due to peroxynitrite generation.

Ischaemia–reperfusion injury can be modelled in the laboratory by exposing primary neuronal cultures to glutamate or its agonists, or to various reactive oxygen species, NO donors, peroxynitrite, or by combined oxygen–glucose deprivation. In cerebellar granule cells, glutamate induces a rapid increase in poly (ADP-ribose) immunoreactivity (73). PARP inhibitors are protective in these models of brain injury induced by glutamate or compounds which generate NO. The rank order of potency of different classes of PARP inhibitors correlates with the degree of protection. Moreover, the protection afforded by PARP inhibition is not associated with changes in calcium influx induced by glutamate. In addition, primary cortical cultures from $PARP^{-/-}$ mice are resistant to toxicity from NMDA, compounds which generate NO, and combined oxygen–glucose deprivation (48, 49, 73–79). These systems only model a component of the complex chain of events initiated *in vivo* following an ischaemic insult or stroke. Nevertheless, the pathophysiological relevance of these observations is supported by the observation that increased poly (ADP-ribosyl)ation has been demonstrated in the reperfused brain (80, 81). In $PARP^{-/-}$ mice a markedly reduced infarct volume is observed after transient middle-cerebral artery occlusion (49, 82). The reduction in infarct volume was observed both in $PARP^{-/-}$ mice that had either a genetic background identical to the wild-type strain (82) or a mixed 129/C57B6 genetic background (49). Thus, the reduction in infarct volume was due to the absence of the *PARP* gene product and not to other genetic variables. Following focal ischaemia in the ipsilateral hemisphere, PARP activation was examined by evaluating (ADP-ribose) polymer formation or levels of NAD^+. Poly (ADP-ribose) formation was increased and NAD^+ levels were decreased following focal ischaemia in wild-type tissue, while no poly (ADP-ribose) formation was observed in $PARP^{-/-}$ tissue (49) and NAD^+ levels did not fall (82). PARP activation is mainly related to NO production by bNOS, because in mice deficient in bNOS, which were subjected to middle cerebral artery occlusion–reperfusion, PARP activation is markedly diminished (80). A novel potent PARP inhibitor, 3, 4-dihydro-5-[4–1(1-piperidynil) butoxy]-1(2H)-isoquinolinone, also prevents neural damage following vascular stroke in rats; although, for unexplained reasons, the protection provided by this particular agent wanes when the dose of the agent is increased (83). Finally, the PARP inhibitor, INH2BP, protects against astrocyte damage and stroke development induced by peroxynitrite (81).

The protection observed in the PARP$^{-/-}$ mice exceeds the degree of protection reported for any other transgenic model, including the bNOS$^{-/-}$ mice. This observation suggests a common role for PARP activation by other excitotoxic mechanisms, in addition to the production of free radicals and NO (see also Chapters 4 and 7).

It has been suggested that PARP activation contributes to the pathogenesis of other forms of brain injury and to neurodegenerative disorders. For instance, it has been reported that in spinal cord sections taken following traumatic spinal cord injury there is evidence of peroxynitrite formation and activation of PARP (84). PARP inhibitors provide protection in an *in vitro* model of traumatic brain injury (74). Furthermore, inactivation of PARP markedly improves the functional outcome in traumatic brain injury (85).

In addition, PARP activation has been implicated in the pathogenesis of Parkinson's disease, a chronic progressive neurological disorder related to the degeneration of neurons in the substantia nigra which contain melanin; this is another disease where activation of NMDA receptors plays a crucial role in its pathogenesis. The synthetic heroin analogue, 1-methyl-4-phenyl-1,2,3,6-tetrahydropyridine (MPTP) can selectively damage neurons in the nigrostriatal dopaminergic pathway and produce parkinsonism in experimental animals (86). There is evidence for both the production of free radicals and oxidants derived from oxygen (87), and for the overproduction of NO and peroxynitrite (88) in the pathogenesis of MPTP neurotoxicity. In brain injury induced by MPTP, the neuronal NO synthase (bNOS) is the source of cytotoxic NO and peroxynitrite. Accordingly, protection is provided by the bNOS inhibitor 7-nitro-indazole (88, 89). Furthermore, genetically engineered mice which lack the bNOS gene are significantly more resistant to toxicity induced by MPTP than their wild-type littermates (89). Direct evidence for the involvement of PARP in the pathogenesis of toxicity induced by MPTP comes from a mouse model of Parkinson's disease. MPTP treatment reduces striatal dopamine and cortical noradrenaline levels by more than 50% in these animals, while simultaneous treatment with each of five different inhibitors of PARP ameliorates the catecholamine depletion induced by MPTP. The protective potency of benzamide and its derivatives parallels their efficacy as enzyme inhibitors (90). Furthermore, mice lacking a functional *PARP-1* gene are also resistant to MPTP neurotoxicity (91).

The advantages of PARP inhibitors as agents for the treatment of stroke, brain injury, and also other forms of reperfusion injuries include the possibility of effective delayed treatment, since PARP inhibitors target a delayed process of cell death. Most PARP inhibitors, in the therapeutic dose range, tend to have little influence on haemodynamic parameters, which can also be considered an advantage. It is clear that PARP is not the only factor involved in the pathogenesis of cell and organ injury in response to oxidant or nitrosative stress. The relative importance of PARP in mediating oxidant injury is dependent on cell type. Furthermore, the protection wanes when cells are challenged with extremely high concentrations of oxidants that trigger cytotoxic effects independent of PARP, such as direct inhibitory effects on the mitochondrial respiratory chain or inhibitory effects on other intracellular energetic or redox processes. Finally, direct interactions of the oxidants with proteins, lipids, arachidonic acid, and other molecules may also play a significant role in the development of cellular injury. It is likely that there are important synergistic interactions between PARP activation and these other cellular processes of cytotoxicity (Fig. 5.1). Nevertheless, from the evidence presented here it seems that inhibition of PARP alone can 'tip the balance' and significantly influence the outcome of stroke, brain injury, and various other ischaemia–reperfusion states.

5.5   Myocardial ischaemia–reperfusion

The participation of oxidants and free radicals derived from oxygen and nitrogen, is also well established in the pathogenesis of myocardial ischaemia–reperfusion injury (92, 93). Furthermore, activation of PARP may also be responsible, at least in part, for the development of reperfusion injury in the heart. In experimental rabbit myocardial infarction, pharmacological inhibitors of PARP, such as nicotinamide or 3-aminobenzamide (3-AB), given immediately prior to the reperfusion of the ischaemic myocardium substantially reduced the infarct size, while the structurally related, but inactive analogue, 3-aminobenzoic acid did not protect (92). Similarly, in anaesthetized rats 3-AB gives protection after myocardial infarction elicited by occlusion and reperfusion of the left coronary artery (93), as shown by:

(1) decrease in infarct size;
(2) reduction of histological damage;
(3) reduced plasma creatine phosphokinase activity, which means less myocardial necrosis;
(4) reduced cardiac myeloperoxidase content; and
(5) higher myocardial ATP levels. (Fig. 5.2).

These data were confirmed using PARP knockout (–/–) animals. These animals are protected against myocardial ischaemia and reperfusion injury (demonstrated by reduced myocyte necrosis), as shown by reduced histological damage and lower plasma creatine phosphokinase activity after reperfusion. Furthermore, there is a reduced neutrophil infiltration into the

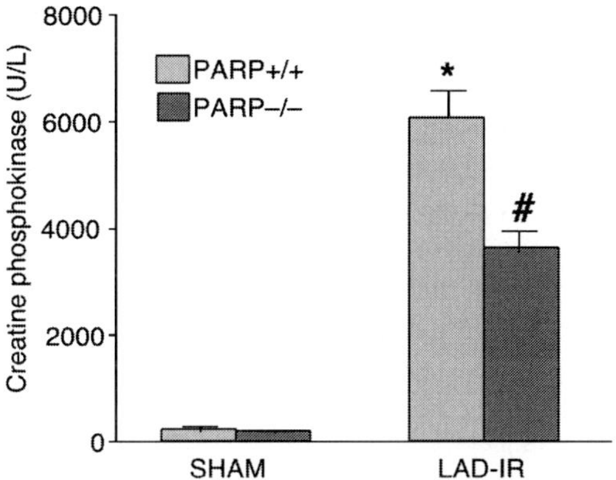

Fig. 5.2 Role of PARP activation in the development of peroxynitrite (ONOO⁻) induced cell necrosis and apoptosis in thymocytes challenged with peroxynitrite. Thymocytes were treated for 4h with 5,10 and 20 μM peroxynitrite or 10 μM dexamethasone. Cells were then stained with Annexin V-FITC and propidium iodide (PI) and two-color analysis was performed by flow cytometry. An increase in the number of apoptotic (Annexin V-FITC single positive) cells was observed in response to 5 μM peroxynitrite treatment, whereas necrotic (stained by both Annexin V-FITC and PI) cells dominated in response to 20 μM peroxynitrite PARS-deficient cells were protected against the necrotic loss of membrane integrity, as indicated by decreased PI uptake, and inhibition or genetic inactivation of PARP increased the number of cells in the normal population, as well as in the apoptotic population. Reprinted, in modified form, with permission, from: Virág, L., Scott, G.S., Salzman, A.L., and Szabó, C. (1998). Peroxynitrite-induced thymocyte apoptosis: the role of caspases and poly-(ADP-ribose) synthetase (PARS) activation. Immunology **94**, 345–355, with permission of Blackwell Science Ltd.

reperfused hearts in the absence of PARP, which is associated with a reduced expression of P-selectin and intercellular adhesion molecule-1 (ICAM-1) (94). The resistance of the PARP knockout mice to myocardial reperfusion injury is also evident 24 hours after the release of the occlusion. Furthermore, the production of inflammatory cytokine mediators, such as TNF-α and free radicals (nitric oxide), is also reduced in the PARP$^{-/-}$ animals, consistent with an overall suppression of the severity of the myocardial inflammation in the absence of PARP (95).

The reduced cardiac myeloperoxidase activity in the presence of 3-AB or in the PARP$^{-/-}$ animals (93–95), indicates that PARP plays a role in controlling the infiltration of neutrophil granulocytes into the reperfused myocardium. Although 3-AB is not a direct scavenger of peroxynitrite and does not inhibit the synthesis or action of its precursors, inhibition or genetic inactivation of PARP also caused a marked reduction in the amount of reactive peroxynitrite formed in the reperfused myocardium (93–95). This latter observation suggests that PARP inhibition, in an indirect way, influences the amounts of reactive peroxynitrite produced. (For a possible explanation of this result see

Chapters 3 and 7). It is most likely that inhibition of PARP interrupts a self-amplifying, positive feed-forward cycle, which involves: (1) neutrophil recruitment $\rightarrow$ (2) peroxynitrite formation $\rightarrow$ (3) endothelial injury and/or upregulation of adhesion receptors $\rightarrow$ (4) more neutrophil recruitment. This cycle can be interrupted by PARP inhibitors at the level of endothelial injury (45) or at the level of adhesion receptor expression (94).

In fibroblasts lacking a functional *PARP* gene, expression of ICAM-1 stimulated by cytokines, is significantly reduced when compared to fibroblasts from animals which express PARP (94). Similarly, in cultured human endothelial cells, the expression of P-selectin and ICAM-1 dependent on oxidants or cytokines was reduced when PARP was inhibited by 3-AB (94). These studies point towards a direct regulation by PARP of the expression of adhesion receptors. PARP is reported to regulate the expression of a number of other genes, which include the major histocompatibility complex class II gene, *ras*, c-*myc*, DNA methyltransferase gene, protein kinase C, collagenase, and the gene for the inducible NO synthase (96, 97). The mechanisms by which PARP regulates gene expression may be related to an effect on activator-dependent transcription (98), and/or via an effect on activation of the transcription factor NF-κB (99). (See Chapters 3 and 7 for further discussion of these ideas.)

5.6   Splanchnic ischaemia–reperfusion

A series of studies in anaesthetized rats suggested that the peroxynitrite–PARP pathway may be significant in the pathophysiological changes seen in the reperfused gut (46). Splanchnic occlusion shock can be induced in rats by clamping both the superior mesenteric artery and the coeliac trunk for 45 min, followed by release of the clamp (reperfusion). There was a marked increase in the oxidation of dihydrorhodamine-123 to rhodamine (a marker of oxidative processes induced by peroxynitrite) in the plasma of these shocked rats after reperfusion, but not during ischaemia alone (46). Immunohistochemical examination showed a marked increase in nitrotyrosine, indicating the presence of peroxynitrite, in the necrotic ileum of the shocked rats. In addition, in *ex vivo* studies of aortic rings from these shocked rats there was a reduction in the contractions caused by noradrenaline and also impaired responsiveness to the relaxant effect of acetylcholine, which means vascular hyporeactivity and endothelial dysfunction, respectively.

Splanchnic artery ischaemia and reperfusion also resulted in a marked increase in epithelial permeability. Inhibition of PARP with 3-AB treatment significantly reduced ischaemia–reperfusion injury in the bowel as evaluated by histological examination. It also significantly increased mean arterial blood pressure, improved contractile responsiveness to noradrenaline, enhanced the endothelium-dependent relaxation, and reduced the reperfusion-induced increase in epithelial permeability (46). 3-AB also prevented the infiltration of neutrophils into the reperfused intestine, as evidenced by reduced myeloperoxidase activity, improved the histological status of the reperfused tissues, reduced the production of peroxynitrite during reperfusion, and improved survival (46). Similarly, in PARP$^{-/-}$ (null) mice the degree of injury and neutrophil infiltration in splanchnic ischaemia–reperfusion injury is decreased (100). These results demonstrate that PARP inhibition exerts multiple protective effects in splanchnic artery occlusion–reperfusion shock, and suggest that peroxynitrite and/or hydroxyl radicals, produced during the reperfusion phase, cause DNA strand breaks, PARP activation, and subsequent cellular dysfunction. Therefore in shock caused by splanchnic occlusion, the vascular endothelium may be an important site of protection by PARP inhibition. The reduced neutrophil infiltration and reduced nitrotyrosine levels seen after inhibition or inactivation of PARP may be related to the interruption of positive feedback cycles, similar to that already described in relation to the reperfused heart.

5.7 Haemorrhagic shock

Severe blood loss followed by fluid resuscitation is considered to be a form of 'whole-body ischaemia–reperfusion injury'. Recent investigations tested the effects of pharmacological inhibitors of PARP in rodents and pigs. In rats suffering severe haemorrhagic shock treated with the PARP inhibitor INH2BP, there was a significant improvement in the survival rate; the inhibitor increased the 50% survival time from 30 min to approximately 70 min. Moreover, treated animals maintained higher mean arterial blood pressure values. Similar to these protective effects, the NOS inhibitor L-NMA as well as a cell-permeable superoxide dismutase mimetic, improved survival times and blood pressure during haemorrhagic shock (101).

Similarly, in pigs resuscitated with Ringer's lactate solution after haemorrhagic shock, inhibition of PARP with 3-AB prevented the deterioration of mean arterial pressure and cardiac output during the resuscitation phase, thereby prolonging survival. These observations suggest that the cardiovascular decompensation in haemorrhagic shock is, at least in part, mediated by the activation of PARP (102). Because the PARP inhibitor neither reduced afterload nor increased heart rate, the basis for the increased cardiac index presumably represents an increase in stroke volume secondary to an augmentation of venous return, or an inotropic effect. The improved myocardial performance in the animals treated with the PARP inhibitor was also associated with a slight effect on the left atrial pressure during resuscitation. Therefore, perhaps PARP (or 3-aminobenzamide) affects vasoconstrictor and venous capacitance. The role of PARP-1 in hemorrhagic shock was

Table 5.3 Role of PARP activation in various forms of ischaemia–reperfusion

Reperfusion model	Mode of PARP inhibition	Effect of PARP inhibition	Ref.
Transient focal cerebral ischaemia/ reperfusion in rodents	3-AB, 3,4-dihydro-5-[4–1 (1-piperidynil)buthoxy]-1(2H)-isoquinolinone	Suppression in infarct size	74, 75
Transient focal cerebral ischaemia/ reperfusion in mice	PARP$^{-/-}$ phenotype	Marked suppression in infarct size	49, 82
Rat splanchnic ischaemia/reperfusion	3-AB	Reduction in epithelial hyperpermeability, improved histology, reduced neutrophil infiltration. Reduction in the apparent reactive peroxynitrite formation.	46, 102
Rat and rabbit myocardial ischaemia/ reperfusion	3-AB; 4-AB; 4-amino-1,8-naphthalimide; 1,5-dihydroisoquinoline	Reduction in infarct size; improved myocardial contractility, reduced plasma creatinine phosphokinase levels, improved myocardial ATP, reduced neutrophil infiltration. Reduction in the apparent reactive peroxynitrite formation.	92, 93, 105
Skeletal muscle ischaemia/reperfusion in the rabbit	3-AB; nicotinamide; 1,5-dihydro-isoquinoline	Reduction in the degree of muscle necrosis	92
Rat model of retinal ischaemia/ reperfusion	3-AB	Prevention of loss of inner retinal thickness; improved histology	106

[a]Reprinted, in modified form, with permission, from Szabó, C. and Dawson, V. L. (1998). Role of poly (ADP-ribose) synthetase activation in inflammation and reperfusion injury. *Trends Pharmacol. Sci.*, **19**, 287–298, with permission of Elsevier Science Ltd.
Abbreviations: AB, aminobenzamide; INH2BP, 5-iodo-6-amino-1,2-benzopyrone; NMDA, N-methyl-D-aspartate.

recently confirmed using PARP-1 deficient mice. The PARP-1 deficient mice respond to hemorrhagic shock with maintained blood pressure, increased survival rate, and improved intestinal function (103).

5.8   Other pathophysiological conditions—conclusions

The effects of PARP inhibition or absence in reperfusion shock are summarized in Table 5.3. PARP activation in reperfusion injury is not confined to the brain, heart, or gut. Recent reports have demonstrated that the inhibition of PARP provides protective effects in the reperfused, skeletal muscle, retina, and other organs (92, 104). However, pharmacological inhibition of PARP is without beneficial effect in the reperfused liver, perhaps consistent with the lack of protection afforded by PARP inhibitors on hepatocytes exposed to oxidants as described earlier.

In addition to the various forms of reperfusion injury, PARP activation has been implicated in the pathophysiology of a variety of local and systemic inflammatory conditions, including arthritis, diabetes, allergic encephalomyelitis, inflammatory bowel disease, and endotoxic shock (96, 97). It is noteworthy that since the publication of that review in 1998, several independent groups have reported that PARP knockout animals do not develop diabetes after treatment with high doses of streptozotocin (105, 107), nor after multiple low doses of streptozotocin (Mabley and Szabó, unpublished observations 1999).

Over the last 5 years, PARP has emerged as a novel pharmacological target for the therapy of ischaemia–reperfusion. Future studies, with novel, potent and selective PARP inhibitors, in carefully selected patient populations suffering from acute reperfusion injury, such as stroke or myocardial infarction, are expected. Such studies should clarify the practical, clinical utility of this approach.

Acknowledgement

This work was supported by grants from the National Institutes of Health, R29GM54773 and R01GM60915-01.

References

1. Berger, N. A., Sims, J. L., Catino, D. M., and Berger, S. J. (1983). Poly (ADP-ribose) mediates the suicide response to massive DNA damage: studies in normal and DNA repair defective cells. In *ADP-ribosylation, DNA repair, and cancer* (ed. M. Miwa, O. Hayaishi, S. Shall, M. Smulson, and T. Sugimura), pp. 219–226. Japan Science Society Press, Tokyo.

2. Berger, N. A. (1985). Poly (ADP-ribose) in the cellular response to DNA damage. *Radiat. Res.* **101**, 4.
3. Berger, N. A. and Berger, S. J. (1986). Metabolic consequences of DNA damage: the role of poly (ADP-ribose) polymerase as mediator of the suicide response. *Basic Life Sci.* **38**, 357.
4. Berger, N. A., Berger, S. J., and Gerson, S. L. (1987). DNA repair, ADP-ribosylation and pyridine nucleotide metabolism as targets for cancer chemotherapy. *Anticancer Drug Des.* **2**, 203.
5. Berger, N. A. (1991). Oxidant-induced cytotoxicity: a challenge for metabolic modulation. *Am. J. Respir. Cell. Mol. Biol.* **4**,.
6. Yamamoto, H., Uchigata, Y., and Okamoto, H. (1981). Streptozotocin and alloxan induce DNA strand breaks and poly(ADP-ribose) synthetase in pancreatic islets. *Nature* **294**, 284.
7. Uchigata, Y., Yamamoto, H., Kawamura, A., and Okamoto, H. (1982). Protection by superoxide dismutase, catalase, and poly(ADP-ribose) synthetase inhibitors against alloxan- and streptozotocin-induced islet DNA strand breaks and against the inhibition of proinsulin synthesis. *J. Biol. Chem.* **257**, 6084.
8. Sandler, S., Welsh, M., and Andersson, A. (1983). Streptozotocin-induced impairment of islet B-cell metabolism and its prevention by a hydroxyl radical scavenger and inhibitors of poly(ADP-ribose) synthetase. *Acta Pharmacol. Toxicol.* **53**, 392.
9. Sims, J. L., Berger, S. J., and Berger, N. A. (1983). Polymerase inhibitors preserve nicotinamide adenine dinucleotide and adenosine 5'-triphosphate pools in DNA-damaged cells: mechanism of stimulation of unscheduled DNA synthesis. *Biochemistry* **22**, 5188.
10. Berger, S. J., Sudar, D. C., and Berger, N. A. (1986). Metabolic consequences of DNA damage: DNA damage induces alterations in glucose metabolism by activation of poly (ADP-ribose) polymerase. *Biochem. Biophys. Res. Comm.* **134**, 227.
11. Mizumoto, K. and Farber, J. L. (1995). Growth inhibition and cell killing by *N*-methyl-*N*-nitrosourea: metabolic alterations that accompany poly(ADP-ribosyl)ation. *Arch. Biochem. Biophys.* **319**, 512.
12. Schraufstatter, I. U., Hinshaw, D. B., Hyslop, P. A., Spragg, R. G., and Cochrane, C. G. (1986). Oxidant injury of cells. DNA strand-breaks activate polyadenosine diphosphate-ribose polymerase and lead to depletion of nicotinamide adenine dinucleotide. *J. Clin. Invest.* **77**, 1312.
13. Schraufstatter, I. U., Hyslop, P. A., Hinshaw, D. B., Spragg, R. G., Sklar, L. A., and Cochrane CG (1986). Hydrogen peroxide-induced injury of cells and its prevention by inhibitors of poly(ADP-ribose) polymerase. *Proc. Natl Acad. Sci. USA* **83**, 4908.
14. Andreoli, S. P. (1989). Mechanisms of endothelial cell ATP depletion after oxidant injury. *Pediatr. Res.* **25**, 97.
15. Agarwal, S., Drysdale, B. E., and Shin, H. S. (1988). Tumor necrosis factor-mediated cytotoxicity involves ADP-ribosylation. *J. Immunol.* **140**, 4187.
16. Lichtenstein, A., Gera, J. F., Andrews, J., Berenson, J. and Ware, C. F. (1991). Inhibitors of ADP-ribose polymerase decrease the resistance of HER2/neu-expressing cancer cells to the cytotoxic effects of tumor necrosis factor. *J. Immunol.* **146**, 2052.

17. Agarwal, S. and Piesco, N. P. (1994). Poly ADP-ribosylation of a 90-kDa protein is involved in TNF-alpha-mediated cytotoxicity. *J. Immunol.* **153**, 473.

18. Kirkland, J. B. (1991). Lipid peroxidation, protein thiol oxidation and DNA damage in hydrogen peroxide-induced injury to endothelial cells: role of activation of poly(ADP-ribose)polymerase. *Biochim. Biophys. Acta* **1092**, 319.

19. Thies, R. L. and Autor, A. P. (1991). Reactive oxygen injury to cultured pulmonary artery endothelial cells: mediation by poly(ADP-ribose) polymerase activation causing NAD^+ depletion and altered energy balance. *Arch. Biochem. Biophys.* **286**, 353.

20. Kossenjans, W., Rymaszewski, Z., Barankiewicz, J., Bobst, A., and Ashraf, M. (1996). Menadione-induced oxidative stress in bovine heart microvascular endothelial cells. *Microcirculation* **3**, 39.

21. Hayward, A. R. and Herberger, M. (1988). Nicotinamide protects target cells from cell-mediated cytolysis. *Cell. Immunol.* **113**, 414.

22. Redegeld, F. A., Chatterjee, S., Berger, N. A., and Sitkovsky, M. V. (1992). Poly-(ADP-ribose) polymerase partially contributes to target cell death triggered by cytolytic T lymphocytes. *J. Immunol.* **149,** 3509.

23. Piela-Smith, T. H., Aune, T., Aneiro, L., Nuveen, E., and Korn, J. H. (1992). Impairment of lymphocyte adhesion to cultured fibroblasts and endothelial cells by gamma-irradiation *J. Immunol.* **148**, 41.

24. Janero, D. R., Hreniuk, D., and Sharif, H. M. (1994). Hydroperoxide-induced oxidative stress impairs heart muscle cell carbohydrate metabolism. *Am. J. Physiol.* **266**, C179.

25. Gilad, E., Zingarelli, B., Salzman, A. L., and Szabó, C. (1997). Protection by inhibition of poly (ADP-ribose) synthetase against oxidant injury in cardiac myoblasts *in vitro. J. Mol. Cell. Cardiol.* **29**, 2585.

26. Bhatnagar, A. (1997). Contribution of ATP to oxidative stress-induced changes in action potential of isolated cardiac myocytes. *Am. J. Physiol.* **272**, H1598.

27. Marini, M., Frabetti, F., Brunelli, M. A., and Raggi, M. A. (1993). Inhibition of poly(ADP-ribose) polymerization preserves the glutathione pool and reverses cytotoxicity in hydrogen peroxide-treated lymphocytes. *Biochem. Pharmacol.* **46**, 2139.

28. Yamamoto, K., Tsukidate, K., and Farber, J. L. (1993). Differing effects of the inhibition of poly(ADP-ribose) polymerase on the course of oxidative cell injury in hepatocytes and fibroblasts. *Biochem. Pharmacol.* **46**, 483.

29. Aalto, T. K. and Raivio, K. O. (1993). Nucleotide depletion due to reactive oxygen metabolites in endothelial cells: effects of antioxidants and 3-aminobenzamide. *Pediatr. Res.* **34**, 572.

30. Blundell, G., Jones, B. G., Rose, F. A., and Tudball, N. (1996). Homocysteine mediated endothelial cell toxicity and its amelioration.*Atherosclerosis* **122**, 163.

31. Hoshino, J., Beckmann, G., and Kroger, H. (1993). 3-Aminobenzamide protects the mouse thymocytes *in vitro* from dexamethasone-mediated apoptotic cell death and cytolysis without changing DNA strand breakage. *J. Steroid Biochem. Mol. Biol.* **44**, 113.

32. Cosi, C. Suzuki, H., Milani, D., Facci, L., Menegazzi, M., Vantini, G., and Kanai, C. (1994). Poly(ADP-ribose) polymerase: early involvement in glutamate-induced neurotoxicity in cultured cerebellar granule cells. *J. Neurosci. Res.* **39**, 38.

33. Burkart, V. and Kolb, H. (1993). Protection of islet cells from inflammatory cell death *in vitro*. *Clin. Exp. Immunol.* **93**, 273.

34. Radons, J., Heller, B., Burkle, A., Hartmann, B., Rodriguez, M. L., Kroncke, K. D., Burkart, V., and Kolb, H. (1994). Nitric oxide toxicity in islet cells involves poly (ADP-ribose) polymerase activation and concomitant NAD$^+$ depletion. *Biochem. Biophys. Res. Commun.* **199**, 1270.

35. Heller, B., Burkle, A., Radons, J., Fengler, E., Jalowy, A., Muller, M., Burkart, V., and Kolb, H. (1994). Analysis of oxygen radical toxicity in pancreatic islets at the single cell level. *Biol. Chem. Hoppe-Seyler* **375**, 597.

36. Inada, C., Yamada, K., Takane, N., and Nonaka, K. (1995). Poly(ADP-ribose) synthesis induced by nitric oxide in a mouse beta-cell line. *Life Sci.* **56**, 1467.

37. Heller, B., Wang, Z. Q., Wagner, E. F., Radons, J., Burkle, A., Fehsel, K., Burkart, V., and Kolb, H. (1995). Inactivation of the poly(ADP-ribose) polymerase gene affects oxygen radical and nitric oxide toxicity in islet cells. *J. Biol. Chem.* **270**, 11176.

38. Hoyt, D. G. and Lazo, J. S. (1994). Acute pneumocyte injury, poly(ADP-ribose) polymerase activity, and pyridine nucleotide levels after *in vitro* exposure of murine lung slices to cyclophosphamide. *Biochem. Pharmacol.* **48**, 1757.

39. Hudak, B. B., Tufariello, J., Sokolowski, J., Maloney, C., and Holm, B. A. (1995). Inhibition of poly(ADP-ribose) polymerase preserves surfactant synthesis after hydrogen peroxide exposure. *Am. J. Physiol.* **269**, L59.

40. Watson, A. J., Askew, J. N., and Benson, R. S. (1995). Poly(adenosine diphosphate ribose) polymerase inhibition prevents necrosis induced by H2O2 but not apoptosis. *Gastroenterology* **109**, 472.

41. Meier, H. L. (1996). The time-dependent effect of 2, 2'-dichlorodiethyl sulfide (sulfur mustard, HD, 1, 1'-thiobis [2-chloroethane]#) on the lymphocyte viability and the kinetics of protection by poly(ADP-ribose) polymerase inhibitors. *Cell Biol. Toxicol.* **12**, 147.

42. Szabó, C., Zingarelli, B., O'Connor, M., and Salzman, A. L. (1996). DNA strand breakage, activation of poly-ADP ribosyl synthetase, and cellular energy depletion are involved in the cytotoxicity in macrophages and smooth muscle cells exposed to peroxynitrite. *Proc. Natl. Acad. Sci. USA* **93**, 1753.

43. Zingarelli, B., O'Connor, M., Wong, H., Salzman, A. L., and Szabó, C. (1996). Peroxynitrite-mediated DNA strand breakage activates poly-ADP ribosyl synthetase and causes cellular energy depletion in macrophages stimulated with bacterial lipopolysaccharide. *J. Immunol.* **156**, 350.

44. Szabó, C., Zingarelli, B., and Salzman, A. L. (1996). Role of poly-ADP ribosyltransferase activation in the vascular contractile and energetic failure elicited by exogenous and endogenous nitric oxide and peroxynitrite. *Circ. Res.* **78**, 1051.

45. Szabó, C., Cuzzocrea, S., Zingarelli, B., O'Connor, M., and Salzman, A. L. (1997). Endothelial dysfunction in a rat model of endotoxic shock. Importance of the activation of poly (ADP-ribose) synthetase by peroxynitrite. *J. Clin. Invest.* **100**, 723.

46. Cuzzocrea, S., Zingarelli, B., O'Connor, M., Salzman, A. L., and Szabó, C. (1997). Role of peroxynitrite and activation of poly (ADP-ribose) synthetase in the vascular failure induced by zymosan-activated plasma. *Br. J. Pharmacol.* **122**, 493.

47. Filep, J. G., Lapierre, C., Lachance, S., and Chan, J. S. (1997). Nitric oxide co-operates with hydrogen peroxide in inducing DNA fragmentation and cell lysis in murine lymphoma cells. *Biochem. J.* **321**, 897.

48. Zhang, J., Dawson, V. L., Dawson, T. M., and Snyder, S. H. (1994). Nitric oxide activation of poly (ADP-ribose) synthetase in neurotoxicity. *Science* **263**, 687.

49. Eliasson, M. J. L, Sampei, K., Mandir, A. S., Hurn, P. D., Traystman, R. J., Bao, J., Pieper, A., Wang, Z. Q., Dawson, T. M., Snyder, S. H., and Dawson, V. L. (1997). Poly (ADP-ribose) polymerase gene disruption renders mice resistant to cerebral ischaemia. *Nature Med.* **3**, 1089.

50. Kamoshima, W., Kitamura, Y., Nomura, Y., and Taniguchi, T. (1997). Possible involvement of ADP-ribosylation of particular enzymes in cell death induced by nitric oxide-donors in human neuroblastoma cells. *Neurochem. Int.* **30**, 305.

51. Szabó, C., Saunders, C., O'Connor, M., and Salzman, A. L. (1997). Peroxynitrite causes energy depletion and increases permeability via activation of poly (ADP-ribose) synthetase in pulmonary epithelial cells. *Am. J. Respir. Cell. Mol. Biol.* **16**, 105.

52. Kennedy, M., Szabó, C., and Salzman, A. L. (1998). Activation of polyADP ribosyl synthetase (PARS) mediates cytotoxicity induced by peroxynitrite in human intestinal epithelial cells. *Gastroenterology* **114**, 510.

53. Virág, L., Scott, G. S., Salzman, A. L. and Szabó, C. (1998). Peroxynitrite-induced thymocyte apoptosis: the role of caspases and poly-(ADP-ribose) synthetase (PARS) activation. *Immunology* **94**, 345.

54. Virág, L., Salzman, A. L., and Szabó, C. (1998). Poly (ADP-ribose) synthetase activation mediates mitochondrial injury during oxidant-induced cell death. *J. Immunol.* **161**, 3753.

55. Palomba, L., Sestili, P., Cattabeni, F., Azzi, A., and Cantoni, O. (1996). Prevention of necrosis and activation of apoptosis in oxidatively injured human myeloid leukemia U937 cells. *FEBS Lett.* **390**, 91.

56. Zamzami, N., Hirsch, T., Dallaporta, B., Petit, P. X., and Kroemer, G. (1997). Mitochondrial implication in accidental and programmed cell death: apoptosis and necrosis. *J. Bioenerg. Biomembr.* **29**, 185.

57. Rosser, B. G. and Gores, G. J. (1995). Liver cell necrosis: cellular mechanisms and clinical implications. *Gastroenterology* **108**, 252.

58. Gow, A. J., Thom, S. R., and Ischiropoulos, H. (1998). Nitric oxide and peroxynitrite-mediated pulmonary cell death. *Am. J. Physiol.* **274**, L112.

59. Southan, G. J. and Szabó, C. (1995). Selective pharmacological inhibition of distinct nitric oxide synthase isoforms. *Biochem. Pharmacol.* **51**, 383.

60. Ignarro, L. J. (1991). Signal transduction mechanisms involving nitric oxide. *Biochem. Pharmacol.* **41**, 485.

61. Geller, D. A. and Billiar, T. R. (1998). Molecular biology of nitric oxide synthases. *Cancer Metastasis Rev.* **17**, 7.

62. Szabó, C. (1996). The role of peroxynitrite in the pathophysiology of shock, inflammation and ischaemia–reperfusion injury. *Shock* **6**, 79.

63. Beckman, J. S. and Koppenol, W. H. (1996). Nitric oxide, superoxide, and peroxynitrite: the good, the bad, and ugly. *Am. J. Physiol.* **271**, C1424.

64. Szabó, C. and Ohshima, H. (1997). DNA injury induced by peroxynitrite. *Nitric Oxide Biol. Chem.* **1**, 323.

65. Szabó, C., Virág, L., Cuzzocrea, S., Scott, G. J., Hake, P., O'Connor, M. P.,

Zingarelli, B., Salzman, A. L., and Kun, E. (1998). Protection against peroxynitrite-induced fibroblast injury and arthritis development by inhibition of poly (ADP-ribose) synthetase. *Proc. Natl. Acad. Sci. USA*, **95**, 3867.

66. Burkart, V., Gross-Eick, A., Bellmann, K., Radons, K. J., and Kolb, H. (1995). Suppression of nitric oxide toxicity in islet cells by alpha-tocopherol. *FEBS Lett.* **364**, 259.

67. Xia, Y., Dawson, V. L., Dawson, T. M., Snyder, S. H., and Zweier, J. L. (1996). Nitric oxide synthase generates superoxide and nitric oxide in arginine-depleted cells leading to peroxynitrite-mediated cellular injury. *Proc. Natl. Acad. Sci. USA*, 93, 6770.

68. Fagni, L., Lafon-Cazal, M., Roundouin, G., Manzoni, O., Lerner-Natoli, M., and Bockaert, J. (1994). The role of free radicals in NMDA-dependent neurotoxicity. *Progr. Brain Res.* **103**, 381.

69. Lafon-Cazal, M., Culcasi, M., Gaven, F., Pietri, S., and Bockaert, J. (1993). Nitric oxide, superoxide and peroxynitrite: putative mediators of NMDA-induced cell death in cerebellar granule cells. *Neuropharmacology* **32**, 1259.

70. Dawson, V. L. (1995). Nitric oxide: role in neurotoxicity. *Clin. Exp. Pharmacol. Physiol.* **22**, 305.

71. Beckman, J. S. (1991). The double-edged role of nitric oxide in brain function and superoxide-mediated injury. *J. Dev. Physiol.* **15**, 53.

72. Hammer, B., Parker, W. D. Jr, and Bennett, J. P. Jr. (1993). NMDA receptors increase OH radicals *in vivo* by using nitric oxide synthase and protein kinase C. *Neuroreport* **5**, 72.

73. Cosi, C., Suzuki, H., Milani, D., Facci, L., Menegazzi, M., Vantini, G., Kanai, Y., and Skaper, S. D. (1994). Poly(ADP-ribose) polymerase: early involvement in glutamate-induced neurotoxicity in cultured cerebellar granule cells. *J. Neurosci. Res.* **39**, 38.

74. Wallis, R. A., Panizzon, K. L., and Girard, J. M. (1996). Traumatic neuroprotection with inhibitors of nitric oxide and ADP-ribosylation. *Brain Res.* **710**, 169.

75. Wallis, R. A., Panizzon, K. L., Hanry, D., and Wasterlain, C. G. (1993). Neuroprotection against nitric oxide injury with inhibitors of ADP-ribosylation. *Neuroreport* **5**, 245.

76. Zhang, J. and Steiner, J. P. (1995). Nitric oxide synthase, immunophyllins and poly (ADP-ribose) synthetase: novel targets for the development of neuroprotective drugs. *Neurol. Res.* **17**, 285.

77. Zhang, J., Pieper, A., and Snyder, S. H. (1995). Poly(ADP-ribose) synthetase activation: an early indicator of neurotoxic DNA damage. *J. Neurochem.* **65**, 1411.

78. Snyder, S. H. (1996). No NO prevents parkinsonism. *Nature Med.* **2**, 65.

79. Didier, M., Bursztajn, S., Adamec, E., Passani, L., Nixon, R. A., Coyle J. T, Wei, J. Y., and Berman, S. A. (1996). DNA strand breaks induced by sustained glutamate excitotoxicity in primary neuronal cultures. *J. Neurosci.* **16**, 2238.

80. Endres, M., Scott, G. S., Namura, S., Salzman, A. L., Moskowitz, M., and Szabó, C. (1998). Role of peroxynitrite and neuronal nitric oxide synthase in the activation of poly (ADP-ribose) synthetase after cerebral ischaemia–reperfusion. *Neurosci. Lett.* **248**, 29.

81. Endres, M., Scott, G. S., Salzman, A. L., Kun, E., Moskowitz, M., and Szabó, C. (1998). Protective effects of 5-iodo-6-amino-1, 2-benzopyrone, a potent

inhibitor of poly (ADP) ribose synthetase against peroxynitrite-induced astrocyte damage and stroke development. *Eur. J. Pharmacol.* **351**, 377.

82. Endres, M., Wang, Z-Q., Namura, S., Waeber, C., and Moskowitz, M. A. (1997). Ischemic brain injury is mediated by the activation of poly(ADP-ribose)polymerase. *J. Cerebr. Blood. Flow. Metab.* **17**, 1143.

83. Takahashi, K., Greenberg, J. H., Jackson, P., Maclin, K., and Zhang, J. (1997). Neuroprotective effects of inhibiting poly (ADP-ribose) synthetase on focal cerebral ischaemia in rats. *J. Cerebr. Blood Flow Metab.* **17**, 1137.

84. Scott, G. S., Jakeman, L., Stokes, B., and Szabó, C. (1999). Peroxynitrite production and activation of poly (adenosine diphosphate-ribose) synthetase in spinal cord injury. *Ann. Neurol.* **45**, 120.

85. Whalen, M. J., Clark, R. S. B., Dixon, C. E., Robichaud, P., Marion, D. W., Vagni, V., Graham, S., Virág, L., Haskó, G., Stachlewitz, R., Szabó, C., and Kochanek, P. (1999). Reduction in deficits of memory and motor function after traumatic brain injury in mice deficient in poly (ADP-ribose) polymerase. *J. Cerebr. Blood Flow Metabol.* **19**, 835.

86. Kopin, I. J., and Markey, S. P. (1988). MPTP toxicity: implications for research in Parkinson's disease. *Ann. Rev. Neurosci.* **11**, 81.

87. Ebadi, M., Srinivasan, S. K., and Baxi, M. D. (1996). Oxidative stress and antioxidant therapy in Parkinson's disease. *Prog. Neurobiol.* **48**, 1.

88. Schulz, J. B., Matthews, R. T., Muqit, M. M., Browne, S. E. and Beal, M. F. (1995). Inhibition of neuronal nitric oxide synthase by 7-nitroindazole protects against MPTP-induced neurotoxicity in mice. *J. Neurochem.* **64**, 936.

89. Przedborski, S., Jackson-Lewis, V., Yokoyama, R., Shibata, T., Dawson, V. L. and Dawson, T. M. (1996). Role of neuronal nitric oxide in 1-methyl-4-phenyl-1,2,3,6- tetrahydropyridine (MPTP)-induced dopaminergic neurotoxicity. *Proc. Natl. Acad. Sci. USA* **93**, 4565.

90. Cosi, C., Colpaert, F., Koek, W., Degryse, A. and Marien, M. (1996). Poly (ADP-ribose) polymerase inhibitors protect against MPTP-induced depletions of striatal dopamine and cortical noradrenaline in C57B1/6 mice. *Brain Res.* **729**, 264.

91. Mandir, A. S., Przedborski, S., Jackson-Lewis, V., Wang, Z. Q., Simbulan-Rosenthal, D., Smulson, M. E., Hoffman, B. E., Guastella, D. B., Dawson, V. L. and Dawson, T. M. (1999). Poly (ADP-ribose) polymerase activation mediates 1-methyl-4-phenyl-1,2,3,6- tetrahydropyridine (MPTP)-induced parkinsonism. *Proc. Natl. Acad. Sci. USA* **96**, 5774.

92. Thiemermann, C., Bowes, J., Myint, F. P., and Vane, J. R. (1997). Inhibition of the activity of poly (ADP ribose) synthetase reduces ischaemia–reperfusion injury in the heart and skeletal muscle. *Proc. Natl. Acad. Sci. USA* **94**, 679.

93. Zingarelli, B., Cuzzocrea, S., Zsengeller, Z., Salzman, A. L., and Szabó, C. (1997). Beneficial effect of inhibition of poly-ADP ribose synthetase activity in myocardial ischaemia–reperfusion injury. *Cardiovasc. Res.* **36**, 205.

94. Zingarelli, B., Salzman, A. L., and Szabó, C. (1998). Genetic disruption of poly (ADP ribose) synthetase inhibits the expression of P-selectin and intercellular adhesion molecule-1 in myocardial ischaemia–reperfusion injury. *Circ. Res.* **83**, 85.

95. Yang, Z., Zingarelli, B., and Szabó, C. (2000). Role of poly (ADP-ribose) synthetase in the delayed myocardial ischemia–reperfusion injury. *Shock* **13**, 126.

96. Szabó, C. (1998). Role of poly (ADP-ribose) synthetase in inflammation. *Eur. J. Pharmacol.* **350**, 1.
97. Szabó, C. and Dawson, V. L. (1998). Role of poly (ADP-ribose) synthetase activation in inflammation and reperfusion injury. *Trends Pharmacol. Sci.* **19**, 287.
98. Meisterernst, M., Stelzer, G., and Roeder, R. G. (1997). Poly (ADP-ribose) polymerase enhances activator-dependent transcription *in vitro*. *Proc. Natl. Acad. Sci. USA* **94**, 2261.
99. Le Page, C., Sanceau, J., Drapier, J. C., and Wietzerbin, J. (1998). Inhibitors of ADP-ribosylation impair inducible nitric oxide synthase gene transcription through inhibition of NF kappa B activation. *Biochem. Biophys. Res. Comm.* **243**, 451.
100. Szabó, A., Salzman, A. L., and Szabó, C. (1998). Poly (ADP-ribose) synthetase mediates neutrophil sequestration and lipid peroxidation after mesenteric ischaemia. *Shock* **9** (Suppl. 1), 55.
101. Szabó, C. (1998). Potential role of the peroxynitrite-poly (ADP-ribose) synthetase pathway in a rat model of severe hemorrhagic shock. *Shock* **9**, 341.
102. Szabó, A., Hake, P., Salzman, A. L., and Szabó, C. (1998). Inhibition of poly (ADP-ribose) synthetase exerts protective effects in a porcine model of hemorrhagic shock. *Shock* **10**, 347.
103. Liaudet L., Soriano, F. G., Habley, J. G., Szabo, E., Virag, L., Salzman, A. L., Szabo, C. (2000) Protection against hemorrhagic shock in mice genetically deficient in poly (ADP-ribose) polymerase. *Proc Natl. Acad. Sci. USA* **97**, 10203.
104. Lam, T. T. (1997) The effect of 3-aminobenzamide, an inhibitor of poly-ADP-ribose polymerase, on ischemia/reperfusion damage in rat retina. *Res. Commun. Mol. Pathol. Pharmacol.* **95**, 241.
105. Burkart, V., Wang, Z. Q., Radons, J., Heller, B., Herceg, Z., Stingl, L., Wagner, E. F., and Kolb, H. (1999). Mice lacking the poly(ADP-ribose) polymerase gene are resistant to pancreatic beta-cell destruction and diabetes development induced by streptozocin. *Nature Med.* **5**, 314.
106. Masutani, M., Suzuki, H., Kamada, N., Watanabe, M., Ueda, O., Nozaki, T., Jishage, K., Watanabe, T., Sugimoto, T., Nakagama, H., Ochiya, T., and Sugimura, T. P. (1999). Poly(ADP-ribose) polymerase gene disruption conferred mice resistant to streptozotocin-induced diabetes *Proc. Natl. Acad. Sci. USA* **9**, 2301.
107. Pieper, A. A., Brat, D. J., Krug, D. K., Watkins, C. C., Gupta, A., Blackshaw, S., Verma, A., Wang, Z. Q., and Snyder, S. H. (1999). Poly(ADP-ribose) polymerase-deficient mice are protected from streptozotocin-induced diabetes. *Proc. Natl. Acad. Sci. USA* **96**, 3059.

6

New poly (ADP-ribose) polymerase inhibitors for chemo- and radiotherapy of cancer

Nicola J. Curtin, Bernard T. Golding, Roger J. Griffin, David R. Newell, Michael J. Roberts, Sheila Srinivasan, and Alex W. White

6.1 Introduction

The earliest indication that an NAD^+-utilizing enzyme participates in the cellular response to DNA damage was the demonstration that cytotoxic DNA alkylating agents inhibit glycolysis by virtue of NAD^+ depletion (1). The novel nucleic acid, poly (ADP-ribose) was discovered in the 1960s (2, 3). These observations were reconciled by the finding that alkylating agents cause the cellular activation of the enzyme, poly (ADP-ribose) polymerase (PARP), which catalyses the synthesis of poly (ADP) ribose from NAD^+ (4). Subsequent work has shown that PARP is an abundant nuclear enzyme which is activated ($\approx$500-fold) by strand breaks in DNA. PARP has been implicated in a variety of processes including DNA repair, recombination, and the regulation of gene expression (for reviews see refs 5–9). That mice with a disrupted ('knockout') *PARP* gene are both viable and fertile casts some doubt on the essential nature of PARP for normal growth and development (10–12). Nevertheless, there is compelling evidence for the role of PARP in the response of cells to genotoxic stress. The activation of PARP by DNA strand breaks is very rapid, apparently precedes the resealing of DNA lesions, and imposes a high energy cost on the cell. These factors suggest that poly (ADP-ribose) synthesis is important in the immediate cellular response to DNA damage.

The role of PARP in DNA repair has not been clearly defined. Direct modification of histones may occur causing their electrostatic repulsion from DNA. Alternatively, the automodification of PARP may cause the positively charged histones to dissociate from DNA and associate with the negatively charged poly (ADP-ribose) (13). The net result of histone dissociation from DNA would be to relax the chromatin structure, which in turn may improve the access of repair enzymes to DNA. Thus, PARP may facilitate the repair of the tightly packaged non-transcribed regions of the genome, but not necessarily the more loosely packaged transcribed regions. Pertinent to this is the observation that there is no difference in the rate of base excision repair (in which PARP is implicated) between transcribed and non-transcribed regions of the genome, whereas nucleotide excision repair (which is not associated with PARP activity) proceeds more rapidly in active genes (14). Moreover, the repair of damage induced by *N*-methyl-*N*-nitrosourea (MNU) in non-transcribed DNA regions, but not in transcribed sequences, was inhibited in PARP-deficient V79 cells (15). Another postulated role of PARP in DNA repair is that poly (ADP-ribose) may act as a signal to other nuclear proteins and recruit repair enzymes to the site of the DNA break (9). The physical association of PARP with proteins involved in DNA repair, namely XRCC1 (and hence also to DNA polβ and DNA ligase III) (16) as well as those involved in DNA replication; DNA polα (17) and other enzymes of the DNA synthesome (18) have been demonstrated recently and are reviewed in Chapters 2 and 3. Whatever the precise function of PARP is in DNA repair, its inhibition can have profound effects on the survival of cells following exposure to DNA damaging agents. PARP inhibitors may therefore be useful adjuncts to the chemo- or radiotherapy of cancer.

The development of 3-substituted benzamides as inhibitors of PARP by Purnell and Whish (19, see Section 6.2.1) enabled the investigation of the effects of DNA damaging agents in PARP-inhibited cells. Most studies have investigated the potentiation of monofunctional alkylating agents and ionizing radiation, as these are the most potent activators of PARP, although some other agents—for example, bleomycin, hydrogen peroxide, cisplatin, 1,3-bis(2-chloroethyl)-1-nitrosourea (BCNU), etoposide, and (±)-*anti*-benzo[*a*]pyrene diolepoxide—have also been studied (reviewed by Griffin *et al.* in ref. 20). The conclusion from most of these findings is that PARP participates in base excision repair but not nucleotide excision repair.

Investigations of the role of PARP conducted in PARP null mice and cells (10–12) demonstrate an increase in sister-chromatid exchanges after treatment with genotoxic agents, indicating the importance of PARP for genomic stability. However, the response to ionizing radiation and DNA alkylating agents of the two reported strains of PARP$^{-/-}$ mice and cells are not absolutely identical, as reviewed by LeRhun *et al.* (21). The PARP knockout models are described more fully in Chapter 3. In addition, PARP$^{-/-}$ cells synthesize ADP-ribose polymers (albeit to a lesser extent than do +/+ cells) in response to 1-methyl-3-nitro-1-nitrosoguanidine (MNNG) exposure (22), indicating the existence of other enzyme(s) with PARP activity that are activated by DNA damage. Such PARP homologues, described in Chapter 2, may complicate the interpretation of data obtained from PARP null mice.

The use of PARP inhibitors to obstruct DNA repair and hence increase the cytotoxicity of certain classes of anticancer agents remains a viable strategy, because ADP-ribose polymer formation by the alternative PARP in the PARP null cells was also inhibited by benzamide (22). Furthermore, inhibited PARP and a lack of PARP are not exactly equivalent. The difference between inhibited PARP and a PARP deficiency has been most clearly illustrated by Satoh and Lindahl (23) who demonstrated that cell extracts depleted of PARP (analogous to PARP$^{-/-}$ mice or cells) could repair nicked-plasmid DNA with similar efficiency to those containing PARP. In contrast, DNA repair was inhibited in PARP-containing cell extracts in the absence of NAD$^+$, or in the presence of 3-aminobenzamide (3-AB). Thus, in the context of DNA repair, the inhibition of PARP has a deleterious effect, whereas the absence of PARP is neutral. Inhibition of DNA repair in the presence of a PARP inhibitor, or absence of NAD$^+$, would be explained if PARP remains bound to the DNA and continues to shield the nick from the DNA repair machinery, thus preventing rejoining of the DNA strand break.

Much of the understanding of the function of PARP has been based on early studies with benzamide inhibitors. However, the benzamides lack potency and specificity, such that at the millimolar concentrations needed to achieve PARP inhibition and cytotoxic potentiation *in vitro,* they can also inhibit *de novo* purine biosynthesis (24) and cytochrome P450 (25). It is not surprising, therefore, that the benzamides have been shown to modulate the cytotoxicity of some agents by PARP-independent mechanisms (26–28). Nevertheless, the role of benzamides as inhibitors of the repair of some types of DNA damage through an interaction with PARP has been vindicated on two counts. First, it has been shown that in a series of PARP inhibitors a positive correlation exists between the potency of PARP inhibition and radiopotentiation, whereas there is no correlation with the inhibition of glucose metabolism (29). Second, it has been observed that the *trans*-dominant inhibition of PARP by molecular modification also potentiates cytotoxic agents (30, 31).

Collectively, the studies described above have demonstrated a potential utility for PARP inhibitors to augment cytotoxic cancer chemo- and radiotherapy. However, until recently it has not been possible to realize this potential because of the lack of potency of available inhibitors.

6.2 Development of PARP inhibitors

6.2.1 Early inhibitors of PARP

The first inhibitors of PARP described were nicotinamide and 5-methylnicotinamide, which were shown to function as competitive inhibitors (32). However, it was quickly realized that these compounds were non-specific. Nicotinamide is a substrate for nicotinamide-*N*-methyltransferase deaminase and phosphoribosyl transferase (33), while it is also an inhibitor of microsomal NADase, mono (ADP-ribosyl) transferases, and adenosine 3',5'-cyclic monophosphate phosphodiesterase (34).

In 1975, Shall reported the nicotinamide isosteres benzamide and pyrazinamide as PARP inhibitors (35). Purnell and Whish investigated analogues of benzamide in an attempt to find more specific PARP inhibitors (19). The introduction of polar substituents on to the benzamide ring increased the aqueous solubility of the compounds compared to benzamide, without compromising inhibitory activity (19). In addition, 3-aminobenzamide (3-AB) and 3-methoxybenzamide (3-MB) were shown to be better competitive inhibitors of PARP than benzamide and came to be regarded as 'benchmark' inhibitors. Replacement of the carboxamide with a carboxylate group in 3-substituted benzamides greatly reduced PARP-inhibitory activity, as did alkylation of the carboxamide (19). In contrast, the acylated analogue 3-acetoamidobenzamide retained inhibitory activity. The absence of a ring nitrogen in the benzamide analogues compared to the nicotinamide derivatives prevents their metabolism by NAD^+-metabolizing enzymes.

A systematic analysis of PARP inhibitors was conducted by Sims *et al.* (36), who confirmed the findings of Purnell and Whish (19). Sims *et al.* tested 33 analogues of the nicotinamide and adenine portions of NAD^+ using a permeabilized cell assay. Benzamide, 3-AB, and 3-MB were found to be the most potent PARP inhibitors, causing 96% inhibition of the enzyme. Pyridine derivatives with a carboxamide group at the 2- or 4-position (picolinamide and isonicotinamide) were found to exhibit inhibitory activity, although these compounds were less potent than nicotinamide (36). Pyrazinamide was only slightly less potent than nicotinamide (36). It was determined that *N*-alkylation of the ring nitrogen of nicotinamide resulted in a substantial decrease of inhibitory activity, as did reduction of the pyridine ring.

Cantoni and co-workers investigated the inhibitory activity of sulfur-containing benzamide derivatives (37). They carefully selected compounds which had a gradual increase in dipole moments, hydrogen-bonding strengths, and steric hindrance of the amide functional group (sulfonamide > thioamide > carboxamide). The sulfonamide and thioamide analogues of benzamide were found to be much less active than benzamide, as was thiophene-3-carboxamide, although this compound is regarded as a classical bioisostere of benzamide (37).

The specificity of the nicotinamide and benzamide inhibitors has been questioned by various authors (26, 38), because ADP-ribosylation reactions are carried out not only by PARP, but also by mono (ADP-ribosyl) transferases and NAD^+ glycohydrolases. This led Rankin *et al.* (39) to evaluate the selectivity of compounds, previously identified as PARP inhibitors, on all three classes of enzyme. The results indicated that inhibition of PARP occurred at micromolar concentrations, whereas the effects on mono (ADP-ribosyl) transferases and NAD^+ glycohydrolases were insignificant at these concentrations (39). However, Sestili *et al.* demonstrated that the *in vitro* activities of inhibitors did not necessarily correlate with their *in vivo* activities (40).

6.2.2 Mechanism-based PARP inhibitors

A series of compounds designed to interfere with the ADP-ribosylation catalysed by PARP was synthesized by Slama and Simmons (41). They replaced the β-D-ribosyl ring of the nicotinamide ribonucleoside moiety of NAD^+ with a 2,3-dihydroxycyclopentane ring. The resulting compound, 'carba-NAD^+', closely resembles NAD^+ in shape and overall charge. It functions as a mechanism-based inhibitor, as replacement of the oxygen atom renders the 'nicotinamide-ribose' bond resistant to cleavage by PARP. Owing to the resemblance between carba-NAD^+ and NAD^+ it is an efficient cofactor for alcohol dehydrogenases from yeast and horse liver (41). The diastereoisomer, ψ-carba-NAD^+, has the incorrect geometry for binding and catalysis, and is neither a cofactor nor an inhibitor of the alcohol dehydrogenases. The diastereoisomeric mixture of carba-NAD^+ and ψ-carba-NAD^+ was shown to be an effective inhibitor of the mono (ADP-ribosyl) transferases. Surprisingly, after the mixture was separated, the inhibitory activity was attributed to ψ-carba-NAD^+, the 'unnatural' diastereoisomer (41). The carba-NAD^+ analogues were found to be poor PARP inhibitors and limited in their effectiveness owing to their abilities to interfere with other NAD^+-metabolizing enzymes.

Recently, Watson *et al.* (42) have described 3-substituted benzamides and 5-substituted isoquinolin-1(2*H*)-ones bearing electrophilic groups (e.g. an epoxide) designed to interact with the carboxylate of an acceptor protein. None of the compounds were significantly more potent than the corresponding non-alkylating analogue (3-substituted benzamide or 5-substituted isoquinolines).

6.2.3 Conformationally restricted inhibitors of PARP

Banasik *et al.* identified a class of PARP inhibitors possessing a unifying structural feature: the carboxamide group conjugated to the aromatic ring was constrained within a ring system (43, 44). Compounds evaluated included 4-amino-1,8-naphthalimide (inhibitor 1, Fig. 6.1), 2-nitro-6-[5*H*]-phenanthridinone (inhibitor 2a), 1,5-dihydroxyisoquinoline (inhibitor 3), and 2-methylquinazolin-4-[3*H*]-one (inhibitor 4a, Fig. 6.1) which had IC_{50} values of 0.18, 0.35, 0.39, and 5.6 µM, respectively. Compound 1 exhibited a more than 100-fold greater potency than benzamide ($IC_{50} = 22$ µM) or 3-AB ($IC_{50} = 33$ µM), and was by far one of the most active inhibitors evaluated at the time (43, 44). Banasik *et al.* found that the majority of the very potent inhibitors were selective for PARP compared with the mono (ADP-ribosyl) transferases. Molecular orbital calculations indicated that the controlling factor for inhibitory activity was the ability of the oxygen atom of the carboxamide group to donate π-electrons to a positively charged region within

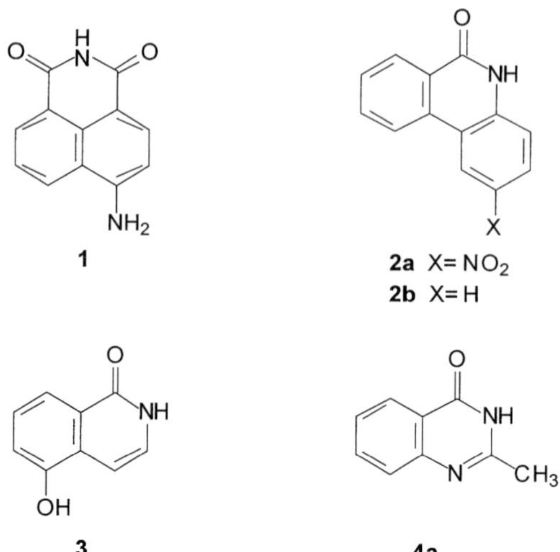

Fig. 6.1   Conformationally restricted inhibitors of PARP.

the enzyme active-site (43). It was suggested that the oxygen atom served as an electron donor to a putative hydrogen-bond acceptor within the catalytic domain (43). Since the activity of inhibitors bearing *N*-substituted carboxamides was poor, it was also inferred that an unsubstituted carboxamide group is required to act as an electron acceptor to a putative hydrogen-bond donor.

Further information regarding the conformation of the carboxamide group of NAD^+ in the enzyme active-site came from the work of Hong and Goldstein (45). Their *ab initio* calculations were based on the premise that the carboxamide group of NAD^+ adopts one of two specific orientations relative to the catalytic groups within the enzyme's active site, as depicted in Fig. 6.2. It was demonstrated that the biologically active conformation was that with the carbonyl group orientated *anti* to the 2,3-bond of the heteroaromatic ring. Evidence in support of these findings was provided by Suto *et al.*, who independently investigated conformationally restricted compounds (46). Using the 3-substituted benzamides as a template, they synthesized compounds containing an ethano bridge linking the benzene ring to the amide nitrogen (see Fig. 6.3) (46). It was found that the 5-substituted dihydroisoquinolines (Fig. 6.3, inhibitor **5**) were extremely potent inhibitors of PARP, which were 50- to 70-fold more potent than the 7-substituted dihydroisoquinolines (Fig. 6.3, inhibitor **6**). Suto *et al.* concluded that the positioning of the aromatic substituent, along with the restriction of the carboxamide rotation into the *anti* configuration was critical for potent activity, as a decrease in activity was observed when the ring substituent was placed in the 6-, 7-, or 8-position (46). The most potent inhibitor identified in the dihydroisoquino-

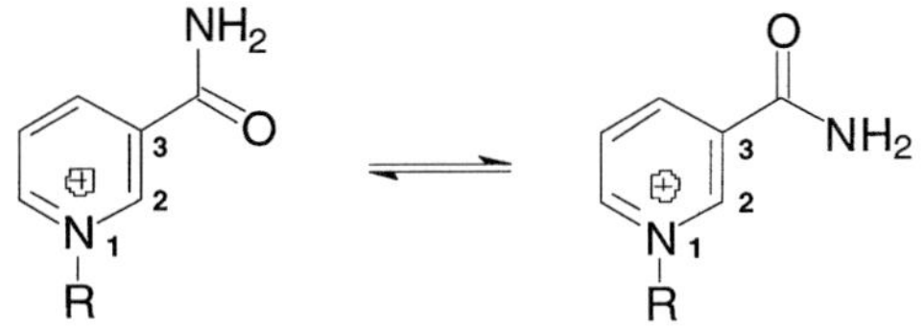

Fig. 6.2 Possible orientations of the carboxamide group of NAD^+. (From ref. 45.)

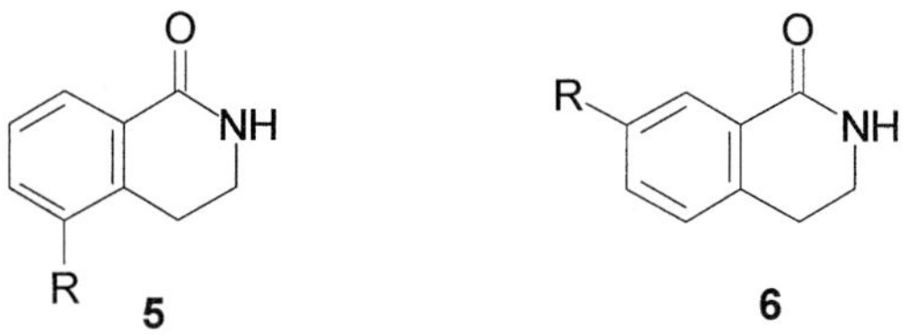

5a PD128763

$(R = CH_3, IC_{50} = 0.4\ \mu M)$

Fig. 6.3 Conformationally restricted benzamide analogues. (From ref. 46.)

line series was PD128763 (i.e. 3,4-dihydro-5-methylisoquinolin-1(2*H*)-one; compound **5a**, Fig. 6.3), which proved to be $\geqslant 50$ times more potent than 3-AB against calf thymus PARP. (IC_{50} values were 0.13 μM and 8 μM for PD128763 and 3-AB, respectively) (47) and murine PARP in permeabilised <1210 cells (IC_{50} values were 0.4 μM and 19 μM for PD128763 and 3AB respectively) (71). It was suggested that owing to its excellent PARP inhibitory activity, compound **5a** (R = Me) could act as a radiosensitizer and thus be of clinical value for use in conjunction with radiotherapy (48).

6.2.4 Design and synthesis of heterocyclic inhibitors of PARP

Sestili and associates (40) investigated the potency of a number of close benzamide analogues, both against PARP and as inhibitors of the rejoining of DNA strand breaks produced by the monofunctional DNA methylating agent methyl methanesulfonate (MMS). These studies identified certain structural features that were desirable for, or detrimental to, PARP inhibition. On the basis of these structural features and the data on conformationally restricted inhibitors, a number of novel PARP inhibitors have been developed as part of the Anticancer Drug Discovery Initiative programme here at Newcastle.

A novel approach was used for restricting the carboxamide moiety into the active *anti* position (see Fig. 6.2). The benzoxazole-4-carboxamides (Fig. 6.4) are electron-rich heterocycles and, as such, should enhance the donor properties of the carbonyl oxygen. An intramolecular hydrogen bond between the amide and oxazole nitrogen constrains the carboxamide group into the necessary conformation for activity against PARP (49, 50). The benzoxazole-4-carboxamides (inhibitors 7a–c, Fig. 6.4) with IC_{50} values for PARP inhibition ranging from 2 to 10 µM were found to be more potent PARP inhibitors than 3-substituted benzamides, e.g. 3-AB (IC_{50} = 19 µM) under the same assay conditions (49, 50). Replacement of the carboxamide group of 2-methyl benzoxazole-4-carboxamide by a carboxylate resulted in loss of inhibitory activity. *N*-methylation of the carboxamide group also resulted in a loss of activity, confirming the importance of an unsubstituted amide group (50).

The structural features required for PARP inhibition (as discussed in Section 6.2.3) are present in the basic quinazolinone architecture (see Fig. 6.5). However, until recently, quinazolinone derivatives were neglected as PARP inhibitors, despite reports of 2-methylquinazolin-4-[3*H*]-one exhibiting good activity against the enzyme (44, 50). During the synthesis of a series of benzoxazole-4-carboxamides, an unexpected rearrangement was observed which resulted in the formation of quinazolinone derivatives (50, 51). These were assayed for inhibitory activity against PARP and were found to be extremely potent. The quinazolinones identified from this rearrangement, 8-hydroxy-2-methylquinazolin-4-[3*H*]-one, NU1025 (inhibitor **4b**, Fig. 6.5), and 8-hydroxy-2-(4'-nitrophenyl)quinazolin-4-[3*H*]-one, NU1057 (inhibitor 4c, Fig. 6.5) have IC_{50} values of 0.4 and 0.2 µM, respectively (50, 51).

Following the synthesis and biological evaluation of the benzoxazole-4-carboxamides (51), a series of benzimidazole-4-carboxamides (Fig. 6.6), exhibiting extremely potent PARP inhibitory activity was discovered (52).

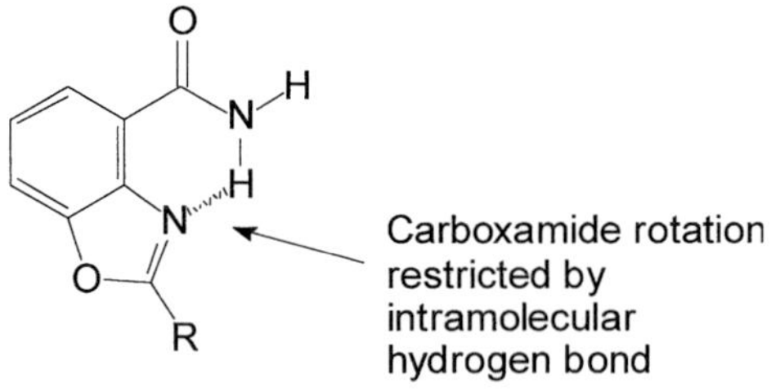

7a NU1056 [R= CH₃] IC_{50}= 9.8 µM

7b NU1040 [R= C(CH₃)₃] IC_{50}= 8.4 µM

7c NU1051 [R= Ph] IC_{50}= 2.1 µM

Fig. 6.4 Benzoxazole derivatives.

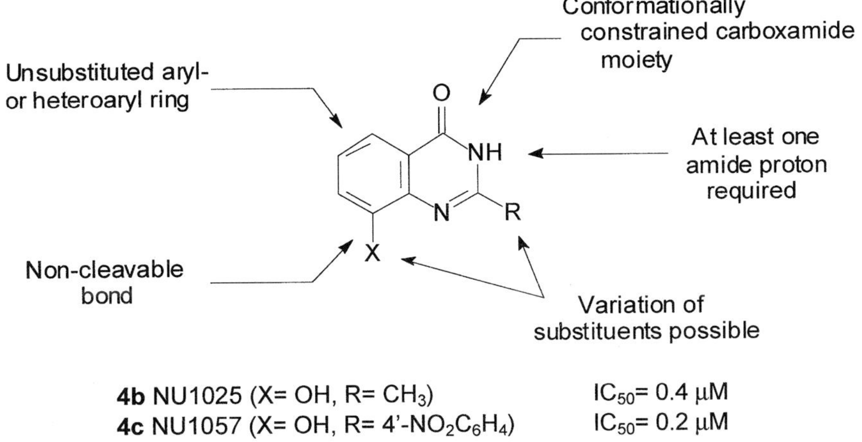

4b NU1025 (X= OH, R= CH$_3$)   IC$_{50}$= 0.4 µM
4c NU1057 (X= OH, R= 4'-NO$_2$C$_6$H$_4$)   IC$_{50}$= 0.2 µM

Fig. 6.5   Structural features of quinazolinone inhibitors.

Their carboxamide group was again constrained *via* an intramolecular hydrogen bond in an analogous manner to the benzoxazoles, but the presence of the N^1 imidazole nitrogen allowed an additional site on the compounds to be elaborated. The benzimidazoles are similar in electronic configuration to the benzoxazoles and possess the structure–activity requirements for optimal PARP inhibition (see Section 6.2.3 and Fig. 6.6) (52), subsequently confirmed by crystal structure data on their complexes with the catalytic domain of chicken PARP (see Section 6.2.6).

A series of 2-substituted benzimidazole derivatives was synthesized with alkyl, aryl, and substituted-aryl functionalities as the 2-substituents (52, 53). The series was extended to include di- and tri-substituted arylbenzimidazoles (54). These derivatives were found to be significantly more potent than the corresponding benzoxazoles. Compounds bearing both electron-withdrawing and electron-donating groups in the 4'-position of the 2-aryl substituent were found to exhibit excellent PARP inhibitory activity in the nanomolar concentration range (52). The 2-alkyl benzimidazole, 2-methyl-1*H*-benzimidazole-4-carboxamide, NU1064, and the more potent 2-aryl derivative, 2-(4'-hydroxyphenyl)-1-*H*-benzimidazole-4-carboxamide, NU1085 (inhibitors **8a** and **8d**, Fig. 6.6) were selected for use in potentiation studies and were shown to enhance the cytotoxicity of the alkylating agent temozolomide, see Section 6.3.1 (55, 56). Data from kinetic experiments indicated that NU1064 acts as a competitive inhibitor (55).

Shinkwin *et al.* (57) recently prepared thieno[3,4-*c*]pyridin-4(5*H*)-ones and thieno[3,4-*d*]pyrimidin-4(3*H*)-ones, in which a thiophene ring replaces the benzenoid ring of known PARP inhibitors of the isoquinoline (e.g. inhibitor **3**; see Fig. 6.1) and quinazolinone (e.g. inhibitor **4a**; see Fig. 6.1) classes. The

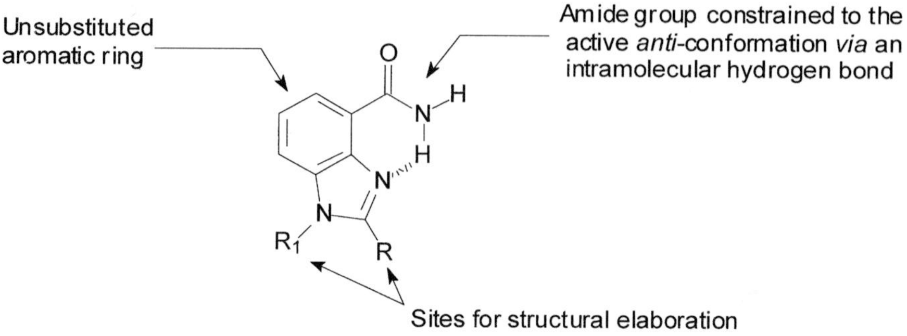

8a NU1064 (R = CH$_3$, R$_1$ = H)   IC$_{50}$ = 1.10 μM
8b NU1070 (R = Ph, R$_1$ = H)   IC$_{50}$ = 0.10 μM
8c NU1076 (R = 4′-CH$_3$OC$_6$H$_4$, R$_1$ = H)   IC$_{50}$ = 0.06 μM
8d NU1085 (R = 4′-HOC$_6$H$_4$, R$_1$ = H)   IC$_{50}$ = 0.08 μM

Fig. 6.6   Benzimidazole structure–activity relationships. (From ref. 52.)

most potent compound, 6-methyl-7-nitrotieno[3,4-*c*]pyridin-4(5*H*)one, at 10 μM inhibited PARP activity by approximately 90%.

6.2.5   Summary of structure–activity requirements

Thus far, it has been established that analogues of the natural substrate, NAD$^+$, are effective inhibitors of PARP. Potent enzyme-inhibitory activity appears to be associated with the following features:

1. An unsubstituted aromatic or polyaromatic heterocyclic system (saturated analogues show poor activity).
2. The presence of a carboxamide group (carboxylate and thiocarboxylate analogues have no activity).
3. Restriction of the carboxamide moiety into the *anti* conformation (derivatives with unrestricted carboxamides are less active).
4. The presence of at least one amide proton for putative hydrogen-bonding (*N*-alkylation of the carboxamide moiety results in loss of activity).
5. A non-cleavable bond at the position corresponding to the 3-position of the benzamides

These features are summarized in Fig. 6.7.

With regard to the design of novel inhibitors, other factors must also be taken into consideration. Ideally, the inhibitor should possess functional sites which allow for elaboration of the molecule, in order to probe the enzyme-active site. If the inhibitor is to be of clinical value as a therapeutic agent, then

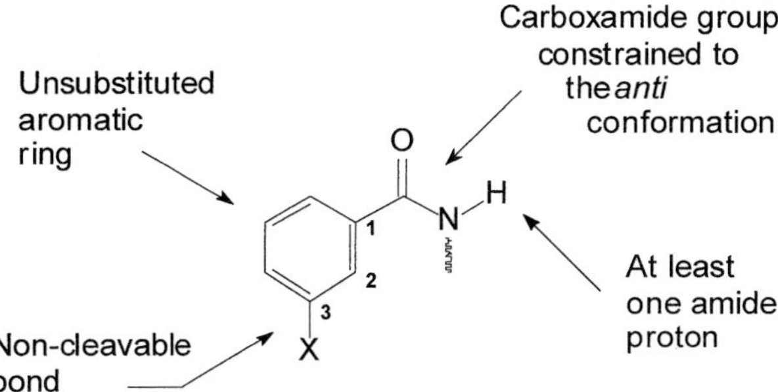

Fig. 6.7 Summary of structure–activity relationships.

solubility problems must be addressed and a water-solubilizing substituent may be necessary.

6.2.6   Crystal structure of the enzyme-active site

The crystal structure of the catalytic fragment of chicken PARP bound to PD128763 (inhibitor 5a, Fig. 6.3) provided conclusive information on the nature of enzyme-inhibitor binding (58). Since the catalytic domains of human and chicken PARP share 100% homology in amino-acid sequence, the information gained from this crystal structure is of potential value in the design of novel PARP inhibitors. Interestingly, the catalytic fragment of chicken PARP (PARP-CF) was found to be similar in structure to the catalytic domains of microbial toxins, although the amino-acid sequence homology is relatively low (58).

PARP-CF was crystallized with and without PD128763 (inhibitor 5a). The inhibitor was bound to PARP-CF by two hydrogen bonds from the carboxamide group to the peptide backbone of Gly863, and a third hydrogen bond from the carbonyl oxygen to the side chain of Ser904 (58, 59). It was assumed that the inhibitor binds to PARP-CF in an analogous manner to the nicotinamide portion of NAD^+. Hydrophobic interactions between PD128763 and the adjacent Tyr907 were also identified. Glu988 is a residue involved in catalysis (60, 61), and was located at a distance of only 4 Å from a methyl carbon atom (C-9) of PD128763. This glutamate may act as a general base to activate the 2′-hydroxyl group of an acceptor NAD^+ molecule for nucleophilic attack at C-1 of the ribose of an acceptor NAD^+ molecule. Alternatively, it could stabilize the intermediate oxocarbenium ion formed from the cleavage of the ribose–nicotinamide bond of NAD^+ during the course of the normal catalytic reaction. Figure 6.8 shows a schematic representation of the enzyme–inhibitor interactions between PARP-CF and PD128763 (58). The earlier predictions regarding the structure–activity requirements of PARP inhibitors (see Section 6.2.3) are therefore confirmed.

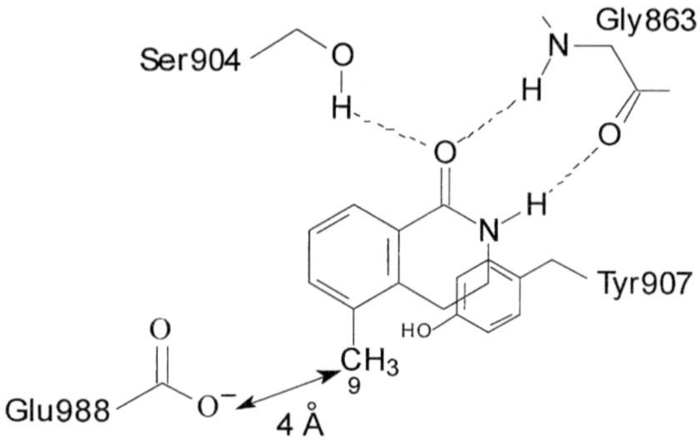

Fig. 6.8 Enzyme–inhibitor interactions between PARP-CF and PD128763. (From ref. 58.)

More recently, the structures of PARP-CF bound to the quinazolinone (NU1025; inhibitor 4b, Fig. 6.5), 3-MB, and 4-amino-1,8-naphthalimide (inhibitor 1, Fig. 6.1) were determined (59). For all these inhibitors the binding of the carboxamide group to the protein mimics the motif found for PD128763 (inhibitor 5a, Fig. 6.3). The positions of binding of all the inhibitors are similar, i.e. in the nicotinamide subsite of the NAD^+-binding region of the protein. Almassy *et al.* (54) have determined the crystal structures of several quinazolinones and benzimidazoles bound to PARP-CF. These show the same binding features around the benzene-carboxamide moiety as those described for PARP-CF with PD128763 (inhibitor 5a). In addition, the 2-aryl group occupies a relatively spacious pocket, as shown in Fig. 6.9 for NU1085, with still vacant regions evident, which might be satisfied by further substitution on the aromatic ring.

6.3 Biological activity of PARP inhibitors

6.3.1 Novel PARP inhibitors as potentiators of the in vitro activity of chemo- and radiotherapy

As described in Section 6.2, a number of studies have addressed the need for PARP inhibitors that are significantly more potent than the 'benchmark' 3-substituted benzamides. The approach of Suto *et al.* (46) and the work of the Newcastle-upon-Tyne group (50, 53) has been to develop new PARP inhibitors distinct from 3-substituted benzamide. These studies were based both on structure–activity relationships for the benzamides and the proposed interactions of NAD^+ within the PARP catalytic site, and have led to the development of potent compounds with IC_{50} values for PARP inhibition of less than 1.0 μM (see Sections 6.2.3 and 6.2.4).

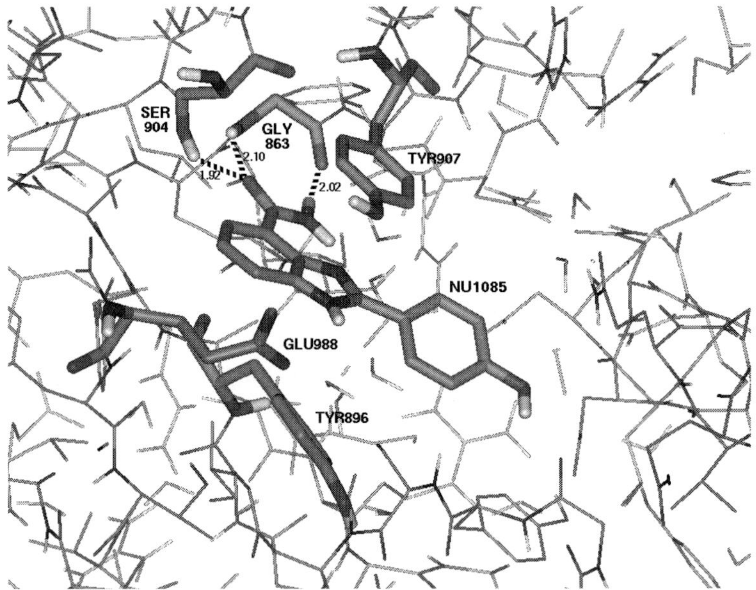

Fig. 6.9 NU1085 (inhibitor 8d, see Fig. 6.6) complexed with the catalytic domain of chicken PARP. Three hydrogen bonds (yellow) form the principal interactions of the inhibitor with PARP (distances are in Angstroms). (From ref. 54.)

The potentiation of both alkylating agents and ionizing radiation by PARP inhibition has been confirmed by molecular studies. For example, *trans*-dominant inhibition of PARP, through overexpression of the DNA binding domain, sensitizes cells to the methylating agent MNNG and to γ-irradiation (30, 31). (For a detailed discussion of current data see Chapter 4). However, conclusions based on the use of benzamides regarding the role of PARP in response to other cytotoxic agents is less certain. Studies with the new, more potent PARP inhibitors are helping to clarify both the cytotoxic agents and types of DNA damage that involve PARP-dependent repair mechanisms. For example, there are conflicting data on the potentiation of the cytotoxicity of DNA cross-linking agents, e.g. cisplatin, by the benzamide inhibitors. Enhancement of cisplatin cytotoxicity has been observed in ovarian and cervical cancer cells (62), whereas no enhancement was seen in a study of mammary carcinoma cells (63). These apparently conflicting data could be due to cell-specific differences in repair mechanisms and/or the non-specific effects of the benzamides. An investigation using the more potent inhibitors identified by Banasik *et al.* (44), i.e. 4-amino-1,8-naphthalimide (inhibitor 1, see Fig. 6.1) (5 μM), 2-nitro-6-[5*H*]-phenanthridinone (inhibitor 2a, see Fig. 6.1) (10 μM) and 1,5-dihydroxyisoquinoline (inhibitor 3, see Fig. 6.1) (15 μM) as well as 4-hydroxyquinazoline (1 mM) and 3-aminobenzamide (1 mM), failed to detect any potentiation of cisplatin in a panel of ovarian cancer cell lines. In contrast, all the novel PARP inhibitors potentiated the cytotoxicity of the monofunctional agent MNNG, but only 3-aminobenzamide potentiated the cytotoxicity of the bifunctional agent, BCNU (64).

Recently, radiopotentiation by 4-amino-1,8-naphthalimide (inhibitor 1, see Fig. 6.1) has been investigated in a panel of rodent and human cell lines. The level of enhancement was directly proportional to inhibitor concentration, and sensitizer enhancement ratios of 1.3 to 1.5 were obtained at 20 μM 4-amino-1,8-naphthalimide (65).

Using 6-[5*H*]-phenanthridinone (inhibitor 2b, see Fig. 6.1) Weltin and co-workers (66) demonstrated an 80% reduction in PARP activity in murine lymphoma cells. The compound was 500 times more potent than 3-methoxybenzamide, and at 20 μM 6-[5*H*]-phenanthridinone caused a synergistic enhancement of nitrogen mustard cytotoxicity (66). Transient exposure to 100 μM 6-[5*H*]-phenanthridinone before and after γ-irradiation caused a significant reduction in cell proliferation and an increase in apoptosis compared to irradiation alone (67). Subsequent investigations by this group examined the effect of 6-[5*H*]-phenanthridinone on a variety of anticancer agents in a panel of human and murine tumour cell lines (68). 6-[5*H*]-Phenanthridinone enhanced the antiproliferative activity of SN38 (topoisomerase I inhibitor), taxotere, BCNU, and cisplatin, but antagonized the effect of adriamycin. In some cells 6-[5*H*]-phenanthridinone increased the cytotoxicity of bleomycin, but in others it had no effect. Thus, potentiation by 6-[5*H*]-phenanthridinone is dependent on both the cell type and the mechanism of cytotoxic drug action.

Shah and colleagues (69) have used 1,5-dihydroxyisoquinoline (inhibitor 3, see Fig. 6.1) to investigate the function of PARP in recovery from oxidative stress, generated by using a xanthine–xanthine oxidase system. Using an isolated system, the IC_{50} for PARP inhibition was 0.3 μM with nearly complete inhibition at 10 and 100 μM. In whole cells, 10 and 100 μM concentrations of 1,5-dihydroxyisoquinoline were required to inhibit oxidant-induced PARP activity by 95 and 98%, respectively. PARP activity in these cells was still greater than in cells not exposed to oxidative stress, and a concentration of 1 mM of 1,5-dihydroxyisoquinoline was required for complete suppression of oxidant-induced PARP activation. Prevention of recovery from oxidative damage was only observed at 1 mM 1,5-dihydroxyisoquinoline and not at concentrations of 10 or 100 μM, implying that even very low levels of PARP activity (< 5% of control) are sufficient for recovery from sublethal oxidative damage and that complete PARP inactivation is necessary for inhibition of recovery.

A number of potent dihydroisoquinolinone PARP inhibitors designed and synthesized by Suto and associates (46) increased the X-ray sensitivity of V79 (Chinese hamster lung) cells, by affecting both the shoulder and the slope of the survival curve. One of the most potent compounds was PD128763 (inhibitor 5a, R = Me, see Fig. 6.3), IC_{50} for PARP inhibition = 0.4 μM, and this compound has been subject to further investigation. The effect of PD128763 on cellular recovery from X-irradiation was compared with that of 3-aminobenzamide, which at 10 mM completely blocked recovery from potentially lethal damage but only partially blocked recovery from sublethal damage. A 20-fold lower concentration of PD128763 (0.5 mM) not only blocked recovery but actually increased cell killing induced by X-irradiation (48). Striking potentiation of the cytotoxicity of the monofunctional alkylating agent streptozotocin and the nitroimidazole RSU 1069 (7- and 36-fold, respectively) by 0.5 mM PD128763 was observed in L1210 cells (70), whereas the potentiation of these agents by 5 mM 3-aminobenzamide was only two- to threefold. The cytotoxicity of the chloroethylating agents BCNU and CCNU was not increased by PD128763, in agreement with the observation that several other potent inhibitors failed to potentiate the cytotoxicity of BCNU in ovarian cancer cells (64).

A study comparing the effects of PD128763 and an equipotent novel PARP inhibitor NU1025, 8-hydroxy-2-methylquinazolin-4[3*H*]-one (inhibitor 4b, see Fig. 6.5), with the 40- and 50-fold weaker classical inhibitors benzamide and 3-aminobenzamide, has been performed in L1210 cells (71). This study revealed an excellent correlation between the potency of the compounds as PARP inhibitors and their ability to enhance the DNA strand-breakage and cytotoxicity of the methylating agent temozolomide. Concentrations of only 50–100 μM NU1025 or PD128763 were sufficient for a four- to sevenfold potentiation of temozolomide cytotoxicity, whereas 1 mM benzamide and 5 mM 3-aminobenzamide were needed to achieve a four-fold potentiation. Similarly, 100 μM PD128763 completely prevented temo-zolomide-induced NAD^+ depletion, while 1 mM of 3-AB was required for the same effect. Further studies in L1210 cells treated with a variety of anticancer agents in combination with NU1025 (200 μM) have demonstrated: a fourfold potentiation of 5-methyltriazenoimidazole-4-carboxamide (MTIC), the methylating species derived from temozolomide; a 2.6-fold potentiation of the topoisomerase I inhibitor camptothecin; and a 1.4-fold potentiation of γ-irradiation and bleomycin (55, 72). Under the same experimental conditions there was no potentiation of the topoisomerase II inhibitor etoposide or the thymidylate synthase inhibitors CB3717 and nolatrexed. The structurally dif-ferent PARP inhibitor NU1064, 2-methyl-1*H*-benzimidazole-4-carboxamide (inhibitor 8a, Fig. 6.6, IC_{50} for PARP inhibition = 1.1 μM) also potentiated the cytotoxicity of temozolomide in a concentration-dependent manner, with maximum potentiation achieved at a concentration of 200 μM NU1064 (55). The more potent inhibitor NU1085, 2-(4′-hydroxyphenyl)-1*H*-benzimida-zole-4-carboxamide (inhibitor 8d, Fig. 6.6, IC_{50} for PARP inhibition = 0.08 μM) was investigated in a panel of 12 human cancer cell lines (56). A concen-tration of 10 μM NU1085 was sufficient to enhance the growth inhibitory effect of temozolomide 1.5- to 4-fold and to cause a 2.5- to 6-fold enhancement of topotecan-induced growth inhibition. The ability of NU1085 to potentiate growth inhibition and cytotoxicity showed no evidence of tissue specificity and was not influenced by the reported p53 status of the cell lines (56).

In the majority of *in vitro* chemo- and radiopotentiation studies, PARP inhibitors have been used at concentrations that do not affect cell growth or survival (48, 55, 65, 66, 71, 72), but their intrinsic cytotoxic properties have not been reported. However, in the study by Boulton *et al.* (71) the cytotoxicity of benzamide, 3-aminobenzamide, PD128763, and NU1025 was investigated. There was a $\geq$ 10-fold difference in the concentration of PD128763 and NU1025 required to produce maximal potentiation of temozolomide cytotoxicity and the LC_{50} of the inhibitors alone i.e. 50–100 μM was required for maximum potentiation and LC_{50} values of 0.99 and 1.6 mM were obtained for PD128763 and NU1025, respectively (71). The difference in the concentrations required for maximum potentiation and intrinsic cytotoxicity were less for 3-aminobenzamide and benzamide: 5 mM and 1 mM were required for potentiation, with the LC_{50} values being 14 and 6 mM for 3-aminobenzamide and benzamide, respectively. This may be due to the more pronounced non-specific effects of the benzamides as described earlier (24, 25). Maximum potentiation of temozolomide cytotoxicity was observed at a concentration of 200 μM NU1064, which itself inhibited survival by 50% (55). In the study by Weltin *et al.* (67) continuous exposure to 50 μM phenanthridinone was completely cytostatic, but the transient exposure to 100 μM used in the radiopotentiation studies only reduced cell survival by 15%. PARP null cells grow more slowly than their wild-type counterparts (12), and if PARP is necessary for optimum cell growth then this may contribute to the growth inhibitory effects of the PARP inhibitors. However, the possibility that at high concentrations the novel inhibitors may act on other vital cellular targets, as has been demonstrated for the benzamides (24, 25), cannot be excluded. Investigations using PARP null cells will facilitate the distinction of PARP-specific from non-specific toxicities of the inhibitors.

All the PARP inhibitors described above are competitive inhibitors with respect to NAD$^+$. Buki and co-workers (73) have adopted a different approach in their use of 6-nitroso-1,2-benzpyrone (inhibitor 9a, Fig. 6.10) to inactivate PARP (K_i = 40 μM). The proposed mechanism of action of 6-nitroso-1,2-benzpyrone is via ejection of Zn^{2+} from one of the two PARP zinc fingers. However, this does not appear to prevent PARP binding to DNA as measured by the obstruction of DNA synthesis by the Klenow fragment. To date, 6-nitroso-1,2-benzpyrone and the analogue, 5-iodo-6-amino-1,2-benzpyrone (inhibitor 9b, Fig. 6.10), have been investigated for their protective effect and direct cytotoxicity (discussed below), rather than as resistance modulators in conjunction with DNA damaging agents.

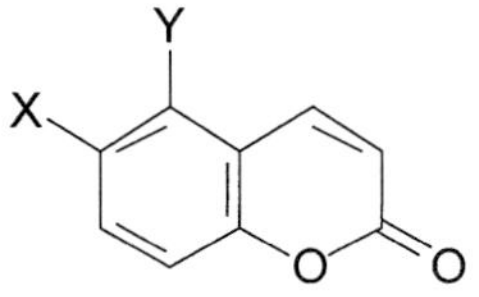

9a X= NO, Y= H

9b X= NH$_2$, Y= I

Fig. 6.10   Benzopyrone PARP inhibitors. (From ref. 73.)

6.3.2   Protective effect of PARP inhibitors

In contrast to the increased toxicity of anticancer drugs when PARP is inhibited, PARP inhibitors may in certain cases exert a protective effect by preventing a suicidal NAD$^+$- and, hence, ATP-depletion. PARP activation has been implicated in tissue damage following ischaemia and reperfusion and in the inflammatory response (see Chapter 5). For example, nitric oxide produced in inflammatory responses may cause DNA damage and cell lysis. Radons *et al.* (74) observed that nitric oxide (involved in cell lysis by macrophages) activated PARP in pancreatic islet cells. These authors showed that the activation of PARP and cell lysis could be inhibited by 0.54 μM 4-amino-1,8-naphthalimide (inhibitor 1, Fig. 6.1) or 9 mM 3-aminobenzamide, but not by mono (ADP-ribosylation) inhibitors. Similarly, Zhang *et al.* (75) observed that a number of PARP inhibitors, including benzamides, 4-amino-1,8-naphthalimide, and 1,5-dihydroxyisoquinoline (inhibitor 3, Fig. 6.1) could protect rat cerebral cortical cultures from neurotoxicity caused by *N*-methyl-D-aspartate and nitric oxide. It was concluded that neurotoxicity was mediated by energy depletion following PARP activation. Similarly, several PARP inhibitors: benzamide, 3-AB, and 6-[5*H*]-phenanthridinone (inhibitor 2b), but not the inactive analogue, benzoic acid, prevented cell death from peroxynitrite cytotoxicity in the NSC34 spinal-cord cell line (76). The zinc-finger interacting PARP inhibitor, 5-iodo-6-amino-1,2-benzpyrone (inhibitor 9b), also protects against peroxynitrite cytotoxicity in glioma cells *in vitro* and in a murine stroke model (77) and delays the onset of collagen-induced arthritis, in which peroxynitrite is implicated (78). The role of PARP inhibitors in the protection from nitric oxide-induced tissue damage in a variety of conditions is reviewed in ref. 79.

6.3.3   PARP inhibitors and apoptosis

Apoptosis represents a common, final cell-death pathway triggered by a variety of stimuli including hormones and cytotoxic drugs. Not surprisingly, the genes that control apoptosis are frequently mutated in cancer cells (for reviews see refs 80–82). Restoration of apoptotic pathways in malignant cells which have a reduced response to apoptotic signals would obviously have potential in the therapy of cancer, and a body of evidence is accumulating that implicates PARP in the apoptotic process. (See Chapter 4 for a comprehensive discussion of this topic.)

Kaufmann *et al.* (83) were the first to demonstrate the cleavage of PARP during the induction of apoptosis by a variety of chemotherapeutic agents. PARP cleavage has now been recognized as a common feature of programmed cell death, irrespective of the triggering event, and is frequently used as a marker of apoptosis. Cleavage results in a small amino-terminal fragment (25 kDa) containing the DNA binding domain and a larger C-terminal domain comprising the automodification and catalytic domains. The cleavage of PARP is mediated by a protease related to interleukin-1β-converting enzyme and the cell-death gene (*CED-3*) in *Caenorhabditis elegans* (84) now termed caspase-3. Evidence that PARP cleavage is an important component of the apoptotic pathway comes from studies with PARP null cells transfected with an uncleavable, but catalytically active, PARP. In these cells there was a delay in Fas-mediated apoptosis (85). The role of PARP cleavage in apoptosis is described in Chapter 4.

In addition, the commitment step in apoptosis is characterized by chromatin condensation coincident with the activation of an endogenous Ca^{2+}/Mg^{2+}-dependent endonuclease without gene transcription or protein synthesis, suggesting activation may involve derepression of a constitutively expressed protein (86, 87). PARP has been shown to inhibit a Ca^{2+}/Mg^{2+} endonuclease (88, 89) and the apoptotic endonuclease may be maintained in the repressed state by poly (ADP-ribosyl)ation, with derepression being achieved by inactivation of PARP.

It is therefore tempting to speculate that PARP inhibitors might augment apoptosis, particularly in malignant cells if reduced apoptotic capacity were due to a deficiency in caspase-3 activity. For example, inhibition of PARP by a small molecule could mimic cleaved PARP and stimulate apoptosis by derepressing the apoptotic endonuclease activity. In addition, if the amino-terminal fragment of apoptotically cleaved PARP plays an active part in blocking DNA ends from repair, inhibited PARP might perform the same function. The availability of more potent PARP inhibitors means that these hypotheses can be investigated. Such studies may reveal that PARP inhibitors alone have a therapeutic utility.

To investigate the potential utility of PARP inhibitors as therapeutic agents in their own right, Rice *et al.* (90) studied 6-nitroso-1,2-benzpyrone (inhibitor 9a, Fig. 6.10) and 3-nitrosobenzamide, which are believed to inactivate PARP at one zinc-finger site. The nitroso compounds induced apoptosis in 855–2 and HL-60 human leukaemic cells, but not in human peripheral-blood stem cells. Apoptosis was accompanied by a threefold increase in Ca^{2+}/Mg^{2+} endonuclease activity and was not dependent on protein synthesis. These authors proposed that the endonuclease is maintained in a latent form by poly (ADP-ribosyl)ation and that inactivation of PARP by the C-nitroso compounds results in derepression of endonuclease activity. Subsequent studies by Mendeleyev *et al.* (91) have used 4-iodo-3-nitrobenzamide which, on intracellular reduction to 4-iodo-3-nitrosobenzamide, inhibits PARP through ejection of Zn^{2+} from the first zinc finger. The cytotoxicity of 4-iodo-3-nitrobenzamide was determined in a panel of human and animal cells, and toxicity was reported to correlate with the rate of reduction to the nitroso compound.

The effect on apoptosis of 3-aminobenzamide is curious; in some cases it can protect cells from undergoing glucocorticoid- or topoisomerase inhibitor-induced apoptosis (92, 93), while in others it can enhance the effect of glucocorticoids (94). Furthermore, the induction of apoptosis by 3-aminobenzamide alone has also been demonstrated (93), an effect which could be blocked by aphidicolin, suggesting that apoptosis induced by PARP inhibition is associated with DNA replication.

To investigate the ability of PARP inhibitors to induce apoptosis, the effects of the novel potent PARP inhibitor, NU1025 (inhibitor 4b, Fig. 6.5), have been studied in a panel of human tumour cell lines. The extent and temporal sequence of the induction of apoptosis by 1 mM NU1025 differed between the different cell types, and was not related to the reported p53 status of the cell lines (95).

6.3.4   PARP inhibitors as potentiators of chemo- and radiotherapy in vivo

Relatively high concentrations of even the most potent PARP inhibitors are required for the *in vitro* potentiation of some cytotoxic agents (e.g. oxidative stress), suggesting that residual PARP activity must be reduced to less than 5% of the total stimulatable activity to prevent cellular recovery (69). However, it may be possible to use lower concentrations of inhibitor *in vivo* to achieve sufficient depletion of PARP activity. It has been shown that the activity of the enzyme in cultured cells is higher than that in cells obtained directly from the *in vivo* tumour (96). In addition, most of the new inhibitors are competitive with respect to NAD^+. Lower intracellular NAD^+ concentrations are observed in tumours compared to normal tissues (97, 98). The $NAD^+/NADH$ ratio is lowered in hypoxic tumour cells, with NAD^+ levels reduced at least threefold (99), and indeed NADH is itself a PARP inhibitor (100) with a K_i of 5 μM. Together these data suggest that the inhibitors will be more effective in tumour cells and that some selective potentiation of cytotoxic agents is theoretically possible.

Nicotinamide and benzamides have been extensively studied for their ability to sensitize tumour cells *in vivo* to both single and fractionated doses of ionizing radiation Furthermore, these compounds preferentially sensitize the tumours compared to the normal tissues (reviewed in refs 20 and 101). The precise mechanism for the potentiation is unclear; for although they are PARP inhibitors, they also increase tumour blood flow, which could have the effect of reducing hypoxia-mediated radioresistance. However, there is recent evidence to suggest that nicotinamide, at doses of 100 mg/kg or greater, may also contribute to radiosensitization by inhibiting DNA repair, possibly *via* PARP inhibition (102). Clinical trials of nicotinamide with carbogen breathing to improve tumour oxygenation have been conducted in combination with radiotherapy for head and neck cancer (103) and gliomas (104). However, some toxic side-effects were observed, including renal and hepatic toxicity.

Several *in vivo* studies have been conducted to investigate the potentiation of chemotherapy by the benzamide inhibitors. However, despite the recognized role of PARP in DNA strand-break repair, not all the cytotoxic drugs studied *in vivo* are thought to act solely, or even primarily, via DNA strand-break formation (reviewed in ref. 20). Moreover, the PARP inhibitors used in *in vivo* studies generally lack the specificity which would allow PARP inhibition to be unequivocally implicated as a mechanism of chemo- or radiosensitization. For example, the ability of pyrazinamide to potentiate the activity of cyclophosphamide against the RIF-1 and Lewis lung murine tumours (105), or of metoclopramide to increase the activity of BCNU and cisplatin (106, 107), may not arise through PARP inhibition, and the two compounds are almost certainly acting via different mechanisms.

The PARP inhibitor most frequently used *in vivo* is 3-AB, which has been demonstrated to enhance the antitumour activity of bleomycin (108), cisplatin (109), chlorambucil (110), and cyclophosphamide (111) against Ehrlich ascites, cisplatin against the sarcoma 180 (109), and DTIC against the Greene melanoma and L1210 leukaemia (111). A possible explanation for *in vivo* chemopotentiation by the benzamides is the observed effects on body temperature and drug clearance elicited by these compounds. For example, Horsman and co-workers (112) demonstrated that potentiation of melphalan was only achieved at doses of 3-aminobenzamide that were greater than 100 mg/kg, and at these doses the compound caused significant hypothermia and reduced plasma clearance of the cytotoxic drug.

It is likely that the role of PARP inhibitors as chemo- and radiosensitizers will be more clearly defined by the use of the novel inhibitors. Treatment of mice bearing SCC7 sarcomas with the combination of PD128763 (at the maximum tolerated dose of 100 mg/kg) and 2.5 Gy X-irradiation caused tumour regressions with significant tumour growth delay, with a more than threefold increase of the therapeutic effect over X-rays alone. Similarly, PD128763 enhanced the X-ray activity against RIF-1 and KHT sarcomas. Notably, there was no increase in host toxicity in these studies (47).

Martin and colleagues (113) have investigated the ability of 1,5-dihydroxyisoquinoline to enhance the efficacy of a combination of *N*-phospho-noacetyl-L-aspartic acid (PALA) + 6-methylmercaptopurine riboside + 6-aminonicotinamide against murine breast cancer *in vivo*. The frequency of tumour regressions was significantly increased from 48% to 66%. However, there was an indication that toxicity was also increased as the inclusion of 1,5-dihydroxyisoquinoline caused a further decrease in body weight.

Preliminary *in vivo* evaluation of the benzimidazole PARP inhibitors have been undertaken as a prelude to antitumour chemosensitization studies. Pharmacokinetic studies in tumour-bearing mice show rapid absorption of NU1085, with peak plasma levels of 100 μM observed 30 min after a single intraperitoneal dose of 40 mg/kg. However, the drug was eliminated rapidly, and at 60-min postinjection the concentration in the tumour had declined to 7 μM (114).

The availability of PARP knockout mice will enable the dissection of effects due specifically to PARP inhibition from side-effects that are unrelated to PARP. In collaboration with Drs J. Menissier de Murcia and G. de Murcia we have investigated the toxicity of PD128763 in PARP knockout and wild-type mice of the same strain. We observed that PD128763 induced a dose-dependent reduction in body temperature which was almost identical in both +/+ and –/– mice, demonstrating that hypothermia is not due to PARP inhibition but instead the interaction of PD128763 with some other physiological target.

6.4 Conclusions and future directions

The data accumulated over the last two decades, initially from studies using the classical benzamide inhibitors and latterly through molecular studies and the use of more potent inhibitors, demonstrate that PARP inhibitors can increase the potency of certain cytotoxic agents. Thus, PARP inhibitors have the potential to improve chemotherapy and radiotherapy in the clinic, and the new generation of inhibitors that are 2–3 orders of magnitude more potent than the benzamides may be sufficiently potent to be active at pharmacologically achievable concentrations. The limited *in vivo* data that exists (47, 113) indicates this is so. Further *in vivo* studies need to be conducted to determine the toxicity and antitumour activity of these compounds in combination with cytotoxic therapies. Moreover, investigation of the selectivity of the novel PARP inhibitors to tumours, rather than enhancing the undesirable toxicity to normal tissues, needs to be addressed.

There are other important developments that should enable more potent and specific inhibitors of PARP to be designed for clinical use. First, the crystal structure of the catalytic domain of chicken PARP has been elucidated both with and without co-crystallized PD128763, 4-amino-1,8-naphthalimide, 3-methoxybenzamide, NU1025 (58, 59), and NU1085 (54). The study of the binding interactions of these inhibitors with PARP will lead to the design and synthesis of new compounds predicted to have high levels of potency. Second, the availability of cells derived from PARP knockout mice will enable the specificity of these inhibitors for the cellular target to be evaluated. The PARP$^{-/-}$ mice will enable the delineation of toxic side-effects (such as hypothermia) unrelated to PARP that may then be eliminated by the use of structurally different inhibitors. The prospect for the development of a PARP inhibitor for clinical use within the next few years seems extremely promising.

Acknowledgements

We gratefully acknowledge the financial support of the Cancer Research Campaign and Agouron Pharmaceuticals Inc., San Diego, USA, and Dr R. Almassy, Agouron, for the crystallography data.

References

1. Roitt, I. M. (1956). The inhibition of carbohydrate metabolism in ascites-tumour cells by ethyleneimines. *Biochem. J.*, **63**, 300.
2. Fijimura, S., Sugimura, T., and Okabe, K. (1965). NMN-activated poly(A)polymerase in nuclei from rat liver hepatoma cells. *Proc. Annu. Meeting of Jpn Biochem. Soc.*, **134**, 691. (In Japanese.)
3. Chambon, P., Weil, J. D., Doly, J., Strosser, M. T. and Mandel, P. (1966). On the formation of a novel adenylic compound by enzymatic extracts of liver nuclei. *Biochem. Biophys. Res. Commun.*, **25**, 638.
4. Whish, W. J. D., Davies, M. I., and Shall, S. (1975). Stimulation of poly(ADP-ribose) polymerase activity by the antitumour antibiotic, streptozotocin. *Biochem. Biophys. Res. Commun.*, **65**, 722.
5. Shall, S. (1984). ADP-ribose in DNA repair: a new component of DNA excision repair. *Adv. Radiat. Biol.*, **11**, 1.
6. Boulikas, T. (1991). Relation between carcinogenesis, chromatin structure and poly(ADP-ribosylation). *Anticancer Res.*, **11**, 489.
7. Sugimura, T. and Miwa, M. (1994). Poly(ADP-ribose): a historical perspective. *Mol. Cell. Biochem.*, **138**, 5.
8. de Murcia, G. and Menissier de Murcia, J. (1994). Poly(ADP-ribose) polymerase: a molecular nick sensor. *TIBS*, **19**, 172.
9. Lindahl, T., Satoh, M. S., Poirier, G. G., and Klungland, A. (1995). Post-translational modification of poly(ADP-ribose) polymerase induced by DNA strand breaks. *TIBS*, **20**, 405.
10. Wang, Z-Q., Auer, B., Stingl, L., Berghammer, H., Haidacher, D., Schweiger, M., and Wagner, E. R. (1995). Mice lacking ADPRT and poly(ADP-ribosyl)ation develop normally but are susceptible to skin disease. *Genes Dev.*, **9**, 509.
11. Menissier de Murcia, J., Neidergang, C., Trucco, C., Ricoul, M., Dutrillaux, B.,

Mark, M., Oliver, F. J., Masson, M., Dierich, A., LeMeur, M., Walztinger, C., Chambon, P., and deMurcia, G. (1997). Requirement of poly(ADP-ribose) polymerase in recovery from DNA damage in mice and in cells. *Proc. Natl Acad. Sci. USA*, **94**, 7303.

12. Wang, Z-Q., Stingl, L., Morrison, C., Jantsch, M., Los, M., Schulze-Osthoff, K., and Wagner, E. F. (1997). PARP is important for genomic stability but dispensible in apoptosis. *Genes Dev.*, **11**, 2347.

13. Althaus, F. R., Hoffer, L., Kleczkowska, H. E., Malamga, M., Naegeli, H., Panzer, P. L., and Realini, C. A. (1994). Histone shuttling by poly ADP-ribosylation. *Mol. Cell. Biochem.*, **138**, 53.

14. Scicchitano, D. A. and Hanawalt, P. C. (1989). Repair of methylpurines in specific DNA sequences in Chinese hamster ovary cells; absence of strand specificity in the dihydrofolate reductase gene. *Proc. Natl Acad. Sci. USA*, **86**, 3050.

15. Ray, L. S., Chatterjee, S., Berger, N. A., Grishko, V. I., LeDoux, S. P., and Wilson, G. L. (1996). Catalytic activity of poly(ADP-ribose) polymerase is necessary for repair of *N*-methylpurines in non-transcribed but not in transcribed, nuclear DNA sequences. *Mutat. Res.*, **363**, 105.

16. Mason, M., Niedergang, C., Schreiber, V., Muller, S., Menissier deMurcia, J., and deMurcia, G. (1998). XRCC1 is specifically associated with poly(ADP-ribose) polymerase and negatively regulates its activity following DNA damage. *Mol. Cell. Biol.*, **18**, 3563.

17. Dantzer, F., Nasheuer, H-P., Vanescg, J-L., deMurcia, G., and Menissier deMurcia, J. (1998). Functional association of poly (ADP-ribose) polymerase with DNA polymerase α-primase complex: a link between DNA strand break detection and DNA replication. *Nucleic Acids Res.*, **26**, 1891.

18. Simbulan-Rosenthal, C. M., Rosenthal, D. S., Boulares, A. H., Hickey, R. J., Malkas, L. H., Coll, J. M., and Smulson, M. E. (1998). Regulation of the expression or recruitment of the components of the DNA synthesome by poly(ADP-ribose) polymerase. *Biochemistry*, **37**, 9363.

19. Purnell, M. R. and Whish, W. J. D. (1980). Novel inhibitors of poly(ADP-ribose) synthetase. *Biochem. J.*, **185**, 775.

20. Griffin, R. J., Curtin, N. J., Newell, D. R. Golding, B. T., Durkacz, B. W., and Calvert, A. H. (1995). The role of inhibitors of poly(ADP-ribose) polymerase as resistance-modifying agents in cancer therapy. *Biochimie*, **77**, 408.

21. LeRhun, Y., Kirkland, J. B., and Shah, G. M. (1998). Cellular responses to DNA damage in the absence of poly(ADP-ribose) polymerase. *Biochem. Biophys. Res. Comm.*, **245**, 1.

22. Shieh, W. M., Amé, J-C., Wilson, M. V., Wang, Z-Q., Koh, D. W., Jacobson, M. K., and Jacobson, E. L. (1998). Poly(ADP-ribose) polymerase null mouse cells synthesise ADP-ribose polymers. *J. Biol. Chem.*, **273**, 30069.

23. Satoh, M. S. and Lindahl, T. (1992). Role of poly(ADP-ribose) formation in DNA repair. *Nature*, **356**, 356.

24. Milam, K. M., Thomas, G. H., and Cleaver, J. E. (1986). Disturbances in DNA precursor metabolism associated with exposure to an inhibitor of poly(ADP-ribose) synthetase. *Exp. Cell Res.*, **165**, 260.

25. Eriksson, C., Busk, L., and Brittebo, E. B. (1996). 3-Aminobenzamide: effects on cytochrome P450-dependent metabolism of chemicals and on the toxicity of dichlobenil in the olfactory mucosa. *Toxicol. Appl. Pharmacol.*, **136**, 324.

26. Moses, K., Harris, A. L., and Durkacz, B. W. (1988). Adenosine-diphosphoribosyltransferase inhibitors can protect against or potentiate the cytotoxicity of S-phase acting drugs. *Biochem. Pharmacol.*, **37**, 2155.

27. Moses, K., Harris, A. L., and Durkacz, B. W. (1988). Synergistic enhancement of 6-thioguanine by ADP-ribosyltransferase inhibitors. *Cancer Res.*, **48**, 5650.

28. Moses, K., Willmore, E., Harris, A. L., and Durkacz, B. W. (1990). Correlation of enhanced 6-mercaptopurine cytotoxicity with increased phosphoribosylpyrophosphate levels in Chinese hamster ovary cells treated with 3-aminobenzamide. *Cancer Res.*, **50**, 1992.

29. BenHur, E., Chen, C. C., and Elkind, M. M. (1985). Inhibitors of poly(adenosinediphosphoribose) synthetase, examination of metabolic perturbations and enhancement of radiation response in Chinese hamster cells. *Cancer Res.*, **45**, 2123.

30. Molinete, M., Vermeulen, W., Burkle, A., Menissier de Murcia, J., Kupper, J. H., Hoeijmakers, J. H. J., and de Murcia, G. (1993). Overproduction of the poly(ADP-ribose) polymerase DNA-binding domain blocks alkylation-induced DNA repair synthesis in mammalian cells. *EMBO J.*, **12**, 2109.

31. Kupper, J. H., van Gool, L., and Burkle, A. (1995). Molecular genetic systems to study the role of poly(ADP-ribosyl)ation in the cellular response to DNA damage. *Biochimie*, **77**, 450.

32. Clark, J. B., Ferris, G. M., and Pinder, S. (1971). Inhibition of nuclear NAD nucleoside and poly ADP-ribose polymerase activity from rat liver by nicotinamide and 5-methylnicotinamide. *Biochim. Biophys. Acta*, **238**, 82.

33. Banasik, M. and Ueda, K. (1994). Inhibitors and activators of ADP-ribosylation reactions. *Mol. Cell. Biochem.*, **138**, 185.

34. Shimoyama, M., Kawai, M., Nasu, S., Shioji, K., and Hoshi, Y. (1975). Inhibition of adenosine 3'-, 5'-monophosphate phosphodiesterase by nicotinamide and its homologs *in vitro*. *Physiol. Chem. Physics*, **7**, 125.

35. Shall, S. (1975). Experimental manipulation of the specific activity of poly(ADP-ribose) polymerase. *J. Biochem.*, **77**, 2.

36. Sims, J. L., Sikorski, G. W., Catino, D. M., Berger, S. J., and Berger, N. A. (1982). Poly(adenosine diphosphate-ribose) polymerase inhibitors stimulate unscheduled deoxyribonucleic acid synthesis in normal human lymphocytes. *Biochemistry*, **21**, 1813.

37. Cantoni, O., Sestili, P., Spadoni, G., Balsamini, C., Cucchiarini, L., and Cattabeni, F. (1987). Analogues of benzamide containing a sulfur atom as poly(ADP-ribose) transferase inhibitors. *Biochem. Int.*, **15**, 329.

38. Hunting, D. J., Gowans, B. J., and Henderson, J. F. (1985). Specificity of inhibitors of poly(ADP-ribose) synthesis. Effects on nucleotide metabolism in cultured cells. *Mol. Pharmacol.*, **28**, 200.

39. Rankin, P. W., Jacobson, E. L., and Benjamin, R. C. (1989). Quantitative studies of inhibitors of ADP-ribosylation *in vitro* and *in vivo*. *J. Biol. Chem.*, **264**, 4312.

40. Sestili, P., Spadoni, G., Balsamini, G., Scovassi, I., Cattabeni, F., Duranti, E., Cantoni, O., Higgins, D., and Thomson, C. (1990). Structural requirements for inhibitors of poly(ADP-ribose) polymerase. *J. Cancer Res. Clin. Oncol.*, **116**, 615.

41. Slama, J. T. and Simmons, A. M. (1988). Carbonicotinamide adenine dinu-

cleotide: synthesis and enzymological properties of the carbocyclic analogue of the oxidized nicotinamide adenine dinucleotide. *Biochemistry*, **27**, 183.

42. Watson, C. Y., Whish, W. J. D., and Threadgill, M. D. (1998). Synthesis of 3-substituted benzamides and 5-substituted isoquinolin-1(2*H*)-ones and preliminary evaluation as inhibitors of poly(ADP-ribose)polymerase (PARP). *Bioorg. Med. Chem.*, **6**, 721.

43. Banasik, M., Komura, H., Saito, I., Abed, N. A. N., and Ueda, K. (1989). New inhibitors of poly(ADP-ribose) synthetase. In *ADP-ribose transfer reactions: mechanisms and biological significance* (ed. M. K. Jacobson and E. L. Jacobson), p. 130. Springer-Verlag, New York.

44. Banasik, M., Komura, H., Shimoyama, M., and Ueda, K. (1992). Specific inhibitors of poly(ADP-ribose) synthetase and mono(ADP-ribosyl)transferase. *J. Biol. Chem.*, **267**, 1569.

45. Hong, L. and Goldstein, B. M. (1992). Carboxamide group conformation in the nicotinamide and thiazole-4-carboxamide rings: implications for enzyme binding. *J. Med. Chem.*, **35**, 3560.

46. Suto, M. J., Turner, W. R., Arundel-Suto, C. M., Werbel, L. M., and Sebolt-Leopold, J. S. (1991). Dihydroisoquinolines: the design and synthesis of a new series of potent inhibitors of poly(ADP-ribose) polymerase. *Anticancer Drug Des.*, **7**, 107.

47. Leopold, W. R. and Sebolt-Leopold, J. S. (1990). Chemical approaches to improved radiotherapy. In *Cytotoxic anticancer drugs: models and concepts for drug discovery and development* (ed. F. A. Valeriote, T. H. Corbett, and L. H. Baker), p. 179. Kluwer, Boston, MA.

48. Arundel-Suto, C. M., Scavone, S. V., Turner, W. R., Suto, M. J., and Sebolt-Leoplod, J. S. (1991). Effects of PD128763, a new potent inhibitor of poly(ADP-ribose) polymerase, on X-ray induced cellular recovery processes in Chinese hamster V79 cells. *Radiat. Res.*, **126**, 367.

49. Bleasdale, C., Calvert, A. H., Curtin, N. J., Durkacz, B. W., Golding, B. T., Griffin, R. J., Newell, D. R., Pemberton, L., and Rhodes, D. (1994). Inhibitors of poly(adenine diphosphate ribose) polymerase to potentiate DNA-reactive drugs. *Br. J. Cancer*, **69** (Suppl. XXI), 16.

50. Griffin, R. J., Pemberton, L. C., Rhodes, D., Bleasdale, C., Bowman, K., Calvert, A. H. Curtin, N. J., Durkacz, B. W., Newell, D. R., Porteous, J. K., and Golding, B. T. (1995). Novel potent inhibitors of the DNA repair enzyme poly(ADP-ribose) polymerase (PARP). *Anticancer Drug Des.*, **10**, 507.

51. Pemberton, L. C. (1994). Novel inhibitors of poly(adenine diphosphate ribose) polymerase to potentiate DNA reactive drugs. PhD thesis, University of Newcastle-upon-Tyne.

52. White, A. W. (1996). The design of novel inhibitors of poly(ADP-ribose) polymerase to potentiate cytotoxic drugs. PhD thesis, University of Newcastle-upon-Tyne.

53. Griffin, R. J., Srinivasan, S., White, A. W., Bowman, K., Calvert, A. H., Curtin, N. J., Newell, D. R., and Golding, B. T. (1996). Novel benzimidazole and quinazolinone inhibitors of the DNA repair enzyme poly(ADP-ribose) polymerase. *Pharm. Sci.*, **2**, 43.

54. Almassy, R., Bowman, K., Calvert, A. H., Curtin, N. J., Golding, B. T., Hostomska, Zu., Hostomsky, Zd., Griffin, R. J., Maegley, K., Newell, D. R.,

Srinivasan. S., White, A. W., and Webber, S. (1998). 2-Phenylbenzimidazole-4-carboxamides are potent inhibitors of the DNA repair enzyme poly(ADP-ribose) polymerase (PARP). *Ann. Oncol.*, **9** (Suppl. 2), 30.

55. Bowman, K. J., Curtin, N. J., Golding, B. T., Griffin, R. J., and White, A. (1998). Potentiation of anticancer agent cytotoxicity by the potent poly(ADP-ribose) polymerase inhibitors, NU1025 and NU1064. *Br. J. Cancer*, **78**, 1269.

56. Delaney, C. A., Wang, L-Z., Kyle, S., Srinivasan, S., White, A. W., Curtin, N. J., Calvert, A. H., Durkacz, B. W., and Newell, D. R. (1999). Potentiation of temozolomide and topotecan growth inhibition and cytotoxicity by poly(ADP-ribose) polymerase (PARP) inhibitors in a panel of human cancer cell lines. *Proc. Am. Assoc. Cancer Res.*, **40**, 402.

57. Shinkwin, A. E., Whish, W. J. D., and Threadgill, M. D. (1999). Synthesis of thiophenecarboxamides, thieno[3,4-*c*]pyridin-4(5*H*)-ones and thieno[3,4,-*d*]pyrimidin-4(3*H*)-ones and preliminary evaluation as inhibitors of poly(ADP-ribose)polymerase (PARP). *Bioorg. Med Chem.*, **7**, 297.

58. Ruf, A., Menissier de Murcia, J., de Murcia, G., and Schulz, G. E. (1996). Structure of the catalytic fragment of poly(ADP-ribose) polymerase from chicken. *Proc. Natl Acad. Sci. USA*, **93**, 7481.

59. Ruf, A., de Murcia, G., and Schulz, G. E. (1998). Inhibitor and NAD$^+$ binding to poly(ADP-ribose) polymerase as derived from crystal structures and homology modelling. *Biochemistry*, **37**, 3893.

60. Marsischky, G. T., Wilson, B. A., and Collier, R. J. (1995). Role of glutamic acid 988 of human poly(ADP-ribose) polymerase in polymer formation. Evidence for active site similarities to the ADP-ribosylating toxins. *J. Biol. Chem.*, **270**, 3247.

61. Ruf, A., Rolli, V., de Murcia, G., and Schulz, G. E. (1998). The mechanism of the elongation and branching reaction of poly(ADP-ribose) polymerase as derived from crystal structures and mutagenesis. *J. Mol. Biol.*, **278**, 57.

62. Boike, G. M., Petru, E., Sevin, B. U., Averett, H., Chou, T. C., Penalver, M., Donato, D., Schiano, M., Hilsenbeck, S. G., and Perras, J. (1990). Chemical enhancement of cisplatin cytotoxicity in a human ovarian and cervical cancer cell line. *Gynecol. Oncol.*, **38**, 315.

63. Alaoui-Jamali, M., Loubaba, B-B., Robyn, S., Tapiero, H., and Batist, G. (1994). Effect of DNA-repair-enzyme modulators on cytotoxicity of L-phenylalanine mustard and *cis*-diaminedichloroplatinum(II) in mammary carcinoma cells resistant to alkylating drugs. *Cancer Chemother. Pharmacol.*, **34**, 153.

64. Bernges, F. and Zeller, W. J. (1996). Combination effects of poly(ADP-ribose) polymerase inhibitors and DNA damaging agents in ovarian tumor cell lines—with special reference to cisplatin. *J. Cancer Res. Clin. Oncol.*, **122**, 665.

65. Sclicker, A., Peschke, P., Burkle, A., Hahn, E. W., and Kim, J. H. (1999). 4-Amino-1,8-naphthalimide: a novel inhibitor of poly(ADP-ribose) polymerase and radiation sensitizer. *Int. J. Radiat. Biol. Relat. Stud. Phys. Chem. Med.*, **75**, 91.

66. Weltin, D., Marchal, J., Dufour, P., Potworowski, E., Oth, D., and Bischoff, P. (1994). Effect of 6(5H)-phenanthridinone, an inhibitor of poly(ADP-ribose) polymerase, on cultured tumour cells. *Oncol. Res.*, **6**, 399.

67. Weltin, D., Holl, V., Hyun, J. W., Dufour, P., Marchal, J., and Bischoff, P. (1997). Effects of 6(5H)-phenanthridinone, a poly(ADP-ribose) polymerase inhibitor, and ionising radiation on the growth of cultured lymphoma cells. *Int. J. Radiat. Biol.*, **72**, 685.

68. Weltin, D., Holl, V., Hyun, J. W., Marchal, J., Bischoff, P., and Dufour, P. (1997). Effects of treatments combining 6(5H)-phenanthridinone, a new poly(ADP-ribose) polymerase inhibitor, and anticancer agents upon murine and human tumour cell lines. *Proc. Am. Assoc. Cancer Res.*, **38**, 1485.

69. Shah, G. M., Poirier, D., Desnoyers, S., Saint-Martin, S., Hoflack, J. C., Rong, P., Simon, M., Kirkland, J. B., and Poirier, G. G. (1996). Complete inhibition of poly(ADP-ribose) polymerase activity prevents the recovery of C3H10T$^1/_2$ cells from oxidative stress. *Biochim. Biophys. Acta.*, **1312**, 1.

70. Sebolt-Leopold, J. S. and Scavone, S. V. (1992). Enhancement of alkylating agent activity *in vitro* by PD 128763, a potent poly(ADP-ribose) polymerase inhibitor. *Int. J. Rad. Oncol. Biol. Phys.*, **22**, 619.

71. Boulton, S., Pemberton, L. C., Porteous, J. K., Curtin, N. J., Griffin, R. J., Golding, B. T., and Durkacz, B. W. (1995). Potentiation of temozolomide cyto-toxicity: a comparative study of the biological effects of poly(ADP-ribose) poly-merase inhibitors. *Br. J. Cancer*, **72**, 849.

72. Bowman, K., Calvert, A. H., Curtin, N. J., Golding, B. T., Griffin, R. J., Newell, D. R., Srinivasan, S., and White, A. W. (1996). Effect of novel poly(ADP-ribose)polymerase inhibitors on the cytotoxicity of anticancer agents. *Br. J. Cancer*, **73** (Suppl. 16), 13.

73. Buki, K. G., Bauer, P. I., Mendeleyev, J., Hakam, A., and Kun, E. (1991). Destabilisation of Zn^{2+} co-ordination in ADP-ribose transferase (polymerising) by 6-nitroso-1,2-benzopyrone coincidental with inactivation of the polymerase but not the DNA binding function. *FEBS Lett.*, **290**, 181.

74. Radons, J., Heller, B., Burkle, A., Hartmann, B., Rodriguez, M-L., Kroncke, K-D., Burkart, V., and Kolb, H. (1994). Nitric oxide toxicity in islet cells involves poly(ADP-ribose) polymerase activation and concomitant NAD^+ depletion. *Biochem. Biophys. Res. Commun.*, **199**, 1270.

75. Zhang, J., Dawson, V. L., Dawson, T., and Snyder, S. H. (1994). Nitric oxide activation of poly(ADP-ribose) synthetase in neurotoxicity. *Science*, **263**, 687.

76. Cookson, M. R., Ince, P. G., and Shaw, P. J. (1998). Peroxynitrite and hydrogen peroxide induced cell death in the NSC34 neuroblastoma ? spinal cord cell line: role of poly(ADP-ribose)polymerase. *J. Neurochem.*, **70**, 501.

77. Endres, M., Scott, G. S., Salzman, A. L., Kun, E., Moskowitz, M. A., and Szabo, C. (1998). Protective effects of 5-iodo-6-amino-1,2-benzpyrone, an inhibitor of poly(ADP-ribose) synthetase against peroxynitrite-induced glial damage. *Eur. J. Pharmacol.*, **351**, 377.

78. Szabo, C., Virag, L., Cuzzocrea, S., Scott, G. S., Hake, P., O'Connor, M. P., Zingarelli, B., Salzman, A., and Kun, E. (1998). Protection against peroxyni-trite-induced fibroblast injury and arthritis development by inhibition of poly(ADP-ribose) polymerase. *Proc. Natl Acad. Sci. USA*, **95**, 3867.

79. Pieper, A. A., Verma, A., Zhang, J., and Syder, S. H. (1999). Poly(ADP-ribose) polymerase, nitric oxide and cell death. *TIPS*, **20**, 171.

80. Arends, M. J. and Wylie, A. H. (1994). Apoptosis: mechanisms and roles in pathology. *Int. Rev. Exp. Pathol.*, **32**, 223.

81. Dive, C. and Hickman, J. A. (1991). Drug-target interactions: only the first step in the commitment to a programmed cell death? *Br. J. Cancer*, **64**, 192.

82. Martin, S. J., Green, D. R., and Cotter, T. G. (1994). Dicing with death: dissect-ing the components of the apoptosis machinery. *TIBS*, **19**, 26.

83. Kaufmann, S. H., Desnoyers, S., Ottaviano, Y., Davidson, N. E., and Poirier, G. G. (1993). Specific proteolytic cleavage of poly(ADP-ribose) polymerase: an early marker of chemotherapy-induced apoptosis. *Cancer Res.*, **53**, 3976.

84. Nicholson, D. W., Ali, A., Thornberry, N. A., Vaillancourt, J. P., Ding, C. K., Gallant, M., Gareau, Y., Griffin, P. R., Labelle, M., Lazebnik, Y. A., Munday, N. A., Raju, S. M., Smulson, M. E., Yamin, T-T., Yu, V. L., and Miller, D. K. (1995). Identification and inhibition of the ICE/CED-3 protease necessary for mammalian apoptosis. *Nature*, **367**, 37.

85. Oliver, F. J., de la Rubia, G., Rolli, V., Ruiz-Ruiz, M. C., de Murcia, G., and Menissier-de Murcia, J. (1998) Importance of poly(ADP-ribose) polymerase and its cleavage in apoptosis: lessons from an uncleavable mutant. *J. Biol. Chem.*, **273**, 33533.

86. Wyllie, A. H., Beattie, G. J., and Hargreaves, A. D. (1981). Chromatin changes in apoptosis. *Histochem. J.*, **13**, 681.

87. Bursch, W., Kleine, L., and Tenniswood, M. (1990). The biochemistry of cell death by apoptosis. *Biochem. Cell Biol.*, **68**, 1071.

88. Yoshihara, K., Tanigawa, Y., Burzio, L., and Kiode, S. S. (1975). Evidence for adenosine diphosphate ribosylation of $Ca^{2+}Mg^{2+}$-dependent endonuclease. *Proc. Natl Acad. Sci. USA*, **72**, 289.

89. Tanaka, Y., Yoshihara, K., Itaya, A., Kamiya, T., and Koide, S. S. (1984). Mechanism of the inhibition of $Ca^{2+}Mg^{2+}$-dependent endonuclease of bull seminal plasma induced by ADP-ribosylation. *J. Biol. Chem.*, **259**, 6579.

90. Rice, W. G., Hillyer, C. D., Harten, B., Schaeffer, C. A., Dorminy, M., Lackey III, D. A., Kirsten, E., Mendeleyev, J., Buki, K. G., Hakam, A., and Kun, E. (1992). Induction of endonuclease-mediated apoptosis in tumour cells by C-nitroso-substituted ligands of poly(ADP-ribose) polymerase. *Proc. Natl Acad. Sci. USA*, **89**, 7703.

91. Mendeleyev, J., Kirsten, E., Hakam, A., Buki, K. G., and Kun, E. (1995). Potential chemotherapeutic activity of 4-iodo-3-nitrobenzamide: metabolic reduction to the 3-nitroso derivative and induction of cell death in tumour cells in culture. *Biochem. Pharmacol.*, **50**, 705.

92. Hoshino, J., Beckman, G., and Kroger, H. (1993). 3-Aminobenzamide protects the mouse thymocytes *in vitro* from dexamethasone-mediated apoptotic cell death and cytolysis without changing DNA strand breakage. *J. Steroid Biochem. Mol. Biol.*, **44**, 113.

93. Bertrand, R., Solary, E., Jenkins, J., and Pommier, Y. (1993). Apoptosis and its modulation in human promyelocytic HL-60 cells treated with topoisomerase I and II inhibitors. *Exp. Cell Res.*, **207**, 388.

94. Meyer, A. S., Schlechte, J. A., and Schmidt, T. J. (1990). Potentiation of glucocorticoid-induced cytolysis in sensitive human leukemic cells by an inhibitor of ADP-ribosylation. *Leuk. Res.*, **14**, 909.

95. Roberts, M. J., Curtin, N. J., Durkacz, B. W., Golding, B. T., Griffin, R. J., Hall, A. G., Newell, D. R., and Srinivasan, S. (1997). Induction of apoptosis in a panel of human tumour cell lines using NU1025, a novel potent inhibitor of poly(ADP-ribose) polymerase (PARP). *Proc. Am. Assoc. Cancer Res.*, **38**, 1301.

96. Chabert, M. G., Kopp, P. C., Bischoff, P. L., and Mandel, P. (1993). Cell culture of tumours alters endogenous poly(ADPR) polymerase expression and activity. *Int. J. Cancer*, **53**, 837.

97. Jedeiken, L. A. and Weinhouse, S. (1955). Metabolism of neoplastic tissue. VI. Assay of oxidised and reduced diphosphopyridine nucleotide in normal and neoplastic tissues. *J. Biol. Chem.*, **213**, 271.

98. Glock, G. E. and McLean, P. (1957). Levels of oxidised and reduced diphosphopyridine nucleotide and triphosphopyridine nucleotide in tumours. *Biochem. J.*, **65**, 413.

99. Wilson, D. F., Erecinska, M., and Brown, C. (1977). Effect of oxygen tension on cellular energetics. *Am. J. Physiol.*, **233**, 135.

100. Ueda, K., Kawaichi, M., and Hayaishi, O. (1982). Poly(ADP-ribose) synthetase. In *ADP-ribosylation reactions* (ed. O. Hayaishi and K. Ueda), p. 117. Academic Press: New York.

101. Horsman, M. R. (1995), Nicotinamide and other benzamide analogues as agents for overcoming hypoxic cell radiation resistance in tumours (a review). *Acta Oncol.*, **34**, 571.

102. Olsson, A. R., Sheng, Y., Pero, R. W., Chaplin, D., and Horsman, M. R. (1996). DNA damage and repair in tumour and non-tumour tissues of mice induced by nicotinamide. *Br. J. Cancer*, **74**, 368.

103. Kaanders, J. H., Pop, L. A., Marres, H. A., Van der Maazen, R. W. M., Van der Kogel, A. J., and Van Daal, W. A. J. (1995). Radiotherapy with carbogen breathing and nicotinamide in head and neck cancer: feasibility and toxicity. *Radiother. Oncol.*, **37**, 190.

104. Van der Maazen, R. W. M., Thijssen, H. O. M., Kaanders, J. H. A. M., Dekoster, A., Keyser, A., Prick, M. J. J., Grotenhuis, J. A., and Wesseling, P. (1995). Conventional radiotherapy combined with carbogen breathing and nicotinamide for malignant gliomas. *Radiother. Oncol.*, **35**, 118.

105. Horsman, M. R. and Chaplin, D. J. (1994). Enhancement of cyclophosphamide cytotoxicity *in vivo* by the benzamide analogue pyrazinamide. *Br. J. Cancer*, **69**, 648.

106. Salford, L. G., Pero, R. W., Aas, A. T., and Brun, A. (1992). Metoclopramide as a sensitizer of 1,3-bis(2-chloroethyl)-1-nitrosourea treatment of brain tumours in rats. *Anticancer Drugs*, **3**, 267.

107. Lyback, S., Wennerberg, J., Kjellen, E., and Pero, R. W. (1991). Dose schedule evaluation of metoclopramide as a potentiator of cisplatin and carboplatin treatment of xenografted squamous cell carcinoma of the head and neck. *Anticancer Drugs*, **2**, 375.

108. Kato, T., Suzumura, Y., and Fukushima, M. (1988). Enhancement of bleomycin activity by 3-aminobenzamide, a poly(ADP-ribose) synthesis inhibitor, *in vitro* and *in vivo*. *Anticancer Res.*, **8**, 239.

109. Chen, G. and Pan, Q. C. (1988). Potentiation of the antitumour activity of cisplatin in mice by 3-aminobenzamide. *Cancer Chemother. Pharmacol.*, **22**, 303.

110. Petrou, C., Mourelatos, D., Miolou, E., Dozi-Vassiliads, J., and Catsoulacos, P. (1990). Effects of alkylating antineoplastics alone or in combination with 3-aminobenzamide on genotoxicity, antitumour activity and NAD levels in human lymphocytes *in vitro* and Ehrlich ascites tumour cells *in vivo*. *Teratogenesis Carcinog. Mutagen.*, **10**, 321.

111. Darroudi, F., Zwanenberg, T. S. B., Natarajan, A. T., Driessen, O., and van Langevelde, A. (1989). Reduced tumour progression *in vivo* by inhibitor of poly(ADP-ribose) synthetase (3-aminobenzamide) in combination with X-rays

or the cytostatic drug DTIC. In *ADP-ribose transfer reactions: mechanisms and biological significance* (ed. M. K. Jackobson and E. L. Jacobson), p. 390. Springer Verlag, New York.

112. Horsman, M. R., Brown, D. M., Hirst, D. G., and Brown, J. M. (1986). Changes in the response of the RIF-1 tumour to melphalan *in vivo* induced by inhibitors of nuclear ADP-ribosyl transferase. *Br. J. Cancer*, **53**, 247.

113. Martin, D. S., Stolfi, R. L., Nord, L. D., and Colofiore, J. R. (1996). Enhancement of chemotherapeutically-induced apoptosis *in vivo* by biochemical modulation of poly(ADP-ribose) polymerase. *Oncol. Rep.*, **3**, 317.

114. Calabrese, C. R., Thomas, H. D., Srinivasan, S., White, A. W., Boritzki, T., Canan-Koch, S., and Zhang, K. (1999). Pharmacokinetics and tissue distribution of novel potent inhibitors of poly (ADP-ribose) polymerase (PARP). *Proc. Am. Assoc. Cancer Res.*, **40**, 389.

--------------------------------- **7** ---------------------------------

Perspectives

Sydney Shall

7.1 Introduction

There is currently increasing interest in the hypothesis that PARP participates in a signalling cascade. It is very well established that DNA damage and/or DNA repair is the source of signals to other cellular processes. After DNA damage cell-cycle checkpoints both in G_1 and in G_2 may be activated. Furthermore, both DNA replication and RNA transcription are usually inhibited. With high levels of DNA damage, apoptosis may ensue. These consequences imply signalling systems. Most interest in investigations aimed at elucidating these signalling systems have concentrated on the question, how is p53 activated? There have been suggestions that PARP directly modulates p53 functions, either by covalent modification or by non-covalent binding of poly (ADP-ribose) to p53. So far no compelling evidence for such regulation in intact cells or animals has been published. The problem is difficult because p53 may bind to damaged DNA and may then signal downstream events without the assistance of other molecules. Recently, Arnold Levine has suggested (1) that perhaps both p53 and another DNA repair protein simultaneously bind to the damaged site, and interactively transmit a signal. This suggestion clearly does provide a potential place for PARP in a signalling cascade.

Other molecules that may participate in these signalling cascades include DNA-PK and ATM. DNA-PK both binds to DNA ends and has enzymatic properties, and therefore is a good candidate for a signalling molecule. However, it seems that DNA-PK is only involved in the repair of double-stranded (ds)-DNA breaks, and in V(D)J and T-cell receptor recombination reactions (2). It does not seem to be involved in the repair of single-strand (ss) breaks. It seems to me that DNA-PK and PARP function in parallel pathways, approximating to ds breaks and ss breaks, respectively. ATM is a much more complex molecule that has been very widely regarded as a purely signalling protein. Recently, however, there has been a re-evaluation of the evidence and it has been suggested that perhaps ATM also has a direct function in some aspects of DNA repair (3, 4).

7.2. PARP and DNA damage signalling

DNA damage is known to signal information to other cellular processes. There are several processes that are especially relevant here. DNA damage leads, first, to cessation of current DNA replication and RNA transcription. Second, initiation of DNA replication of new replicons in those cells that are in S-phase is inhibited. Third, for those cells that are in a pre-S-phase or pre-mitotic state, there is a cell-cycle block, which prevents cell cycle-progression. Finally, in some cells, there is the initiation of the process of programmed cell death or apoptosis; the mechanisms involved in this initiation and the criteria which trigger this mechanism, are totally obscure at the moment. In addition, DNA damage generally induces a variety of new transcriptional events; I shall not deal with this response, except for the special case of the transcription factor NF-κB.

In the case of DNA base excision repair the mechanism of inhibition of DNA and RNA synthesis has not been clearly established. Possibly, very local effects achieve these inhibitions. The recent data indicating (at least in cell extracts) that PARP interacts both physically and functionally with a growing number of transcriptional co-activators (see Chapter 3), may be relevant to the inhibition of RNA synthesis. It has been shown that PARP is associated with the DNA replication machinery (5–7), which might be relevant to the inhibition of ongoing DNA synthesis, or to the inability to initiate synthesis at new replicons when damage (breaks) is present in the DNA template; indeed, distortions of DNA synthesis in PARP knockout (KO) mice have been recently reported (7, 8). Perhaps, a detailed and careful investigation of DNA synthesis in these animals might be helpful; although it should be kept clearly in mind that these animals have accumulated DNA damage in comparison to wild-type animals, and this in itself would affect DNA synthesis as indicated above. It is plain that more work in these areas is needed to determine if any of the PARP enzymes are relevant to these control mechanisms.

Thus, we come to the most discussed issues, namely the cell-cycle blocks and apoptosis. It may be helpful for our thinking about this subject if we divide the molecules involved into two groups: those directly involved in 'sensing' the presence of DNA damage or of its repair; and those specifically adapted to 'transducing' or 'signalling' this information. Some molecules, like the DNA glycosylases, and perhaps DNA polymerase-β are very well adapted to the function of identifying abnormal DNA structures or DNA repair in progress, but neither of these molecules seem to be adapted to transmit signals. By contrast, both DNA-PK and PARP are very capable of finding evidence of DNA repair, namely double-strand and single-strand DNA breaks, respectively. Furthermore, both these proteins have an enzymatic property that may translate and amplify the signal emanating from the DNA break. They are therefore candidates for signalling messengers. Third, there are proteins like ATM and ATR that have enzymatic properties and which, very evidently, at least for ATM, are signalling molecules; it is more difficult to see them as DNA damage 'sensors', although in fact ATM may in fact well be so (3, 4) (see below). Finally, there is the most famous of all these molecules, p53, with enough potential properties to achieve detection, signalling, and execution of the signal, all by itself, which only makes it that much harder to pin down the blame for each step. Molecules like the Cdk inhibitors seem to be examples of 'pure' signalling molecules.

It is currently unclear whether cell-cycle arrest and apoptosis are two alternative, mutually exclusive, pathways, or whether apoptosis is the consequence of a 'failed' attempt at DNA repair. We know that cell-cycle arrest is activated at low levels of DNA damage, when no apoptosis occurs, which suggests that these two outcomes are independently determined. Of course, it is also possible that the two pathways are parallel, but not necessarily mutually exclusive. What then may be the criteria for initiating apoptosis? Most authors answer that it is merely quantitative, simply a lot of DNA damage; 'saturating the repair capacity' will activate the apoptotic pathway. There is no direct evidence for this assumption, which is a very plausible and likely notion. But, one must not forget altogether the additional, alternative possibility; namely that sometimes it is the quality of the damage that induces the apoptosis. There may be some lesions that are non-repairable and are significant obstacles to cell proliferation. Second, it may be that certain DNA regions are privileged in their ability to activate apoptosis. So, signalling of DNA damage may involve the following components, which I shall consider in turn; the transcription factor NF-κB, DNA-PK, ATM, and p53.

7.2.1 The transcription factor NF-κB

The physiological relationship between PARP and the transcription factor NF-κB may turn out to be the most remarkable property of PARP, and the one most significant for human health. At the time of writing, this interaction is supported by the most convincing evidence of all the available examples that propose that PARP is a component of a signalling pathway and/or that PARP regulates a (or some) transcription factor, and hence indirectly regulates transcription. We already know that PARP is not required for standard RNA synthesis, but the evidence discussed below (and also in Chapter 3) clearly indicates that PARP can regulate (at least indirectly) some transcriptional programmes.

This story started when T. Taniguchi and co-workers (9) described the downregulation of PARP mRNA after the activation of murine macrophages by IFN-γ; they also demonstrated that the mRNA of the MHC class II gene, the Ia-antigen, was increased. When they transfected a PARP cDNA into macrophage cells, they observed that the induction of the Ia-antigen by IFN-γ was much reduced (10). They then expressed PARP under the control of the metallothionein promoter and showed that the expression of PARP was associated with a decreased Ia-antigen induction by interferon (11). They went on to show that when macrophages were treated with either IFN-γ or phorbol ester, the PARP level was downregulated and the HLA-DR and IL-1 proteins were produced. But if PARP was overexpressed from a plasmid, this inhibited the expression of these two immune genes (11). Furthermore, as macrophages develop PARP mRNA and the protein level decreases, the induction of HLA-DR and of IL-1 increases after treatment with IFN-γ or phorbol ester (12). IL-1β acts synergistically with IFN-γ to induce MHC class II genes (13). Addition of noradrenaline to astrocytoma cells increases the level of PARP mRNA, and also decreases the expression of the MHC class II genes (14). This observation may also be relevant to the toxicity caused by 1-methyl-4-phenyl-1,2,3,6-tetrahydropyridine (MPTP) and to the pathogenesis of parkinsonism (15–20). (See also Chapter 5, Section 5.4). This observation is consistent with the notion that PARP suppresses the expression of these immune genes. Expression of an antisense PARP gene resulted in a more than 90% reduction in PARP mRNA. In such cells the induction of MHC class II genes by IFN-γ was amplified (21).

Further evidence which suggests that PARP is significant to the (innate) immune response in some circumstances, came with the observation that poly (ADP-ribosyl)ation was also implicated in the pathways in monocytes and macrophages leading to the expression of inducible nitric oxide synthase (iNOS), IL-6, and TNF-α (22–25). Exposing bone marrow macrophages to Lipopolysaccharide (LPS) induces nitric oxide formation, which was blocked by nicotinamide, 3-aminobenzamide (3-AB) and 3-methoxybenzamide (3-MB) (25). Furthermore, induction of iNOS in L929 cells by TNF-α is prevented by nicotinamide, 3-AB, and 3-methoxybenzamide (23). This subject was transformed by the unexpected explanation of why PARP inhibitors block nitrite production in macrophages after their exposure to either LPS or to IFN-γ. Wietzerbin and co-workers (26) showed that this inhibition of iNOS was mediated by the transcriptional activity of NF-κB. They demonstrated quite clearly that inhibition of PARP impaired the ability of NF-κB to function as a transcriptional activator in the expression of the iNOS gene (26).

The transcription factor NF-κB is crucial to many processes (27–29), including three that are important for our discussion; and these three processes are DNA damage, immune responses, and apoptosis. The transcription factor NF-κB is activated by many agents, including TNF-α, mitogens, phorbol esters, and UV light. The redox state of the cell modulates the activation of this transcription factor. Oxidative stress itself can activate this transcription factor (30–33). In addition to oxidizing agents, NF-κB is also activated by a variety of chemical agents that damage DNA (31, 34–36). In immune responses, NF-κB regulates the expression of TNF-α, the inducible NOS gene, interleukins IL-1β, IL-2, IL-6, and IL-8, as well as the adhesion molecules ICAM-1 and E-selectin. NF-κB is also crucial in preventing the initiation of apoptosis. In fact, NF-κB is a family of five proteins, which can form hetero- and homodimers in many permutations; the different dimers may well have somewhat different properties. The most common combination observed is a p50/p65 heterodimer.

Recently, a very elegant and compelling proof of the functional connection between PARP and NF-κB has been reported. Oliver *et al.* (37) investigated the response of cells from PARP-deficient mice to cytotoxicity induced by TNF-α, and they observed that survival was lower in the PARP-deficient cells. They then demonstrated that the PARP-deficient cells were unable to induce the expression of NF-κB itself. Moreover, this was also true for hydrogen peroxide. Thus, we have here a new DNA repair defect in PARP KO cells, in which the normal induction of the transcription factor NF-κB does not occur. Wietzerbin *et al.* (26), had shown that 3-AB blocked NF-κB expression; but Oliver *et al.* (37) found that while 3-AB blocked expression a little, a much greater inhibition was obtained by the complete absence of the PARP protein. These results may indicate that both the enzyme activity and the PARP protein itself are both able to activate NF-κB expression. This needs to be clarified. It was also shown (37) that in PARP KO cells the translocation of NF-κB to the nucleus occurred as it did in the wild-type cells. Consequently, one must conclude that these authors (37) have discovered a previously unknown level of control of NF-κB, which occurs after it has translocated to the nucleus. The nature of this control is unknown, but the authors make some provocative suggestions relating to the ability of PARP to bind to DNA bends (38, 39) such as those induced in the NF-κB enhanceosome by the high mobility group (HMG)-I(Y) protein (40, 41). They suggest that perhaps PARP might bind to HMG I(Y) and thus influence the conformation of the transcription complex. Perhaps one should also consider the possible participation of the c-ABL protein in this transduction mechanism.

The authors then showed that PARP KO mice are protected to a dramatic extent against the whole-animal toxicity of LPS. Related results were obtained independently by the group of Nicotera, using the Wagner knockout mice (42). Furthermore, in the intact animal iNOS was not induced, nitric oxide was consequently not produced, and the synthesis of both TNF-α and IFN-γ was sharply reduced in PARP KO animals exposed to LPS. Oliver *et al.* (37) very reasonably conclude that PARP is significant in the induction of NF-κB expression after the exposure of cells or animals to LPS, TNF-α, or hydrogen peroxide. This seems to me to be the most convincing example of PARP as a signalling molecule that controls the expression of a transcription factor. In this example, it is still not clear whether PARP is activated by DNA single-strand breaks in every case. Presumably, there are such breaks after exposure to hydrogen peroxide. There is also the possibility that NF-κB is regulated in addition, by the redox changes alone, induced by some DNA damaging agents, such as hydrogen peroxide. We do not know for sure whether PARP functions as a protein independently of its enzyme activity after exposure to LPS and to TNF-α. My prejudice would be that a combination of TNF-α and INF-γ would induce a significant level of DNA strand breaks, but whether this occurs after exposure to LPS is uncertain. Both LPS and the combination of LPS and IFN-γ do cause DNA strand breaks and PARP activation in cultured macrophages (43, 44).

Hassa and Hottiger (45) have independently published essentially similar data to that in the paper by Oliver *et al.* (37). They also demonstrate that poly (ADP-ribose) polymerase is required for specific NF-κB transcriptional activation in intact cells. They show that activation of the HIV-long terminal repeat (LTR) promoter and an artificial promoter dependent on NF-κB is severely reduced in cells from PARP KO mice. This deficiency was restored by transfection of a PARP cDNA. They also show that PARP and NF-κB co-immunoprecipitate from nuclear extracts. Thus, both the papers of Oliver *et al.* (37) and that of Hassa and Hottiger (45) clearly demonstrate that PARP is required for the transcriptional activity of NF-κB under certain circumstances.

7.2.2   The DNA-dependent protein kinase (DNA-PK)

The enzyme DNA-dependent protein kinase (DNA-PK) is a nuclear serine–threonine protein kinase that is activated upon association with DNA. (For a very clear review, see ref. 2). This enzyme is composed of a very large catalytic subunit, called DNA-PKcs, and a regulatory protein called Ku, which is made up of two polypeptides, Ku 70 and Ku 80. This trimeric complex is a crucial component of one pathway of DNA double-strand break repair, as well as of the V(D)J recombination process. In addition, the several components may also participate in other processes that maintain structures like chromatin and telomeres (46, 47).

Because, like PARP, DNA-PK also binds to DNA interruptions we are obliged to consider whether there are physical or functional interactions between DNA-PK and PARP. Because DNA-PK is a protein kinase, I will separately consider whether DNA-PK is a link in the signalling chain from damaged DNA to cell-cycle arrest, or apoptosis, or other outcomes. I will especially draw attention to similarities and differences between the two proteins and their respective physiological role.

The regulation of DNA-PK occurs not only by binding to DNA, but also by interaction with other proteins. One clear example is binding to the protein C1D, which has a very high affinity for DNA. This latter protein interacts with the leucine zipper region of DNA-PKcs (48). C1D bound to DNA can activate DNA-PK in the absence of DNA breaks. The physiological consequence of this activation is not yet clear; but C1D is a component of the nuclear matrix and is induced by ionizing radiation. Perhaps this example might be a model for the activation of PARP without the need for DNA breaks, but which instead is induced by interaction with a particular protein which may bind to a specific structure in DNA.

DNA-PK activity is also regulated by autophosphorylation of the DNA-PKcs subunit. This leads to dissociation from Ku and loss of enzyme activity. This, again, is not unlike the automodification of PARP that leads to the dissociation from the allosterically activating DNA and loss of enzyme activity. DNA-PK is also phosphorylated by the non-receptor protein kinase c-Abl, which is itself induced in response to ionizing radiation.

DNA-PK is regarded as a primary DNA damage sensor rather than an inducible, downstream signalling component. DNA-PK is present in relatively high amounts, up to 1% of HeLa-cell nuclear protein, and the levels of DNA-PK do not seem to be regulated by DNA damaging agents (2). That means, like PARP, the protein is not inducible in general, but that preformed protein is activated by DNA breaks. However, like PARP, DNA-PK is induced when cells enter the cell cycle from the quiescent state (49). Moreover, the level of DNA-PK protein in a variety of species correlates with the life span (2, 50) just as PARP activity does. This may suggest that genomic stability (or, perhaps, specifically DNA double-strand breaks) has a significant consequence for the ageing process.

Contrary to earlier reports, it has recently been demonstrated (51) that the presence of DNA-PK is required if p53 is to bind to its cognate DNA sites after cells are exposed to hydrogen peroxide, mitomycin C, or to ionizing radiation. In addition, DNA-PK phosphorylates p53 after ionizing radiation at serines 15 and 37, and thereby releases the prior inhibition by MDM2 and permits the transactivation of gene expression by p53 (52). Thus it seems that DNA-PK has two functions regarding p53; it activates the DNA binding of p53 and it blocks by phosphorylation the inactivation of p53 by MDM2.

Like PARP, DNA-PK is proteolytically degraded by caspases during apoptosis (53); the probable significance is the same for the two enzymes, namely to prevent the consumption of energy in useless attempts at repair.

Rejoining of DNA double-strand breaks by non-homologous end-rejoining seems to have an absolute requirement for all three subunits of DNA-PK. Furthermore, the complex of XRCC4 with DNA ligase IV, is also essential. This is similar to the requirement for PARP in the base excision–repair pathway, in which the complex of XRCC1 and DNA ligase III is used. Precisely what these two proteins do in these reactions is still unclear. It would perhaps be helpful to determine whether, in a PARP KO mouse or cell strain, PARP is needed for the process of non-homologous DNA end-rejoining.

What information is available regarding the physical or functional interactions between DNA-PK and PARP? Some evidence has been published that DNA-PK and PARP can physically interact (54–57). Ruscetti *et al.* (54) used an antibody column directed against Ku 80 to trap proteins in a HeLa-cell nuclear extract. As expected, Ku 80 and 70 were trapped, as well as DNA-

PKcs. In addition, however, two other proteins were bound and eluted from this column. They were identified as heterogeneous nuclear ribonucleoprotein-U and PARP (54). Furthermore, the authors showed that PARP could be phosphorylated by DNA-PK *in vitro*, but without any apparent consequence. By contrast, they demonstrated that PARP can poly (ADP-ribosylate) DNA-PK, which leads to an activation of the modified protein. It is very unusual to find an example of poly (ADP-ribosyl)ation activating an enzyme; in most previous reports inactivation was observed. Consequently, this report is of special interest. They further showed that the activation of DNA-PK by PARP required the enzyme activity, because addition of NAD was required and the PARP inhibitor 1,5-dihydro-isoquinoline inhibited the activation of DNA-PK. The Ku subunits 70 and 80 were not required for the poly (ADP-ribosyl)ation reaction that needed only the DNA-PKcs. This data implies a functional interaction between the two proteins.

In immunoprecipitation studies Said Aoufouchi (personal communication) showed that the three proteins PARP, Ku 70, and Ku 80 were co-immunoprecipitated, irrespective of whether the precipitation was with an antibody directed against PARP or against Ku 70/80. In both directions, one could bring down all three proteins. Furthermore, Morrison *et al.* (56) reported that PARP and Ku 70/Ku 80 consistently co-immunoprecipitate from crude nuclear extracts.

The interaction between PARP and DNA-PK and/or Ku proteins was confirmed recently in two different contexts—in association with matrix attachment sequences (57), and with the promoter of the HTLV-1 provirus (55)—suggesting a potential function of the molecular complex PARP/DNA-PK in the regulation of chromatin structure and function.

In a very elegant study, a further claim for a functional (genetic) interaction between the two proteins was suggested. Morrison *et al.* (56) crossed PARP KO mice with SCID mice, which were defective in DNA-PK. (However, it should be borne in mind that several laboratories have suggested that there may be residual DNA-PK activity in some *scid* cells, because some of the *scid* mutations seem to be 'leaky', that is to say incomplete. Perhaps, the complete loss of both functions may be lethal). Some animals were successfully bred which lacked both PARP and were defective in DNA-PK, although the vast majority of progeny died just after birth. This argues that PARP and DNA-PK do not function solely in a single pathway (see below). It is not possible from this data to distinguish between the possibility that DNA-PK and PARP function in two essentially independent, parallel pathways, and that they interact in some way. Thus, since both PARP KO animals and SCID mice are more viable, it would seem that survival without both pathways of DNA repair is much more difficult than with either one. Do these two enzyme systems provide back up for each other in intact animals? Or do they do so only in knockout animals in a vital process of adaptation to a serious genetic defect?

The authors also noted that in SCID mice the thymus T cells were unable to develop normally; they displayed neither CD4 nor CD8 surface proteins, and did not show CD3 receptors or T-cell receptor rearrangement. In double-mutant mice (PARP KO and SCID) there were many more cells in the thymus, and there was a large increase in the fraction of cells displaying both CD4 and CD8 surface proteins, although they still did not progress to expression of CD3 or the T-cell receptor rearrangement. When they looked in detail at the molecular events in the T cells, they observed reasonably diverse rearrangements of the *D-to-Jβ* locus in the double-mutant mice, but there was very restricted oligoclonal *V-to-DJβ* rearrangements, and the pattern was quite unlike the wild-type animals. Investigation of the lymph nodes showed that more mature T cells with either CD4 or CD8 surface proteins were present, and that CD3 was expressed and the T-cell receptor rearrangement had occurred. The authors concluded reasonably enough that the absence of PARP (in some way) rescues V(D)J recombination in SCID mice. They then proposed that (in the intact animal, presumably), PARP plays a regulatory role in this specific recombination process, possibly by acting as an antirecombinogenic molecule. However, there is no evidence presented for such a role in the PARP KO mice. Furthermore, there was no comparable rescue in the B cell lineage, which one might have expected.

Other studies have also attempted to discover whether there is a functional interaction between DNA-PK and PARP. Weinfeld *et al.* (58) and D'Silva *et al.* (59) have investigated the relative affinities of DNA-PK and of PARP for various types of DNA strand breaks. First, DNA-PK does not seem to be activated by single-strand nicks in the DNA, which are effective activators for PARP. From this data it seems that PARP might have a greater preference for all types of activating DNA than does DNA-PK. It is notable that the most effective structure reported in these two papers for PARP activation is a blunt-ended DNA. As the authors point out, although nicks are not the most active, they probably predominate quantitatively in most real biological situations.

More evidence of a potential functional relationship between DNA-PK and PARP has been sought by investigating the effect of inhibitors of the two enzymes. Wortmannin is a non-specific, but effective inhibitor of DNA-PK and 8-hydroxy-2-methyl-quinazolin-4-one (NU1025) is a potent inhibitor of PARP. Boulton, Kyle, and Durkacz (60) have studied the effects of these inhibitors following ionizing radiation or exposure to the alkylating agent temozolomide. The cell killing of ionizing radiation, as assessed by colony formation, was potentiated by both enzyme inhibitors; 20 μM Wortmannin gave a dose enhancement factor (at 10% survival) of 4.5 ± 0.6, while with NU1025 it was 1.7 ± 0.2. When used in combination, an enhancement factor of 7.8 ±1.5 was observed. These results suggest that the two enzymes do not function solely in a common pathway, because using both inhibitors gives a greater effect than either one alone. On the other hand, there is no clear evidence here of a synergistic effect, which we would predict if there was significant functional interaction between the two pathways. Similar results were obtained after alkylation damage induced by temozolomide. These results were extended by directly measuring the level of double-strand DNA breaks after the treatments just described. Double-strand breaks were increased 30 minutes after damage by inhibition of either DNA-PK or of PARP. Inhibition of both enzymes gave an additive effect, confirming the implication adduced above. Furthermore, Wortmannin had no effect on the level of DNA single-strand breaks; suggesting that DNA-PK has no significant role in single-strand break repair. As would be predicted from the older literature, NU1025 did increase the levels of single-strand breaks. Although DNA-PK does not seem to be relevant for the repair of single-strand breaks, PARP does seem to contribute to the repair of some double-strand DNA breaks, as indicated by the clear effect of NU1025 on the disappearance of double-strand DNA breaks. It would be of considerable interest to confirm this observation by direct analysis of double-strand DNA break repair in PARP KO cells, because it has been generally assumed for many years that PARP, at least, has very little role in the repair of double-strand breaks.

Finally, there is one further parallel between DNA-PK and PARP. It has been reported that both PARP (61) as well as DNA-PK (62) have some function in the integration of retroviral proviral DNA. Normally, a virally coded enzyme called integrase carries out integration of the proviral DNA, which is an essential step in retroviral replication. This enzyme, at least in subcellular systems, carries out most of the steps by which the free proviral DNA is integrated into the chromosomal DNA. The end product of the integrase reaction leaves the 5' ends of the proviral DNA unligated. Thus, presumably cellular enzymes complete the task as if it were a DNA repair reaction. With this logic in mind, it was clearly demonstrated by Gaken *et al.* (61) that PARP could play a role in this process. Rather less obviously, but of profound importance, it has been convincingly demonstrated that the DNA-PK pathway may also function in the integration of the proviral DNA (62).

I have mentioned before the presence of an alternative pathway of double-strand rejoining that uses a recombination mechanism. We may wonder whether PARP would have any consequence for this pathway. Perhaps in the absence of PARP this recombination pathway is more active (B. W. Durkacz, personal communication).

In conclusion, with the limited experimental data currently available, it is my opinion that the best current guess is that DNA-PK and PARP are components of two parallel pathways in DNA repair, with the DNA-PK being predominantly concerned with the repair of DNA double-strand breaks by non-homologous end-rejoining and with V(D)J and T-cell receptor rearrangement in the immune system. By contrast, while PARP is predominantly concerned with single-strand DNA breaks, which presumably are, at least in part, from base excision pathways, this enzyme system is also concerned with some categories of double-strand DNA repair, as indicated by the recent data of Boulton *et al.* (60). Furthermore, we must wait for further experiments to understand how these two proteins function in chromatin conformation and in telomere physiology. DNA-PK and PARP may turn out to mutually regulate the activity of each other by phosphorylation and by poly (ADP-ribosyl)ation, respectively. This remains to be tested in intact cells. Finally, DNA-PK may transmit (perhaps to ATM, perhaps to p53) signals which it receives from PARP.

7.2.3 ATM

ATM is the gene that is mutated in the hereditary syndrome called ataxia telangiectasia (AT). Patients with this syndrome display a hypersensitivity to ionizing radiation. Cells derived from these patients also show this hypersensitivity. A striking aspect of this condition is that the cells from patients with AT do not show the normal inhibition of DNA synthesis after ionizing radiation. There is clearly a defect in the cellular response to this type of DNA damage. It has been widely concluded that ATM plays a crucial role in the signalling of DNA damage. AT cells that are in the G_2 phase of the cell cycle at the time that the damage is inflicted, fail to arrest at the G_2/M checkpoint. Therefore it has usually been thought that this is due to a checkpoint defect and that AT is a checkpoint mutation (for review see ref. 63). Recently, some workers in this field have suggested that ATM also has a direct function in DNA repair, but the precise role has not been experimentally determined.

The ATM protein contains 3056 residues, with the C-terminal 400 amino acids being highly similar to the catalytic subunit of phosphatidylinositol 3-kinase (PI3-kinase). Several large proteins that are involved in cell-cycle progression or in cellular responses to DNA damage contain a domain like this. For example, I have discussed at length the protein DNA-PK above. In addition, there is a protein related to ATM called AT-related protein (ATR or FRP1). This latter protein is the human homologue to the *Schizosaccharomyces pombe* protein Rad3. All three proteins seem to be protein kinases and are not lipid kinases. ATM is a 370-kDa highly phosphorylated protein. It is constitutively expressed in numerous tissues, and is especially prominent in the testis, spleen, and thymus of adult mice. It seems to be associated with areas of cell division. ATM is found primarily in the nucleus. However, small amounts may be found in the cytoplasm. ATM is essential for meiosis, and in patients with AT there are no germ cells. A similar distribution is found for PARP (see Chapter 2). All three of these large proteins are in fact serine–threonine protein kinases. ATM can phosphorylate p53 on Ser15, which is usually phosphorylated after exposure to ionizing radiation (64, 65). Moreover, ATM has been shown to physically associate with p53 (66). Thus, ATM may be the immediately preceding component to p53 in the pathway that signals the presence of DNA damage.

The non-receptor protein tyrosine kinase called c-Abl also interacts with ATM (67, 68). This protein kinase (c-Abl) binds to a specific motif in ATM that is rich in proline, via an SH3 domain in c-Abl (68). There is also evidence that ATM phosphorylates and activates c-Abl (67). Another protein, which has been reported to interact with ATM, is β-adaptin (69), which is a component of the AP-2 adapter complex that is involved in endocytosis mediated by clathrin. This interaction, which has been observed both in intact cells as well as in subcellular systems, shows a cytoplasmic function of ATM.

The notion that ATM is part of a DNA damage checkpoint is confirmed by the report that the *S. pombe chk1* gene can restore the G_2/M mitotic arrest in AT cells and can overcome the radiosensitivity of AT cells (70). The question therefore arises whether there is any physical or functional interaction between PARP and ATM? Many years ago Edwards and Taylor reported that PARP activity was defective in AT cells (71), but this report was never confirmed. More recently, it has been shown that this result was incorrect (72). There is no evidence of a direct physical or functional interaction between PARP and ATM. It seems likely to me that the two proteins function in parallel pathways. However, there is experimental evidence for what might turn out to be a very significant genetic interaction (Josiane Ménissier-de Murcia, personal communication). I have discussed earlier, at length, the function of PARP in the control of the transcriptional factor NF-κB. It has very recently been demonstrated that the ATM protein is required for the sustained activation of NF-κB following DNA damage (73).

7.2.4   The tumour suppressor p53

The p53 protein is central to the cellular response to DNA damage. This protein is required for the normal response to both endogenous and exogenous DNA damage. It accumulates to a high level in cells exposed to ionizing radiation, ultraviolet (UV) light, or mitomycin C or in cells infected with certain DNA viruses (74–77). p53 is not essential for normal mammalian development, in the same way that PARP is dispensable for normal development. But, equally, a mammalian organism cannot readily survive the absence of either protein. In the case of p53, lack of the protein rapidly leads to multiple malignant tumours, which demonstrates that p53 is an important tumour suppressor. Absence of PARP, on the other hand, means that the organism is severely hypersensitive to ionizing radiation and to genotoxic chemicals. PARP null animals have not been reported to show an excess of malignant tumours. I interpret this result to mean that in the absence of PARP, DNA damage generally kills cells by apoptosis rather than leaving them alive but mutated.

Normally, the level of p53 protein in cells is very low, due to a very short half-life of about 5–20 minutes. The proteases responsible are unknown, but mediation by the ubiquitin pathway has been suggested by some data. In the presence of many types of stress including DNA damage, the levels of p53 in the cell rise considerably. Different types of DNA damage increase the level of p53. Both single- and double-strand breaks caused by ionizing radiation, as well as the intermediates produced after UV irradiation, increase the levels of p53 (74–77). It would be interesting to determine the precise kinetics and extent of p53 change after different kinds of DNA damage, especially after treatment with methylating agents. This has been partially reported for UV and for ionizing radiation by Lu and Lane (78).

The tumour suppressor p53 also has an independent activity as a transcriptional regulator. The number of known target genes is quite large and they include cell-cycle control genes such as *p21, FADD45, MDM2, EGF-receptor, PCNA, cyclin D1, cyclin G, TGF-α*, and the 14–3–3 proteins. A number of genes important in apoptosis may also be regulated by p53. These include *BAX, BCL-Xl, FAS1, FASL, AGF-BP3, PAG608*, and *DR5*. Activation of p53 clearly has many and varied downstream consequences. Two pathways of special interest to our discussion involve the ability to induce a cell-cycle arrest by inhibition of cyclin-dependent kinases (Cdk), and the ability to induce apoptosis. I have remarked before that we do not clearly understand the mechanisms by which cells choose between these alternatives.

Many different protein modifications can influence either the DNA binding or the transcriptional activity of p53. These two different properties of p53 are quite separable, and are not synonymous. The protein modifications experienced by p53 include phosphorylations at serines 6 and 9, and at serines 15 and 33 induced by ionizing and UV radiation, and also at serines 37, 315, 376, 378, and 392. In addition, several lysines may be acetylated, as for example lysines 320 and 352.

It is very important to appreciate that the level of p53 may be dissociated from the activity of p53 as a transcriptional co-activator. It should not be overlooked that the level of p53 and its activity are not well correlated. What we really need to know in any experiment is what changes occur in the relevant, significant activity of p53 after DNA damage. It is well known that the regulation of p53 is complex and that the outcome depends on many possible modulations of its activity. The level of p53 increases after DNA damage (including after UV radiation), mainly due to the increase in the half-life of the protein, but also perhaps because the frequency of translational initiation is increased. Transcriptional control of the p53 level does not seem to be very significant. The increase in the amount of p53 is proportional to the level of damage. But the type of damage and the cell type can also alter the response of p53 to DNA damage.

Because cells use rather different mechanisms for repairing different types of damage, for example ionizing and UV radiation damage, we can reasonably suppose that the pathways that transmit signals from DNA damage to p53 must be rather variable. The precise pathways from DNA damage and its repair to the activation of p53 are not known yet. But it is known that cells defective in the *ATM* gene display a delayed and smaller response of p53. None the less, this result does show that ATM is on the pathway. Because the p53 protein itself can bind to DNA double-strand ends, to DNA, and to repair sites, it may itself be the detector of DNA damage. Another possibility proposed by Levine is that p53 interacts with another DNA repair protein and jointly they send a signal (1).

So now we can examine the evidence that PARP is directly on the pathway to the activation of p53. The central question is whether it is the DNA damage itself that finally activates p53 without the necessary intervention of PARP, or whether PARP is required to achieve the activation of p53 that later leads either to cell-cycle arrest or apoptosis. The involvement of PARP in the p53 pathway was first suggested by the work of Lu and Lane published in 1993 (78). They demonstrated again the accumulation of p53 protein in the nuclei of mammalian cells following DNA damage caused by UV light, X-rays, or a restriction enzyme. In addition, promoters that contain sites for the binding of p53 showed a large increase in transcription in response to DNA damage. These workers then asked whether the p53 response was directly triggered by the DNA damage. To answer this they treated cells with 3-aminobenzamide for various lengths of time. There was no p53 response in undamaged cells treated with 3-AB. In contrast, cells treated with 3-AB immediately after ionizing radiation showed a prolonged p53 response. The p53 response was still very high several hours after irradiation. Single-strand breaks are all repaired rather quickly in this system. Thus, it seems that the main activator of the p53 accumulation was probably double-strand breaks. A detailed analysis of precisely what sort of DNA damage can stimulate p53 accumulation would be useful.

A very influential paper from the Berger laboratory was published 2 years later (79), in which they examined the p53 response in two very different types of mutant mammalian cells (see Chapter 3 for details of these mutant cells). The first cell mutant was deficient in PARP itself (cell lines ADPRT54 and ADPRT351) and the second was unable to maintain the normal high level of NAD (cell line N4), which is the exclusive substrate for PARP. Thus both types of cell mutant would be defective in PARP enzyme activity. The two may also be defective in other biochemical pathways. Both types of mutant cell displayed a reduction in the baseline p53 level. The topoisomerase II inhibitor, etoposide, caused an increase in p53 and subsequent apoptosis in wild-type parental V79 cells. However, in both classes of mutant cell there was no accumulation of p53, nor was there any DNA fragmentation characteristic of apoptosis. The authors therefore concluded that PARP might be involved in the regulation of p53. This work seems to imply that PARP activity is needed for the p53 response.

Reports that chemical inhibition of PARP significantly suppresses the accumulation of p53 in response to ionizing radiation (80, 81) are consistent with this notion. But we must note that these later reports seem to be inconsistent with the earlier paper of Lu and Lane (78). Vaziri *et al.* (80) showed that p53 can associate with PARP and that inhibition of PARP blocks the expression of both *p21* and *mdm2*, which are downstream events of the activation of p53.

Kumari *et al.* (82) have confirmed that p53 can be immunoprecipitated by antibodies to PARP, suggesting that the two proteins may exist as a complex. The C-terminal, 85-kDa portion of PARP mediates the association between the two proteins. They also tested whether p53 was modified by PARP, but they failed to show any covalent poly (ADP-ribosyl)ation of p53 in intact cells. However, they were able to show that in a subcellular system PARP can poly (ADP-ribosylate) p53. The inability of PARP to modify p53 in intact cells has also been reported by another group (83). The ability of PARP to covalently modify p53 within 15 min following γ-irradiation has been recently reported by Smith and Grosovsky (84). Using an anti-topoisomerase I antibody these authors made the interesting observation that the transient but tight complex formed between p53 and topoisomerase I, which takes place only during brief periods of genotoxic stress (85), is poly (ADP-ribosylated) in TK6 human lymphoblastoid cells in response to γ-irradiation. They found that only the p53 species engaged in this complex is heavily poly (ADP-ribosylated) under damage conditions. Topoisomerase I is already known to be a partner of PARP (see Chapter 3); therefore topoisomerase I may effect the association of PARP and p53 during the cellular response to stress when the topological properties of chromatin need to be changed.

A rather different and novel approach has been pioneered by the Althaus laboratory. These authors point out that both PARP and p53 have the property of potentially binding to DNA single-strand breaks. They demonstrate a temporary co-localization of the two proteins at such DNA strand breaks. They have previously shown that poly (ADP-ribose) chains, both free and covalently bound to PARP protein can associate with the p53 protein via its C-terminal domain (86). They also show that these poly (ADP-ribose) chains target three specific regions in the p53 protein, to establish strong but non-covalent interactions. The polymer is bound to two amino-acid sequences in the same region of p53 that binds to DNA in a sequence-specific manner. These amino-acid sequences include the amino acids 231–253 and 326–348 in p53. The binding of polymer chains to p53 blocked both the binding of p53 to sequence-specific DNA and to single-strand breaks. Perhaps this novel approach may represent a new category of polymer interaction. Perhaps also, there are other important proteins with similar sequences to p53, for example the MARCKS protein (87), which may therefore also bind the polymer in this apparently specific way. Perhaps, p53 activity is regulated by this specific, strong non-covalent binding of chains of polymers of ADP-ribose.

An alternative approach has examined the consequences of covalent ADP-ribosylation of p53 (88). Wild-type p53 is apparently not modified in intact cells, but mutant p53 protein is modified. When p53 is specifically bound to DNA it is not modified. Thus, the site of modification in p53 probably overlaps its DNA binding site. The significance of this is not clear.

A very informative experimental approach has been to investigate the behaviour of cells that lack PARP, either because of gene inactivation or by lowering PARP mRNA by an antisense strategy. This methodology has been reviewed in detail in Chapter 3. Here I will note that apoptosis is essentially competent in PARP KO cells and animals. In PARP KO or depleted cells, apoptosis, rather than necrosis, seems to be the preferred mode of cell death. This may be explained by the notion that in the presence of PARP, NAD$^+$ is consumed to a degree that depletes the cellular ATP, which leads to necrosis. This is the cell-suicide hypothesis of Nathan Berger. But if apoptosis still occurs in the absence of PARP, then we can assume that PARP is not essential for the process of apoptosis, although the evidence suggests that p53 is essential for this process. Perhaps, p53 is activated by signals in addition to PARP. So, it appears that PARP is not a necessary activator of p53. None the less, there are several papers that suggest that p53 function is somewhat altered in cells that are lacking or depleted in PARP (89–92). It is unclear whether this is a direct effect of the altered state of the PARP protein, or whether it is an indirect consequence of the inefficient DNA repair that results from the absence of the PARP protein.

These indirect effects may be very common. Many cellular stimuli, especially DNA damage, induce both p53 and NF-κB activity. It has recently been demonstrated that there is a mutual control of these two transcriptional factors (93). The two factors often have opposing effects; p53 is usually required to effect apoptosis, while NF-κB seems to inhibit apoptosis, and presumably in conjunction with p53 directs the cell to cell-cycle arrest and DNA repair. Webster and Perkins (93) show that p53 and NF-κB each inhibit the ability of the other protein to stimulate gene expression. Artificial expression of either protein suppresses the stimulation of transcription by the other protein, when tested with a reporter plasmid in intact cells. Moreover, endogenous TNF-α, which normally will activate NF-κB, will suppress endogenous wild-type p53 activation. The reverse is observed following the exposure of cells to UV light. In this case, p53 blocks the activation of NF-κB. It was shown that both p53 and RelA (p65) interact with the transcriptional co-activator proteins p300 and CREB-binding protein (CBP). The authors suggest that the mechanism of reciprocal regulation may involve competition for a limiting pool of p300 and CBP. PARP may impinge on this cross-regulation either through its direct effect on NF-κB or through an indirect effect on p53. It is also possible that PARP might interact directly with p300 or with CBP. This subject clearly needs further investigation.

7.3 A function of poly (ADP-ribose) in controlling telomere length?

There are two PARP-like enzymes that may be associated with the telomere structures and may therefore influence telomere length. The first is tankyrase, which has been shown to negatively regulate the telomere-specific protein TRF-1 (94). The second relevant enzyme activity is the classical PARP, now termed PARP-1 (see Chapter 2). It has been shown that in PARP-1 KO mice there is an excessive shortening of telomeres compared to that in wild-type mice (95). This excessive shortening is seen in different genetic backgrounds, and in both embryonic and adult mice. The *in vitro* telomerase activity is not altered in these PARP-1 KO mice. As expected, they show severe chromosomal instability, as shown by increased frequencies of chromosome fusions and aneuploidy (for more details on the function of PARP in chromosomal instability, see Chapter 3). A third PARP homologue, VPARP (for Vaults PARP) has been recently identified in a two-hybrid screen, using the cDNA encoding the major Vaults particle (MVP) protein (96) (see Chapter 2). Vaults particles are large ribonucleoprotein complexes supposed to be involved in the nucleocytoplasmic traffic. Interestingly, the third protein component of the Vaults particle, TEP-1, is also a partner of the telomerase complex (97). Thus, it seems that two, and perhaps three, different *PARP* genes are significant in maintaining telomere length. This new data may be part of the explanation of the intriguing data from the Bürkle laboratory, showing a correlation between the life span of animals and the specific activity of PARP. In addition, it has been reported that the protein Ku is found to be associated with telomeres in mammals (46, 47). As mentioned above, this protein associates not only with DNA-PK, but also with PARP. Consequently, it will be necessary to determine whether the complex of PARP and Ku are significant in the control of telomere length. Since telomeres seem to be of considerable importance both in the cell fate following DNA damage, in cellular life span, and in cellular immortalization, there will be some exciting results to uncover in analysing the role of the new *PARP* genes in the genome protection and in telomere maintenance in more detail (98).

7.4 General comments

It is now both appropriate and possible to investigate experimentally the idea that perhaps some PARP functions rely only on the protein and not on its enzymatic properties. It is now possible to create mutants of PARP that lack the ability to bind to DNA breaks, or alternatively that do bind to breaks correctly, but which entirely lack enzymatic activity (For details of these mutants, see Chapter 2). Thus, we can test whether any effect really needs the enzymatic function of PARP.

For many years there has been evidence available that PARP can modulate chromatin structure. This was shown to be a property of the synthesis of the polymer. Now we can test another potential mechanism. Perhaps PARP may alter chromatin structure by simply binding to DNA or to various other proteins that can modulate chromatin structure. The very elegant work of the de Murcia laboratory showing that DNA complexes with PARP have a very specific angled structure has been discussed in Chapter 2 (see Figure 2.5). Perhaps in addition, PARP may alter chromatin structure simply by direct binding to important constituents. In this way, PARP may alter chromatin structure without invoking its enzymatic activity. In addition, the ability of PARP to create these specific angled DNA structures raises the possibility that there may other DNA 'cross-over' structures that PARP may recognize. For example, PARP may bind to replication forks, to hairpin structures, or to recombination complexes. These may be associated with PARP functions that do not involve the synthesis of the polymer. Specifically, data has been published reporting that PARP may be associated with matrix elements (57). Thus, we may wonder whether PARP contributes to the loop structure of chromatin. Furthermore, the relationship between PARP and topoisomerase II has been repeatedly suggested, but is not at all clear. Perhaps this too is a matter of chromatin conformation.

We have compared in some detail the data obtained with enzyme inhibitors and the data obtained from the PARP KO mice and cells (99). We have drawn attention to the similarity in the conclusions derived from these two very different experimental approaches. I would like now to note some differences. First, enzyme inhibitors and experiments with dominant-negative mutants (such as the DNA binding domain (DBD) of PARP), presumably work by blocking the cyclic character of the PARP process. Specifically, they lead to the protein or the DBD being persistently attached to the single-strand break in the DNA, and in this way may block further progress of the repair. By contrast, the PARP KO cells simply lack the enzyme. Consequently, there will be circumstances where the two approaches will give quite different results. The use of enzyme inhibitors or the DBD will tell us what happens when we interrupt the PARP cycle of response to a break in the DNA. On the other hand, the PARP KO cells or mice will tell us how the cells or animals behave without this important enzyme. Here the confounding problem is to determine whether the cell has adapted to this pathological state by altering its normal responses. But there is yet another confounding problem in both types of experiment. Suppose it is true that one or more functions of PARP are independent of its enzymatic activity. In this case the two approaches may give very different results. This idea needs to be experimentally examined further.

I should like to end with some very general observations. It is now well established that PARP is required for an adequate cellular response to certain types of DNA damage, especially in the base excision–repair pathways (see Chapter 3). In the absence of PARP the efficiency of DNA repair is poorer. It may well be the case that PARP is also important in other reactions in which there is no DNA damage; this remains to be established. Second, the consequences of the activation of PARP may be quite varied depending on a number of circumstances. For example, it may make a difference whether transcribed sequences are being repaired, and there is some evidence for this. Then the location of the DNA damage in the genome may be significant. For example, damage to telomeric DNA may generate a different response compared to internal sequences. The repair pathway in this case may involve proteins (for example Ku) that may not be used in other cases. Moreover, the outcome of inefficient repair in non-transcribed sequences may be different to that in transcribed sequences. There is considerable evidence that DNA damage may be relevant to the life span of cells; this may be particularly important in the case of telomeric sequences. Perhaps, cell ageing is predominantly related to DNA damage and its repair in telomeric sequences. Because telomeric sequences are repetitive sequences, I would think it might be useful to specifically examine the repair of other repetitive DNA sequences, such as centromeres.

Another special circumstance that I have discussed in some detail is the availability of energy as a determinant of whether apoptosis or necrosis is the outcome of severe damage. It may be that this dichotomy is also significant in less severe circumstances. Perhaps the status of the energy supply of the cell may influence the success of DNA repair when PARP is involved.

Finally, it is becoming increasingly important to consider the downstream consequences of the activation of PARP. Clearly, PARP is required for the activation of the transcriptional factor NF-κB after DNA damage. There is also suggestive evidence that PARP directly or indirectly modulates the activity of p53. DNA damage certainly does lead to the activation of p53, and somehow modulates the choice between cell-cycle block and DNA repair or cell death by apoptosis or necrosis. Finally, we must recognize the mutual control of p53 and NF-κB. Presumably, this interaction is part of a homeostatic mechanism, which has evolved to restore the regulatory elements of the cell to their metastable state that existed before the DNA damage occurred. These perturbations of the regulatory system may have distant effects on the physiology of the cell. Thus, every new and surprising observation made with regard to PARP should be tempered by the question whether such observations are the distant consequences of the activation of PARP after DNA damage. Such a possibility must always be excluded before one concludes that a new function for PARP has been demonstrated. A specific example that needs further investigation in my view, is the evidence that PARP is involved in DNA replication. There is some evidence to this effect. But it is not at all clear whether these data indicate a distant consequence of the activation of PARP by DNA damage, or a direct involvement of PARP in DNA replication independent of any DNA damage. A related observation is the recent evidence that PARP plays some part in the re-initiation of DNA replication in quiescent cells (8). The mechanistic explanation for this very exciting observation will be fascinating. Is this a totally new function for PARP, or is this a facet of a known property of the enzyme, perhaps again a distant effect of the activation of PARP by DNA damage?

Of course, the new *PARP* genes would be expected to have specific functions of their own. In this regard, it will be necessary not only to identify these functions, but also to ask in each case how these new functions relate to the DNA repair function of the first PARP. There are now known to be at least five closely related *PARP* genes (see Chapter 2) (98). The evolutionary relationships between these genes will be of great interest. When did each gene first appear, and how are their varied functions related? It does seem now as if the polymerase genes have evolved from the mono (ADP-ribosyl) transferase enzymes, which seem to be truly ubiquitous in nature. The transferase proteins were presumably also the source of the ADP-ribosylating toxin genes that a number of viruses and bacteria carry.

Poly (ADP-ribose) polymerase was for many years after its discovery a biochemical curiosity. Our understanding has gradually evolved to an understanding that it is only one example of a widespread form of molecular regulation based on the post-translational modification of important proteins. The family of ADP-ribosylation enzymes is now universally accepted as an important component of biological regulation in an astonishing variety of different biological processes. Consequently, in the future, we will read about PARP proteins in many different contexts. It will be very exciting to uncover all the various mechanisms that these PARP enzymes utilize.

References

1. Levine, A. J. (1997). p53, the cellular gatekeeper for growth and division. *Cell*, **88**, 323.
2. Smith, G. C. M. and Jackson, S. P. (1999). The DNA-dependent protein kinase. *Genes Dev.*, **13**, 916.
3. Smith, G. C. M., Cary, R., Lakin, N., Hahn, B. C., Teo, S-H., Chen, D. J., and Jackson, S. P. (1999). Purification and DNA binding properties of the ataxia–telangiectasia gene product ATM. *Proc. Natl. Acad. Sci. USA*, **96**, 11134.
4. Suzuki, K., Kodama, S., and Watanabe, M. (1999). Recruitment of ATM protein to double strand DNA irradiated with ionizing radiation. *J. Biol. Chem.*, **274**, 25571.
5. Simbulan-Rosenthal, C. M. G., Rosenthal, D. S., Hilz, H., Hickey, R., Malkas, L., Applegren, N., Wu, Y., Bers, G., and Smulson, M. (1996). The expression of poly (ADP-ribose) polymerase during differentiation-linked DNA replication reveals that it is a component of the multi-protein DNA replication complex. *Biochemistry*, **35**, 11622.

6. Simbulan-Rosenthal, C. M. G., Rosenthal, D. S., Boulares, A. H., Hickey, R. J., Malkas, L., Coll, J. M., and Smulson, M. E. (1998). Regulation of the expression or recruitment of components of the DNA synthesome by poly (ADP-ribose) polymerase. *Biochemistry*, **37**, 9363.

7. Dantzer, F., Nasheuer, H. P., Vonesch, J. L., de Murcia, G., and Menissier-de Murcia, J. (1998). Functional association of poly(ADP-ribose) polymerase with DNA polymerase alpha-primase complex: a link between DNA strand break detection and DNA replication.*Nucleic Acids Res.*, **26**, 1891.

8. Simbulan-Rosenthal, C. M., Rosenthal, D. S., Luo, R., and Smulson, M. E. (1999). Poly (ADP-ribose) polymerase upregulates E2F-1 promoter activity and DNA pol alpha expression during early S phase. Oncogene, **18**, 5015.

9. Taniguchi, T., Yamauchi, K., Yamamoto, T., Toyoshima, K., Harada, N., Tanaka, H., Takahashi, S., Yamamoto. H., and Fujimoto, S. (1988). Depression in gene expression for poly (ADP-ribose) synthetase during the interferon-gamma-induced activation process of murine macrophage tumor cells. *Eur. J. Biochem.*, **171**, 571.

10. Taniguchi, T., Takahashi, S., Yamamoto, H., Fujimoto, S., and Okoyama, H. (1991). Requirement of down-regulation of NAD+ ADP-ribosyl transferase for the interferon-gamma-induced activation process of murine macrophage tumor cells. *Eur J. Biochem.*, **195**, 557.

11. Nomura, I., Kurashige, T., and Taniguchi, T. (1991). Inhibitory effect of interferon-gamma-dependent induction of major histocompatibility complex class II antigen by expressing exogenous poly (ADP-ribose) synthetase gene. *Biochem. Biophys. Res. Commun.*, **175**, 685.

12. Tomoda, T., Kurashige, T., and Taniguchi, T. (1992). Inhibition of interferon-gamma- and phorbol ester-induced HLA-DR and interleukin-1 production by the expression of a transfected poly (ADP-ribose) synthetase gene in human leukemia THP-1 cells. *Biochim Biophys Acta*, **1135**, 79.

13. Tomoda, T., Kurashige, T., and Taniguchi, T. (1992). Stimulatory effect of interleukin-1 beta on the interferon-gamma-dependent HLA-DR production. *Immunology*, **76**, 15.

14. Taniguchi, T., Nemoto, Y., Ota, K., Imai, T., and Tobari, J. (1993). Participation of poly (ADP-ribose) synthetase in the process of norepinephrine-induced inhibition of major histocompatibility complex class II antigen expression in human astrocytoma cells. *Biochem Biophys Res Commun.*, **193**, 886.

15. Kopin, I. J. and Markey, S. P. (1988). MPTP toxicity: implications for research in Parkinson's disease *Annu. Rev. Neurosci.*, **11**, 81.

16. Ebadi, M., Srinivasan, S. K., and Baxi, M. D. (1996). Oxidative stress and antioxidant therapy in Parkinson's disease. *Progr. Neurobiol.*, **48**, 1.

17. Schulz, J. B., Matthews, R. T., Muqit, M. M., Browne, S. E., and Beal, M. F. (1995). Inhibition of neuronal nitric oxide synthase by 7-nitroindazole protects against MPTP-induced neurotoxicity in mice. *J. Neurochem.*, **64**, 936.

18. Przedborski, S., Jackson-Lewis, V., Yokoyama, R., Shibata, T., Dawson, V. L., and Dawson, T. M. (1996). Role of neuronal nitric oxide in 1-methyl-4-phenyl-1,2,3,6- tetrahydropyridine (MPTP)-induced dopaminergic neurotoxicity. *Proc. Natl Acad. Sci. USA*, **93**, 4565.

19. Cosi, C., Colpaert, F., Koek, W., Degryse, A., and Marien, M. (1996). Poly (ADP-ribose) polymerase inhibitors protect against MPTP-induced depletions

of striatal dopamine and cortical noradrenaline in C57Bl/6 mice. *Brain Res.*, **729**, 264.

20. Mandir, A. S., Przedborski, S., Jackson-Lewis, V., Wang, Z. Q., Simbulan-Rosenthal, D., Smulson, M. E., Hoffman, B. E., Guastella, D. B., Dawson, V. L., and Dawson, T. M. (1999). Poly (ADP-ribose) polymerase activation mediates 1-methyl-4-phenyl-1,2,3,6-tetrahydropyridine (MPTP)-induced Parkinsonism. *Proc. Natl Acad. Sci. USA*, **96**, 5774.

21. Qu, Z., Fujimoto, S., and Taniguchi, T. (1994). Enhancement of interferon-gamma-induced major histocompatibility complex class II gene expression by expressing an anti-sense RNA of poly (ADP-ribose) synthetase. *J. Biol. Chem.*, **269**, 5543.

22. Hauschildt, S., Scheipers, P., and Bessler, W. G. (1991). Inhibitors of poly (ADP-ribose) polymerase suppress lipopolysaccharide-induced nitrite formation in macrophages. *Biochem. Biophys. Res. Commun.*, **179**, 865.

23. Hauschildt, S., Scheipers, P., Bessler, W. G., and Mulsch, A. (1992). Induction of nitric oxide synthase in L929 cells by tumour-necrosis factor alpha is prevented by inhibitors of poly (ADP-ribose) polymerase. *Biochem. J.*, **288**, 255.

24. Pellat-Deceunynck, C., Wietzerbin, J., and Drapier, J. C. (1994). Nicotinamide inhibits nitric oxide synthase mRNA induction in activated macrophages. *Biochem. J.*, **297**, 53.

25. Heine, H., Ulmer, A. J., Flad, H. D., and Hauschildt, S. (1995). Lipopolysaccharide-induced change of phosphorylation of two cytosolic proteins in human monocytes is prevented by inhibitors of ADP-ribosylation. *J. Immunol.*, **155**, 4899.

26. Le Page, C., Sanceau, J., Drapier, J-C., and Wietzerbin, J. (1998). Inhibitors of ADP-ribosylation impair inducible nitric oxide synthase gene transcription through inhibition of NF Kappa B activation. *Biochem. Biophys. Res. Commun.*, **243**, 451.

27. Thanos, D. and Maniatis, T. (1995). NF-κB: a lesson in family values. *Cell*, **80**, 529.

28. Baeuerle, P. A. and Baltimore, D. (1996). NF-κB: ten years after. *Cell*, **87**, 13.

29. Ghosh, S., May, M. J., and Kopp, E. B. (1998). NF-κB and REL proteins: evolutionarily conserved mediators of immune responses. *Annu. Rev. Immunol.*, **16**, 225.

30. Schreck, R., Rieber, P., and Baeuerle, P. A. (1991). Reactive oxygen intermediates as apparently widely used messengers in the activation of the NF-kappa B transcription factor and HIV-1. *EMBO J.*, **10**, 2247.

31. Anderson, M. T., Staal, F. J., Gitler, C., Herzenberg, L. A., and Herzenberg, L. A. (1994). Separation of oxidant-initiated and redox-regulated steps in the NF-kappa B signal transduction pathway. *Proc. Natl Acad. Sci. USA*, **91**, 11527.

32. Schulze-Osthoff, K., Los, M., and Baeuerle, P. A. (1995). Redox signalling by transcription factors NF-kappa B and AP-1 in lymphocytes. *Biochem. Pharmacol.*, **50**, 735.

33. Piret, B., Legrand-Poels, S., Sappey, C., and Piette, J. (1995). NF-kappa B transcription factor and human immunodeficiency virus type 1 (HIV-1) activation by methylene blue photosensitization. *Eur. J. Biochem.*, **228**, 447.

34. Brach, M. A., Kharbanda, S. M., Herrmann, F., and Kufe, D. W. (1992). Activation of the transcription factor kappa B in human KG-1 myeloid

leukemia cells treated with 1-beta-D-arabinofuranosylcytosine. *Mol. Pharmacol.*, **41**, 60.

35. Quinto, I., Ruocco, M. R., Baldassarre, F., Mallardo, M., Dragonetti, E., and Scala, G. (1993). The human immunodeficiency virus type-1 long terminal repeat is activated by monofunctional and bifunctional DNA alkylating agents in human lymphocytes. *J. Biol. Chem.*, **268**, 26719.

36. Legrand-Poels, S., Bours, V., Piret, B., Pflaum, M., Epe, B., Rentier, B., and Piette, J. (1995). Transcription factor NF-κB is activated by photosensitization generating oxidative damages. *J. Biol. Chem.*, **270**, 6925.

37. Oliver, F. J., Ménissier-de Murcia, J., Nacci, C., Decker, P., Andriantsitohaina, R., Muller, S., de la Rubia, G., Stoclet, J. C., and de Murcia, G. (1999). Resistance to endotoxic shock as a consequence of defective NF-kappaB activation in poly (ADP-ribose) polymerase-1 deficient mice. *EMBO J.*, **18**, 4446.

38. Gradwohl, G., Mazen, A., and de Murcia, G. (1987). Poly (ADP-ribose) polymerase forms loops with DNA. *Biochem. Biophys. Res. Commun.*, **148**, 913.

39. Sastry, S. S. and Kun, E. (1990). The interaction of adenosine diphosphoribosyl transferase (ADPRT) with a cruciform DNA. *Biochem Biophys Res Commun.*, **167**, 842.

40. Thanos, D. and Maniatis, T. (1995) Virus induction of human IFN beta gene expression requires the assembly of an enhanceosome. *Cell*, **83**, 1091.

41. Perrella, M. A., Pellacani, A., Wiesel, P., Chin, M. T., Foster, L. C., Ibanez, M., Hsieh, C. M., Reeves, R., Yet, S. F., and Lee, M. E. (1999). High mobility group-I(Y) protein facilitates nuclear factor-kappa B binding and *trans*-activation of the inducible nitric-oxide synthase promoter/enhancer. *J. Biol. Chem.*, **274**, 9045.

42. Kuhnle, S., Nicotera, P., Wendel, A., and Leist, M. (1999). Prevention of endotoxin-induced lethality, but not of liver apoptosis in poly (ADP-ribose) polymerase-deficient mice. *Biochem Biophys Res Commun.*, **263**, 433.

43. Zingarelli, B., O'Connor, M., Wong, H., Salzman, A. L., and Szabó, C. (1996). Peroxynitrite-mediated DNA strand breakage activates poly-ADP ribosyl synthetase and causes cellular energy depletion in macrophages stimulated with bacterial lipopolysaccharide. *J. Immunol.*, **156**, 350.

44. Szabó, C., Virág, L., Cuzzocrea, S., Scott, G. J., Hake, P., O'Connor, M. P., Zingarelli, B., Salzman, A. L., and Kun, E. (1998). Protection against peroxynitrite-induced fibroblast injury and arthritis development by inhibition of poly (ADP-ribose) synthetase. *Proc. Natl Acad. Sci. USA*, **95**, 3867.

45. Hassa, P. O. and Hottiger, M. O. (1999). A role of poly (ADP-ribose) polymerase in NF-κB transcriptional activation. *Biol. Chem.*, **380**, 953.

46. Hsu, H-L., Gilley, D., Blackburn, E., and Chen, D. (1999). Ku is associated with the telomere in mammals. *Proc. Natl Acad. Sci. USA*, **96**, 12454.

47. Bianchi, A. and de Lange, T. (1999). Ku binds telomeric DNA *in vitro*. *J. Biol. Chem.*, **274**, 21223.

48. Yavuzer, U., Smith, G. C. M., Bliss, T., Werner, D., and Jackson, S. P. (1998). DNA end-independent activation of DNA-PK mediated via association with DNA-binding protein C1D. *Genes Dev.*, **12**, 2188.

49. Lee, S. E., Mitchell, R. A., Cheng, A., and Hendrickson, E. A. (1997). Evidence for DNA-PK-dependent and independent DNA double-strand break repair

pathways in mammalian cells as a function of the cell cycle. *Mol. Cell. Biol.*, **17**, 1425.

50. Anderson, C. W. and Lees-Miller, S. P. (1992). The nuclear serine/threonine protein kinase DNA-PK. *Crit. Rev. Eukaryot. Gene Expr.*, **2**, 283.

51. Woo, R. A., McClure, K. G., Lees-Miller, S. P., Rancourt, D. E., and Lee, P. W. K. (1998). DNA-dependent protein kinase acts upstream of p53 in response to DNA damage. *Nature*, **394**, 700.

52. Shieh, S-Y., Ikeda, M., Taya, Y., and Prives, C. (1997). DNA damage-induced phosphorylation of p53 alleviates inhibition by MDM2. *Cell*, **91**, 325.

53. Casciola-Rosen, L. A., Anhlat, G. J., and Rosen, A. (1995). DNA-dependent protein kinase is one of a subset of autoantigens specifically cleaved early during apoptosis. *J. Exp. Med.*, **182**, 1625.

54. Ruscetti, T., Lehnert, B. E., Halbrook, J., Trong, H. L., Hoekstra, M. F., Chen, D. J., and Peterson, S. R. (1998). Stimulation of the DNA-dependent protein kinase by poly (ADP-ribose) polymerase. *J. Biol. Chem.*, **273**, 14461.

55. Ariumi, Y., Masutani, M., Copeland, T. D., Mimori, T., Sugimura, T., Shimotohno, K., Ueda, K., Hatanaka, M., and Noda, M. (1999). Suppression of the poly (ADP-ribose) polymerase activity by DNA-dependent protein kinase *in vitro*. *Oncogene*, **18**, 4616.

56. Morrison, C., Smith, G. C. M., Stingl, L., Jackson, S. P., Wagner, E. F., and Wang, Zhao-Qi (1997). Genetic interaction between PARP and DNA-PK in V(D)J recombination and tumourigenesis. *Nature Genet.*, **17**, 479.

57. Galande, S. and Kohwi-Shigematsu, T. (1999). Poly (ADP-ribose) polymerase and Ku autoantigen form a complex and synergistically bind to matrix attachment sequences. *J. Biol. Chem.*, **274**, 20521.

58. Weinfeld, M., Chaudhry, M. A., D'Amours, D., Pelletier, J. D., Poirier, G. G., Povirk, L. F., and Lees-Miller, S. P. (1997). Interaction of DNA-dependent protein kinase and poly (ADP-ribose) polymerase with radiation-induced DNA strand breaks. *Radiat. Res.*, **148**, 22.

59. D'Silva, I., Pelletier, J. D., Lagueux, J., D'Amours, D., Chaudry, M. A., Weinfeld, M., Lees-Miller, S. P., and Poirier, G. G. (1999). Relative affinities of poly (ADP-ribose) polymerase and DNA-dependent protein kinase for DNA strand interruptions. *Biochim. Biophys Acta*, **1430**, 119.

60. Boulton, S., Kyle, S., and Durkacz, B. W. (1999). Interactive effects of inhibitors of poly (ADP-ribose) polymerase and DNA-dependent protein kinase on cellular responses to DNA damage. *Carcinogenesis*, **20**, 199.

61. Gaken, J. A., Tavassoli, M., Gan, S. U., Vallian, S., Giddings, I., Darling, D. C., Galea-Lauri, J., Thomas, M. G., Abedi, H., Schreiber, V., Ménissier-de Murcia, J., Collins, M. K., Shall, S., and Farzaneh, F. (1996). Efficient retroviral infection of mammalian cells is blocked by inhibition of poly (ADP-ribose) polymerase activity. *J. Virol.*, **70**, 3992.

62. Daniel, R., Katz, R. A., and Skalka, A-M. (1999). A role for DNA-PK in retroviral DNA integration. *Science*, **284**, 644.

63. Halazonetis, T. D. and Shiloh, Y. (1999). Many faces of ATM. *Biochim. Biophys. Acta*, **1424**, R45.

64. Siliciano, J. D., Canman, C. E., Taya, Y., Sakaguchi, K., Appela, E., and Kastan, M. B. (1997). DNA damage induces phosphorylation of the amino terminus of p53. *Genes Dev.*, **11**, 3471.

65. Shieh, S-Y., Ikeda, M., Taya, Y., and Prives, C. (1997). DNA damage-induced phosphorylation of p53 alleviates inhibition of MDM2. *Cell*, **91**, 325.

66. Khanna K. K., Keating, K. E., Kozlov, S., Scott, S., Gatei, M., Hobson, K., Taya, Y., Gabrielli, B., Chan, D., Lees-Miller, S. P., and Lavin, M. F. (1998). ATM associates with and phosphorylates p53: mapping the region of interaction. *Nature Genet.*, **20**, 398.

67. Baskaran, R., Wood, L. D., Whitaker, L. L., Canman, C. E., Moragan, S. E., Xu, Y., Barlow, C., Baltimore, D., Wynshaw-Boris, A., Kastan, M. B., and Wang, J. Y. (1997). Ataxia–telangiectasia mutated protein activates c-Abl tyrosine kinase in response to ionising radiation. *Nature*, **387**, 516.

68. Shafman, T., Khanna, K. K., Kedar, P., Spring, K., Koslov, S., Yen, T., Hobson, K., Gatei, M., Zhang, N., Watters, D., Egerton, M., Shiloh, Y., Kharbanda, S., Kufe, D., and Lavin, M. D. (1997). Interaction between ATM protein and c-Abl in response to DNA damage. *Nature*, **387, 520.**

69. Lim, D-S., Kirsh, D. G., Canman, C. E., Ahn, J. H., Ziv, Y., Newman, L. S., Darnell, R. B., Shiloh, Y., and Kastan, M. B. (1998). ATM binds to β-adaptin in cytoplasmic vesicles. *Proc. Natl Acad. Sci. USA*, **95**, 10146.

70. Chen, P., Gatei, M., O'Connell, M. J., Khanna, K. K., Bugg, S. J., Scott, S. P., Hobson, K., and Lavin, M. F. (1999). Chk1 complements the G2/M checkpoint defect and radiosensitivity of ataxia–telangiectasia cells. *Oncogene*, **18**, 249.

71. Edwards, M. J. and Taylor, A. M. R. (1980). Unusual levels of (ADP-ribose)n and DNA synthesis in ataxia–telangiectasia cells following γ-ray irradiation. *Nature*, **287**, 745.

72. Dantzer, F., Ménissier-de Murcia, J., Barlow, C., Wynshaw-Boris, A., and de Murcia, G. (1999). Poly(ADP-ribose) polymerase activity is not affected in ataxia–telangiectasia (AT) cells and knockout mice. *Carcinogenesis*, **20**, 177.

73. Piret, B., Schoonbroodt, S., and Piette, J. (1999). The ATM protein is required for sustained activation of NF-κB following DNA damage. *Oncogene*, **18**, 2261.

74. Maltzman, W. and Czyzyk, L. (1984). UV irradiation stimulates levels of p53 cellular tumor antigen in nontransformed mouse cells. *Mol. Cell. Biol.*, **4**, 1689.

75. Kastan, M. B., Onyekwere, O., Sidransky, D., Vogelstein, B., and Craig, R. W. (1991). Participation of p53 protein in the cellular response to DNA damage. *Cancer Res.*, **51**, 6304.

76. Fritsche, M., Haessler, C., and Brandner, G. (1993). Induction of nuclear accumulation of the tumour-suppressor protein p53 by DNA-damaging agents. *Oncogene*, **8**, 307.

77. Hall, P., McKee, P. H., Menage, H. P., Dover, R., and Lane, D. P. (1993). High levels of p53 protein in UV-induced normal human skin. *Oncogene*, **8**, 203.

78. Lu, X. and Lane, D. P. (1993). Differential induction of transcriptionally active p53 following UV or ionising radiation: defects in chromosome instability syndromes? *Cell*, **75**, 765.

79. Whitacre, C. M., Hashimoto, H., Tsai, M-L., Chatterjee, S., Berger, S. J., and Berger, N. A. (1995). Involvement of NAD-poly (ADP-ribose) metabolism in p53 regulation and its consequences. *Cancer Res.*, **55**, 3697.

80. Vaziri, H., West, M. D., Allsopp, R. C., Davison, T. S., Wu, Y. S., Arrowsmith, C. H., Poirier, G. G., and Benchimol, S. (1997). ATM-dependent telomere loss

in aging human diploid fibroblasts and DNA damage lead to the post-translational activation of p53 protein involving poly (ADP-ribose) polymerase. *EMBO J.*, **16**, 6018.

81. Wang, X., Ohnishi, K., Takahashi, A., and Ohnishi, T. (1998). Poly (ADP-ribosyl)ation is required for p53-dependent signal transduction induced by radiation. *Oncogene*, **17**, 2819.

82. Kumari, S. R., Mendoza-Alvarez, H., and Alvarez-Gonzalez, R. (1998). Functional interaction of p53 with poly (ADP-ribose) polymerase (PARP) during apoptosis following DNA damage: covalent poly (ADP-ribosyl)ation of p53 by exogeneous PARP and non-covalent binding of p53 to the Mr 85,000 proteolytic fragment. *Cancer Res.*, **56**, 5075.

83. Wesierska-Gadek, J., Bugajska-Schretter, A., and Cerni, C. (1996). ADP-ribosylation of p53 tumor suppressor protein: mutant but not wild-type p53 is modified. *J. Cell. Biochem.*, **62**, 90.

84. Smith, H. M. and Grosovsky, A. J. (1999). PolyADP-ribose-mediated regulation of p53 complexed with topoisomerase I following ionizing radiation. *Carcinogenesis*, **20**, 1439.

85. Gobert, C., Skladanowski, A., and Larsen, A. (1999). The interaction between p53 and DNA topoisomerase I is regulated differently in cells with wild-type and mutant p53. *Proc. Natl Acad. Sci. USA*, **96**, 10355.

86. Malanga, M., Pleschke, J. M., Kleczkowska, H. E., and Althaus, F. (1998). Poly (ADP-ribose) binds to specific domains of p53 and alters its DNA binding functions. *J. Biol. Chem.*, **273**, 11839.

87. Schmidt, A. A. P., Pleschke, J., Kleczkowska, H. E., Althaus, F. R., and Vergères, G. (1998). Poly ADP-ribose modulates the properties of MARCKS proteins. *Biochemistry*, **37**, 9520.

88. Wesierska-Gadek, J., Schmid, G., and Cerni, C. (1996). ADP-ribosylation of wild-type p53 *in vitro*: binding of p53 protein to specific p53 consensus sequences prevents its modification. *Biochem. Biophys. Res. Commun.*, **224**, 96.

89. Agarwal, M. L., Agarwal, A., Taylor, W. R., Wang, Z-Q, Wagner, E. F., and Stark, G. R. (1997). Defective induction but normal activation and function of p53 in mouse cells lacking poly (ADP-ribose) polymerase. *Oncogene*, **15**, 1035.

90. Simbulan-Rosenthal, C. M., Rosenthal, D. S., Ding, R-C., Bhatia, K., and Smulson, M. E. (1998). Prolongation of the p53 response to DNA strand breaks in cells depleted of PARP by antisense RNA expression. *Biochem. Biophys. Res. Commun.*, **253**, 864.

91. Wesierska-Gadek, J., Wang, Z-Q., and Schmid, G. (1999). Reduced stability of regularly spliced but not alternatively spliced p53 protein in PARP-deficient mouse fibroblasts. *Cancer Res.*, **59**, 28.

92. Schmid, G., Wang, Z-Q., and Wesierska-Gadek, J. (1999). Compensatory expression of p73 in PARP-deficient mouse fibroblasts as response to a reduced level of regularly spliced wild-type p53 protein. *Biochem. Biophys. Res. Commun.*, **255**, 399.

93. Webster, G. A. and Perkins, N. D. (1999). Transcriptional cross talk between NF-κB and p53. *M ol. Cell. Biol.*, **19**, 3485.

94. Smith, S., Giriat, I., Schmitt, A., and de Lange, T. (1998). Tankyrase, a poly (ADP-ribose) polymerase at human telomeres. *Science*, **282**, 1484.

95. D'Adda di Fagagna, F., Hande, M. P., Tong, W. M., Lansdorp, P. M., Wang,

Z. Q., and Jackson, S. P. (1999). Functions of poly (ADP-ribose) polymerase in controlling telomere length and chromosomal stability. *Nature Genet.*, **23**, 76.

96. Kickhoefer, V. A., Siva, A. C., Kedersha, N. L., Inman, E. M., Ruland, C., Streuli, M., and Rome, L. H. (1999). The 193-kD vault protein, VPARP, is a novel poly (ADP-ribose) polymerase. *J. Cell Biol.*, **145**, 917.

97. Kickhoefer, V. A., Stephen, A. G., Harrington, L., Robinson, M. O., and Rome, L. H. (1999). Vaults and telomerase share a common subunit, TEP1. *J. Biol. Chem.*, **274**, 32712.

98. Jacobson, M. K. and Jacobson, E. L. (1999). Discovering new ADP-ribose polymer cycles: protecting the genome and more. *TIBS*, **24**, 415.

99. Shall, S. and de Murcia, G. (2000), Poly (ADP-ribose) polymerase-1; what have we learned from the deficient mouse model? *Mutat. Res.*, **460**, 1.

Index

Myocardial ischemia-reperfusion *see*
ischemia-reperfusion
Neutrophil infiltration 164, 166, 167
Adhesion molecules 165, 211

NAD+
As a substrate 11, 57–9
Carba 56–8, 181
Depletion *see* cell death
Etheno- 6
Measurement 6
Necrosis *see* cell death
Neuro-degenerative disorders 163
Neutrophil infiltration *see* Myocardial
ischemia-reperfusion
Nicotinamide 1, 8, 11, 53–9, 80, 164, 179–81,
187, 195, 210
Free medium 81
Nitric oxide Synthases (NOS) 158, 161–3,
210, 211
Oxygen species 153, 156, 157, 159, 160
Peroxynitrite 98, 137, 139, 158, 160, 162,
163, 165, 166, 167, 193
Nitric oxide (NO) 153, 158, 161–3, 166
Nitrotyrosine 166
NMDA 130, 139, 161–3
NMN 3
Non-homologous DNA end joining 213, 216
XRCC4 51, 213
Nuclear factor kappa 47, 102, 103, 112, 166,
208–12, 218, 222, 225
Nuclear matrix 38, 47, 102, 212
Nucleolin 38, 111
Nucleolus 38
Nuclear Location Signal
Poly(ADP-ribose) polymerase 10, 35,
38–40
Poly(ADP-ribose) glycohydrolase 15
Numatrin 38, 111

Oct-1 39, 47, 52, 102
OGG1 *see* Base excision repair
Oxocarbenium 58, 187

Parkinsonism v, 163, 210
Pertussis toxin *see* ADP-ribosylating cells
Peyer patches 36
Poly(ADP-ribose), pADPR
Antibodies 6, 7
branching point 5, 8
Boronate chromatography 6
Catabolism 14, 126
electron microscopy 8
HPLC 6
Immunofluorescence 7, 37, 90

natural occurrence 7
non-covalent interaction 99, 111, 129, 207,
221
Nuclear spekles, plate 1
Quantification 5, 6
Size 3, 5–7, 11, 49
Structure 3–5
1, N^6-etheno derivatives 6
Poly(ADP-ribose) glycohydrolase
Antibodies 16
Chromosomal location 16
Domain structure 15, 20
Glycosidic linkage 3, 14, 56, 57
Km 14
Proteolytic fragments 15–17, 21
Sequence comparison 18, 19
Subcellular location 15
Poly(ADP-ribose) polymerase
Activators *see* activation of PARP
Abortive NADase 11
Antibodies *see* antibodies against
PARP
Automodification 10, 11, 12, 50, 99, 106,
137, 177
BRCT motif *see* BRCT
Catalytic domain 9, 52, 60, 62, 125, 182,
185, 187, 197
Chromosome location 35
Dimerization 12, 44, 49
Deficient mice *see also* knockout mice
Genotoxic stress 36, 91, 92, 95, 98, 104,
177, 221
Micronuclei formation 95
Domain structure 10
Elongation 11, 49, 52, 53, 56–8
Initiation 11, 49, 52, 57
Interaction with partners 48, 50, 51,
98–101, 112
Km for NAD^+ 52, 55
Leucine zipper 12, 51
mRNA expression 35–7
PARP signature 39, 53
Poly (ADP-ribose) polymerase
homologues 60, 62–6, 112, 178,
223
Plant PARPs 60, 61
Purification 8, 9
sequence comparison 40, 43
Spliced variant (PARPII) 52
Subcellular location 37
tissue distribution 35, 36
Zinc fingers 12, 13, 44, 45, 47, 52, 88–90,
99, 106, 107, 110, 192
Preadipocytes 85
Phytohaemaglutinin 36
p53 11, 37, 38, 82, 129, 136, 192, 207, 209,
213, 218–22
Pyramidal cells 36, 37